The Phases of Quantum Chromodynamics

This book discusses the physical phases of quantum chromodynamics (QCD) in ordinary as well as in extreme environments of high temperatures and high baryon number. A major theme of this book is the idea that, to understand the dynamics of QCD in ordinary circumstances, one needs to master them in extreme environments. QCD is thought to be characterized in ordinary circumstances by quark confinement through the formation of flux tubes and chiral-symmetry breaking. These properties are believed to be lost in environments of extreme conditions and new phases, the quark–gluon plasma and color superconductivity, are thought to exist.

The book is aimed at graduate students and researchers entering the fields of lattice-gauge theory, heavy-ion collisions, nuclear theory, and high-energy phenomenology, as well as astrophysicists interested in the phases of nuclear matter and their impact on our current ideas of the interiors of dense stars. It is suitable for use as a textbook on lattice-gauge theory, effective Lagrangians, and field-theoretical modeling for nonperturbative phenomena in QCD.

JOHN KOGUT obtained his Ph.D. from Stanford University in 1971. He is a Fellow in the American Physical Society and has held Guggenheim (1987–8) and Sloan Foundation (1976–8) Fellowships. He is co-author of over 200 original papers in elementary-particle, high-energy, and condensed-matter physics, and is the author of one previous book. He is currently Professor of Physics at the University of Illinois, Urbana-Champaign.

MIKHAIL STEPHANOV obtained his Ph.D. from the University of Oxford in 1994. He was a Postdoctoral Research Associate at the University of Illinois, Urbana-Champaign from 1994 to 1997 and then spent two years at the State University of New York at Stony Brook. He is at present Associate Professor at the University of Illinois at Chicago and RHIC Fellow at the RIKEN-BNL Center. He has published over 50 original papers on particle and nuclear physics, lattice quantum-field theory, matter under extreme conditions, and heavy-ion collisions. His research is supported by a DOE Outstanding Junior Investigator Award, as well as by an Alfred P. Sloan Foundation Fellowship.

T0236018

CAMBRIDGE MONOGRAPHS ON
PARTICLE PHYSICS
NUCLEAR PHYSICS AND COSMOLOGY
21

General Editors: T. Ericson, P. V. Landshoff

1. K. Winter (ed.): *Neutrino Physics*
2. J. F. Donoghue, E. Golowich and B. R. Holstein: *Dynamics of the Standard Model*
3. E. Leader and E. Predazzi: *An Introduction to Gauge Theories and Modern Particle Physics, Volume 1: Electroweak Interactions, the 'New Particles' and the Parton Model*
4. E. Leader and E. Predazzi: *An Introduction to Gauge Theories and Modern Particle Physics, Volume 2: CP-Violation, QCD and Hard Processes*
5. C. Grupen: *Particle Detectors*
6. H. Grosse and A. Martin: *Particle Physics and the Schrödinger Equation*
7. B. Andersson: *The Lund Model*
8. R. K. Ellis, W. J. Stirling and B. R. Webber: *QCD and Collider Physics*
9. I. I. Bigi and A. I. Sanda: *CP Violation*
10. A. V. Manohar andd M. B. Wise: *Heavy Quark Physics*
11. R. K. Bock, H. Grote, R. Frühwirth and M. Regler: *Data Analysis Techniques for High-Energy Physics, Second edition*
12. D. Green: *The Physics of Particle Detectors*
13. V. N. Gribov and J. Nyiri: *Quantum Electrodynamics*
14. K. Winter (ed.): *Neutrino Physics, Second edition*
15. E. Leader: *Spin in Particle Physics*
16. J. D. Walecka: *Electron Scattering for Nuclear and Nucleon Structure*
17. S. Narison: *QCD as a Theory of Hadrons*
18. J. F. Letessier and J. Rafelski: *Hadrons and Quark–Gluon Plasma*
19. A. Donnachie, H. G. Dosch, P. V. Landshoff and O. Nachtmann: *Pomeron Physics and QCD*
20. A. Hofmann: *The Physics of Synchrotron Radiation*
21. J. B. Kogut and M. A. Stephanov: *The Phases of Quantum Chromodynamics*

The Phases of Quantum Chromodynamics:

From Confinement to Extreme Environments

JOHN B. KOGUT
University of Illinois, Urbana-Champaign

MIKHAIL A. STEPHANOV
University of Illinois, Chicago

CAMBRIDGE UNIVERSITY PRESS
Cambridge, New York, Melbourne, Madrid, Cape Town, Singapore,
São Paulo, Delhi, Dubai, Tokyo

Cambridge University Press
The Edinburgh Building, Cambridge CB2 8RU, UK

Published in the United States of America by Cambridge University Press, New York

www.cambridge.org
Information on this title: www.cambridge.org/9780521143387

First published 2004
This digitally printed version 2010

A catalogue record for this publication is available from the British Library

ISBN 978-0-521-80450-9 Hardback
ISBN 978-0-521-14338-7 Paperback

Contents

1   **Introduction**   **1**

2   **Background in spin systems and critical phenomena**   **10**
2.1   Notation and definitions and critical indices   10
2.2   Correlation-length scaling and universality classes   14
2.3   Properties of the Ising model   17
2.4   The Kosterlitz–Thouless model   19
2.5   Coulomb gas, duality maps, and the phases of the planar model   24
2.6   Asymptotic freedom in two-dimensional spin systems   30
2.7   Instantons in two-dimensional spin systems   36
2.8   Computer experiments and simulation methods   37
2.9   The transfer matrix in field theory and statistical physics   41

3   **Gauge fields on a four-dimensional euclidean lattice**   **53**
3.1   Lattice formulation, local gauge invariance, and the continuum
      action   53
3.2   Confinement and the strong-coupling limit   57
3.3   Confinement mechanisms in two and four dimensions:
      vortex and monopole condensation   63

4   **Fermions and nonperturbative dynamics in QCD**   **74**
4.1   Asymptotic freedom and the continuum limit   74
4.2   Axial symmetries and the vacuum of QCD   76
4.3   Two-dimensional fermionic models of confinement, axial
      symmetries, and θ vacua   77
4.4   Instantons and the scales of QCD   84

5   **Lattice fermions and chiral symmetry**   **93**
5.1   Free fermions on the lattice in one and two dimensions   93
5.2   Fermions and bosons on Euclidean lattices   101
5.3   Staggered Euclidean fermions   104
5.4   Block derivatives and axial symmetries   107
5.5   Staggered fermions and remnants of chiral symmetry   109
5.6   Exact chiral symmetry on the lattice   111
5.7   Chiral-symmetry breaking on the lattice   117
5.8   Simulating dynamical fermions in lattice-gauge theory   126
5.9   The microcanonical ensemble and molecular dynamics   127
5.10   Langevin and hybrid algorithms   132

6   **The Hamiltonian version of lattice-gauge theory**   **136**
6.1   Continuous time and discrete space   136
6.2   Quark confinement in Hamiltonian lattice-gauge theory
        and thin strings   145
6.3   Relativistic thin strings, delocalization, and Casimir forces   146
6.4   Roughening and the restoration of spatial symmetries   150

7   **Phase transitions in lattice-gauge theory at high
        temperatures**   **158**
7.1   Finite-temperature transitions at strong coupling   158
7.2   Simulations at nonzero temperature   162
7.3   Pure gauge-field simulations at nonzero temperature   165
7.4   Restoration of chiral symmetry and high temperature   170
7.5   Hadronic screening lengths   172
7.6   Thermal dilepton rates and experimental signatures for the
        quark–gluon plasma   174
7.7   A tour of the three-flavor QCD phase diagram   176

8   **Physics of QCD at high temperatures and chemical
        potentials**   **182**
8.1   The thermodynamic background   182
8.2   Hadron phenomenology and simple models of the transition to
        the quark–gluon plasma   202
8.3   A tour of the $T-\mu$ phase diagram   207
8.4   The quark–gluon plasma and the energy scales of QCD   223
8.5   The extreme environment at a relativistic heavy-ion collider   226

9   **Large chemical potentials and color superconductivity**   **236**
9.1   Color superconductivity and color–flavor locking   236
9.2   Calculating the gap at asymptotically large μ 239

9.3	Lowest excitations of the CFL phase	260
9.4	Comments and some further developments	278
10	**Effective Lagrangians and models of QCD at nonzero chemical potential**	**280**
10.1	QCD at finite μ and the sign problem	280
10.2	The random-matrix model of QCD	282
10.3	Two-color QCD and effective Lagrangians	295
10.4	QCD at nonzero isospin chemical potential	307
10.5	Pion propagation near and below T_c	316
11	**Lattice-gauge theory at nonzero chemical potential**	**324**
11.1	Propagators and formulating the chemical potential on a Euclidean lattice	324
11.2	Naive fermions at finite density	326
11.3	The three-dimensional four-Fermi model at nonzero T and μ	330
11.4	Four-flavor SU(2) lattice-gauge theory at nonzero μ and T	335
11.5	High-density QCD and static quarks	345
11.6	The Glasgow algorithm	347
11.7	The Fodor–Katz method for high T, low μ	349
11.8	QCD at complex chemical potential	350
12	**Epilogue**	**355**
	References	357
	Index	363

1
Introduction

The phases of QCD have presented a challenge to theoretical physics for many years. Originally it was considered folly to think about topics such as confinement, chiral-symmetry breaking, and different states of matter in relativistic field theories. Anything beyond traditional perturbation theory was met with skepticism. However, the development of semi-classical methods in field theory, the lattice-gauge formulation of the subject, and the generalizations of analytic methods of analysis from two-dimensional systems, such as duality and transitions driven by topological excitations, have changed all that.

This book is a look at the fundamentals of QCD from this perspective. After introducing lattice-gauge theory, beginning with fundamentals and reaching important recent developments, it will emphasize the application of QCD to the study of matter in extreme environments. Effective Lagrangians, which incorporate the constraints of low-energy dynamics and their symmetry realizations, will also be developed to provide a complementary, and insightful, perspective. Application of perturbative methods will also be presented in the regime of their validity.

Why extreme environments? A major theme of this book is the idea that, to understand the dynamics of QCD in ordinary circumstances, one needs to master them in extreme environments.

For example, to appreciate confinement, one can heat the system until the thermal fluctuations prevent the formation of the thin flux tubes which give rise to a linearly confining potential. In doing so, the special ingredients in the theory's dynamics which favor a vacuum pressure that squeezes flux into thin continuous tubes are emphasized. In addition, the essential ingredient of Gauss' law and exact color gauge symmetry will be seen to lie at the heart of the confining features of ordinary QCD.

Another example, which is central to this book, is afforded by chiral-symmetry breaking, the fact that the theory with massless quarks develops

1

dynamical constituent quark masses consistent with its global chiral symmetries. Again, in ordinary environments quark–anti-quark attractive forces, generated by mechanisms such as perturbative gluon exchanges and nonperturbative instanton forces and even those of flux tubes and confinement, favor the appearance of a condensate of quark–anti-quark pairs in the system's vacuum. This condensate then gives the vacuum an indefinite chirality, which supports dynamically generated masses for its constituents as well as its colorless physical states. When such a vacuum is heated, the violent thermal fluctuations will melt the condensate and lead to restoration of chiral symmetry. The corresponding thermodynamic state, or thermal vacuum, however, is anything but simple. In this hot environment affecting the theory of massless quarks and gluons, we expect plasma formation and screening of color. The hot theory is qualitatively different from its cold relative: color screening replaces color confinement, massless quarks and gluons replace dynamical symmetry breaking, and a rich mass spectrum of colorless states with a large level spacing is replaced by a spectrum of screened states with a smaller level spacing.

Another extreme environment that lies on the experimental horizon is that of cold, dense matter. An example is afforded by the interior of a neutron star, where the baryon number density is far higher than that of ordinary nuclear matter. This environment is not well understood in theories like QCD, for reasons which will be discussed at length within this book. Since the discovery of asymptotic freedom in theories like QCD, it has been realized that matter should become weakly interacting if the Fermi energies of the quarks, controlled by the chemical potential μ, are large in comparison with the confinement scale, which is of the order of a few hundred MeV (or $1 \, \text{fm}^{-1}$). It has only recently fully been appreciated that even this weakly coupled state is rich in intricate phenomena due to pairing instabilities similar to the phenomenon of superconductivity in metals. Reliable calculations are possible in this regime controlled by the smallness of the gauge coupling of strong interactions. However, this control is very quickly lost once the densities decrease, and weak-coupling approximations are not justified. From the lower-density side, we can reliably predict that, as the chemical potential of the baryon charge approaches the mass of the lightest baryon, M_N (i.e., the quark chemical potential approaches $M_N/3$), a transition to a new state of matter occurs, since it becomes energetically favorable for the ground state to contain baryons. We also have empirical evidence leading us to believe that this state is the state of nuclear matter, chunks of which form nuclei of heavy elements, and that astronomically large chunks, held together by gravity, form neutron stars. But how does this matter, which we know is not very dense on the scale of QCD (one baryon per $6 \, \text{fm}^3$) transform into asymptotically free quark matter, as the density, or μ, increases? What happens to the baryons, whose very existence seems doubtful in such dense *quark* matter? Is there a transition, and, if there is, at what μ and of what type? A hypothetical peek into the interior of a neutron star would reveal the answers. Such questions are still in the realm of speculation in QCD research at this time. One of the goals of this book is to provide the tools and background for a new generation of researchers to study this problem.

The challenge of QCD in cold, dense environments is a major theme of this book. We believe that there is a new field of science awaiting us there. It is the condensed matter or even the chemistry of QCD. The point is that, as cold QCD is subjected to a chemical potential, there will be the possibility of the creation of new states of extended QCD matter, analogous to the creation of new states of molecular matter in atomic physics. The phases of molecular matter are extraordinarily rich because molecules themselves have intricate three-dimensional symmetries and can interact among themselves via cooperative phenomena like those involving induced dipole–dipole forces, which generate energy scales several orders of magnitude smaller than the fundamental atomic energy scale, the Rydberg. These effects conspire to generate phase diagrams of molecules that are very complex and interesting, and determine the character of our environment. In the case of QCD, the quarks and gluons have internal symmetries described by their color and flavor indices. There are small symmetry breakings arising from the bare mass splittings between the quarks. There is a rich physical mass spectrum of colorless states with a level spacing of several hundred MeV. How will such a physical system respond to a chemical potential comparable to its level spacing? Such an environment will encourage the production of new, nontrivially ordered states of fundamental matter. Exotic phases consisting of meson condensates, superfluids, crystals, etc. have been proposed. From this point of view, it would be naive to expect reliable predictions here very soon, since such states of matter and their stability will depend on grasping the dynamics of QCD to within an accuracy of a few MeV. A goal, ambitious at that, for this book, is to encourage a new generation of physicists to tackle this new field of the extremely condensed matter, or chemistry, of QCD, which may very well determine our environment on the cosmic scale.

Why is so little known about QCD at nonzero μ and what are the present speculations about new states of matter at vanishing temperature and nonvanishing chemical potential? Lattice-gauge theory has made little progress on this subject because its action for the SU(3) color group in the presence of a nonzero baryon chemical potential becomes complex and not amenable to reliable methods of analysis, such as computer simulations. This problem will be dealt with at length in the text. We will review the random-matrix analysis of the failure of the quenched "approximation" of QCD to model it adequately at nonzero chemical potential. We will also see that models with positive fermion determinants can be analyzed by lattice and effective-Lagrangian methods. In fact the SU(2) color case can be studied in detail and new states of strongly interacting superfluids are predicted.

QCD of three colors, but with the chemical potential for the isospin charge I_3, is very similar to the SU(2) color case (and in fact also similar to the quenched approximation to QCD – as we shall see using the random-matrix approach). The pions condense into a pion superfluid at an easily predictable critical chemical potential.

We will also view the SU(3) theory in an environment where the chemical potential is asymptotically large and there is a huge Fermi sphere of weakly interacting colorful quarks. In this environment, which is amenable to the weak-coupling approach, one can identify a QCD analog of Bardeen–Cooper–Schriefer (BCS) fermion pairing and superconductivity – color superconductivity. This QCD phenomenon is much richer than its condensed-matter (QED) counterpart. It also has much more integrity in it, since in QED the dominant forces between electrons are repulsive and an intricate phonon-mediated interaction is needed in order to provide the necessary attraction. One needs more than just electrons and photons for traditional BCS fermion pairing. In QCD quarks and gluons and their mutual interactions are the only necessary ingredients. The attraction provided by gluons is long-ranged (on the typical scale of $1/\mu$ in such quark matter), thus providing a stronger effect than one could expect from a short-ranged attraction.

The phenomenon of color superconductivity is also richer than its QED counterpart due to there being a larger set of global symmetries, the chiral symmetries of QCD. An especially interesting theoretical case is the limit in which all three quarks are light compared with their chemical potential μ. In this regime the BCS instability leads to breaking of global (axial SU(3) and baryon-number U(1)), as well as local (color) symmetries. This phenomenon known as *color–flavor locking* (CFL) is an interesting example of a mechanism by which chiral symmetry can be broken spontaneously within the domain of weak-coupling perturbation theory. Since the pattern of breaking in CFL is somewhat similar to the vacuum of QCD, the lowest excitations – Goldstone bosons – are also similar, since their existence is dictated by the symmetries of the theory and the Goldstone theorem. What is new in CFL is that the properties of the Goldstone particles are calculable. Imagine calculating the pion decay constant from the first principles in QCD. This turns out to be possible in the CFL phase!

The high-temperature and high-chemical-potential environments will emphasize the property of asymptotic freedom of QCD, the fact that it is weakly coupled and perturbative at short distances while being strongly coupled and full of interesting nonperturbative symmetry realizations at large distances. One of the goals of lattice-gauge theory is to attain an understanding, both theoretical and numerical, of the physics on the multiple length scales of QCD: weak-coupling perturbative behavior at short distances, semi-classical nonperturbative behavior at intermediate distances, and strong-coupling confinement and color screening at large distances. We will cover the basics of lattice-gauge theory, with a strong emphasis on models in lower dimensions and spin models of greater simplicity, in order to develop the tools, insight, and experience to grapple with these phenomena in realistic four-dimensional settings. Duality, topological excitations and related order–disorder phase transitions, transfer matrices, and Hamiltonian methods will all be studied, with the goal of providing the reader with the skills to attack modern problems in QCD from several different vantage points.

A major development in modern field theory has been an understanding of the subtleties of fermion fields and their symmetries. This book spends considerable time on this subject since the lattice has played an important role in many of these topics. The most significant accomplishment of lattice-gauge theory was the demonstration that gauge theories could be formulated nonperturbatively. This means that local color symmetries could be defined even in an ultraviolet-regulated spacetime, and local gauge symmetries would remain exact. The important ingredient in this advance was the realization that the gauge group, rather than the generators, would provide the building blocks of the theory. The lattice action, for example, is constructed using a spacetime lattice of points and links, and gauge fields are represented by group elements on links and Fermi fields by Grassmann variables on sites. The notion of exact local gauge invariance, namely that the local color of an excitation tranforms under local color rotations and that the physics should be expressible in terms of invariants, has exact, geometrical expressions in this framework. This feature lies at the heart of the theory's ability to discuss confinement outside of particular calculational schemes and to discuss other associated nonperturbative phenomena quantitatively. Within this context one also discusses fermions and their special symmetries, such as flavor transformations and axial flavor and axial singlet transformations. This leads to nonperturbative formulations and calculations of chiral-symmetry breaking and dynamical mass generation. All of this is important to the phenomenology of QCD and its relation to the very successful constituent-quark model. The mechanisms of chiral-symmetry breaking in ordinary strongly interacting physics as well as in extreme environments of high temperature and chemical potentials have been revealed by lattice and semi-classical studies after decades of little progress when the only tool one had was perturbation theory. The interplay of chiral-symmetry breaking, semi-classical excitations, and confinement itself is a major theme in QCD and this book.

It is particularly interesting that these two successes, exact gauge invariance and chiral symmetries of fermions, lead to a near catastrophe in the subject. Naively it appears that global fermion symmetries can be formulated exactly on spacetime lattices. At the same time local gauge interactions can also be formulated exactly. This then presents the opportunity to "gauge" the fermion chiral symmetries and make models in which any global chiral symmetry of interest and any local gauge symmetry would also be exact, but such a "success" is too much of a good thing. We know from perturbation theory that axial symmetries are broken by the perturbative anomalies of gauge fields. For example, the neutral pion does decay into two photons and this is caused by the existence of the famous triangle graphs coupling a pion to two photons. In more basic terms, this result means that the axial flavor current that describes the neutral pion is no longer conserved in the presence of a quantized electromagnetic field. Classically there is a conservation law but quantum fluctuations destroy it. No one can argue that the neutral pion's dominant decay mode is into a state of two photons and few physicists would want to say that the standard formulation of axial symmetries and their currents docs not describe the properties of light quarks and composite pions. It looks as though lattice-gauge theory can make quantum theories with too much symmetry!

Actually this problem is "solved" by another lattice shortcoming. If one considers the Dirac equation on the lattice, one finds surprises. In particular, there are typically too many species predicted from the discrete-lattice version of the Dirac equation in the theory's continuum limit and these extra species have just the right handedness and number to cancel out the anomaly that leads to the nonconservation of the global chiral current. This is the famous, or infamous, problem of "species doubling" on the lattice. It shows that the continuum limit of the lattice Dirac operator describes more species of fermions with more global symmetries than intended. In the early days of lattice-gauge theory this restriction led to formulations of lattice fermions that eliminated the unwanted extra fermion species by explicitly breaking most of the chiral symmetries. However, the breaking was constructed such that the continuum limit of the model would have the desired fermion species, gauge invariance, and anomalies, so that no contradictions could be encountered. This book will spend considerable space on these topics because they are central to understanding QCD. The two most useful lattice-fermion methods, which compromise chiral symmetries on the lattice but are engineered to have the correct continuum limits, are called "Wilson" and "staggered" fermions. They will be discussed in detail because they are easy to use both analytically and numerically. These formulations of the Dirac equation couple only nearest-neighbor sites on the lattice, and their actions can be "improved" in the sense that next-nearest-neighbor terms can be added to the action, so that the deviations of the lattice action from the continuum action can be systematically reduced. These improvements should prove important in numerical work on the phases of QCD.

The unpleasant feature of these lattice-fermion methods, namely that chiral symmetries are compromised on the cutoff system while gauge symmetries are not, is addressed in more sophisticated lattice-fermion actions. In particular, we will introduce the domain-wall fermion method which extends the lattice to five dimensions. In the new dimension there is a domain wall where massless, chiral fermions can exist. This scheme produces a four-dimensional lattice-gauge theory with full chiral symmetry, the correct anomaly structure and local gauge invariance. Another, closely related approach to this problem in clashing symmetries changes the definition of lattice chiral transformations to accomodate the nonzero lattice-spacing cutoff. In this scheme, Ginsparg–Wilson fermions, the cutoff system has an exact chiral-like symmetry and all of its important physical consequences but the price one pays here is that the continuous symmetry differs from the continuum chiral symmetry by lattice terms and there is apparent nonlocality in the four-dimensional action.

Our interest in these new lattice-fermion developments is restricted to their impact on understanding the nonperturbative, dynamical symmetry-breaking physics of QCD, which is a left–right-symmetric theory. It is hoped that the methods will produce useful nonperturbative formulations of the left–right-asymmetric models, such as the standard model of electroweak interactions, and other chiral gauge theories.

As we discuss these issues, we will introduce lattice-numerical methods with an emphasis on algorithms that can solve fermion field theories nonperturbatively. Recent developments in the field at high temperature and low chemical potential and their relevance to relativistic heavy-ion colliders (RHICs) will be emphasized.

Since lattice-gauge theory is not covered in graduate-physics curricula, we spend considerable time with the fundamentals of this subject. In fact, the chapters on lattice-gauge theory here could be used as the student's first introduction to the subject. The perspective is toward QCD in extreme environments, but all the fundamentals such as local gauge invariance, continuum limits and asymptotic freedom, confinement mechanisms, lattice fermions, computer algorithms, and chiral-symmetry breaking are covered here in an elementary and self-contained fashion. The use of model lattice-gauge-theory systems, field theories in two and three dimensions, occurs throughout the book to illustrate and introduce concepts with a minimum of formalism. Several sections of the book within the lattice-gauge-theory mantra give background to the goals of lattice-gauge theory by working through continuum-model field theories. For example, the importance of confinement, topology, the anomaly and $U(1)_A$ problem, and the existence of θ vacua are introduced and illustrated within $1 + 1$ quantum electrodynamics. This development emphasizes the need for lattice fermions to capture the chiral (flavor-singlet) and gauge symmetries of QCD correctly – if a failure occurred here, extra unphysical massless modes would populate the theory's physical spectrum. The importance of topological excitations in field theories is illustrated through the two-dimensional planar spin model, the Abelian Higgs model, and the $O(3)$ model.

Much of the material presented here is adopted from the senior author's elementary reviews of the subject that were written when lattice-gauge theory was in its infancy [1].

The lattice approach to QCD provides a microscopic, fundamental formulation of the theory of quarks and gluons. The emphasis is on local gauge symmetries, their expression at the microscopic level, and their implications at large distances, as well as global symmetries, such as flavor and chiral symmetries. Eventually lattice QCD will provide concise solutions for the physics on all length scales. At this time, however, the field proceeds more modestly and in this book, for example, we supplement the lattice approach with the study of effective Lagrangians. In this approach one embodies the local and global symmetries of QCD in an effective field theory that states how the low-energy excitations of the solution of QCD must interact on the basis of our knowledge of the microscopic physics and the realizations of its fundamental symmetries. The successes here include an understanding of Goldstone and pseudo-Goldstone physics of pions, kaons, etc. Since light degrees of freedom control the critical behavior of QCD at its second-order phase transitions, this approach has much to say about extreme environments. In particular, at high temperature it predicts the critical behavior of the quark–gluon-plasma transition. Its generalization to nonzero chemical potential is also constrained by symmetries and conservation laws and will be developed in detail. We will see that it is particularly illuminating for model systems, such as SU(2) color models, in which the transition to a diquark condensate occurs at $\mu_c = m_\pi/2$, within the domain of applicability of chiral-perturbation theory. Unfortunately, it has less to predict about SU(3) QCD in cold but baryon-rich environments. A theoretical framework for QCD at baryon chemical potentials of the order of the mass of the nucleon is lacking, as will be discussed at length in the chapters that follow. One of the challenges here is to find a quantitative, nonperturbative framework to describe diquark condensation leading to breaking of color symmetry and color superconductivity.

One of the main themes of the book is to emphasize the use of methods that, starting from the defining principles of QCD as a quantum-field theory, are able to provide systematically controllable results. Historically, only perturbation theory lived up to this claim. Although the small-coupling expansion provides indispensable information, its domain of validity ends before, and often well before, the regime of interest, where nonperturbative phenomena such as confinement, chiral-symmetry breaking, and phase transitions set in. Lattice-field theory is an instrument that allows one to go into these domains with precision controlled by the lattice spacing. Hence, lattice-field theory takes a prominent place in this book. The method of effective Lagrangians is another very powerful tool, which allows one to embody concisely the information about the global symmetries of the theory and of the ground state. A systematically controllable expansion, now organized by powers of the small energies and momenta of the lowest excitations, can then provide nontrivial dynamical information, which in some cases is even exact. Sometimes the synonym *model-independent* is used to distinguish such systematically controlled, or exact, results from the

. results obtained using phenomenologically or theoretically motivated models. Such models have made and continue to make significant contributions to our qualitative and semiquantitative understanding of nonperturbative phenomena. Unfortunately, or perhaps fortunately, for the coming generation of researchers, at present, there is still a large domain of the QCD phase diagram that is not amenable to first-principles approaches. It is our belief that this will change and we hope that this book will help in that.

In writing a book of this sort it is hard to know where to start and where to stop. We assume that the reader has had or is taking an introductory field-theory course, so that path integrals, gauge fields, Feynman diagrams, and the quark model have been seen before. In addition, we use the language and methods of statistical physics, so some preparation in critical phenomena, Landau mean-field theory, and scaling laws would be useful. For example, we will freely move between the languages of field theory and statistical mechanics and we will assume that the reader can easily grasp the formal correspondences between the two subjects. We cover them briefly in our discussions of the transfer matrix for model lattice systems. Throughout the text we refer the reader to textbooks and review articles for additional background. The student might be studying introductory field theory or critical phenomena concurrently with a reading of this more specialized book. The student need not be an expert in these subjects when he or she begins this book, but we hope that they will emerge an expert at the end.

2
Background in spin systems and critical phenomena

2.1 Notation and definitions and critical indices

The language and formalisms of lattice-gauge theory are based on those of traditional statistical mechanics. Let's begin by introducing terminology and ideas in the context of the statistical mechanics of two-dimensional spin systems. This will lay the basis for more elaborate and challenging constructions in four-dimensional lattice-gauge theory.

Consider the two-dimensional Ising model on a square lattice whose sites are labeled by a vector of integers

$$n = (n_1, n_2) \tag{2.1}$$

Place a "spin" variable $s(n)$ at each site and suppose that s can have only two possible values: "up," $s = +1$, and "down," $s = -1$. Nearest-neighbor spins are defined to interact through an energy

$$S = -J \sum_{n,i} s(n)s(n+i) \tag{2.2}$$

where i denotes one of the two unit vectors of the square lattice, as shown in Fig. 2.1. We choose the coupling J to be positive, so that aligned spins are favored. In the context of Euclidean field theory, the energy of the system would be called the "action." This terminology will be clarified as we progress from statistical mechanics, with its partition function, order parameters, and susceptibilities, to Euclidean field theory, with its path integral, condensates, and propagators.

The Ising-model variables and energy have two particularly important features. First, the energy has only short-ranged interactions. In fact, the interactions involve only nearest-neighbor spins. The locality of the interactions will play a crucial role in determining the phase diagram and the critical behavior,

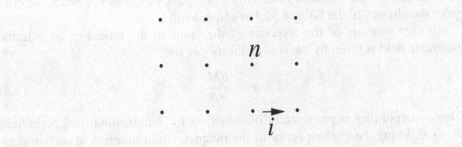

Figure 2.1. Labeling the sites and links of a square two-dimensional lattice.

properties of the system in the immediate vicinity of its phase transition, of the model. Second, the model has a global symmetry: its energy is unchanged if the entire system is flipped upside down. The symmetry group is just Z_2, the global interchange of "up" and "down".

This system of spins can be placed into an external, uniform magnetic field B. The energy acquires another term,

$$S = -J \sum_{n,i} s(n)s(n+i) - B \sum_{n} s(n) \tag{2.3}$$

The external field tends to align all the spins. It distinguishes between the "up" and "down" directions and breaks the global Z_2 symmetry of the original model.

The statistical properties of the Ising model follow from its partition function,

$$Z = \sum_{\text{configs}} \exp(-\beta S) \tag{2.4}$$

where the Boltzmann factor $P = \exp(-\beta S)$ records the relative probability of a particular configuration of spins and $\beta \equiv 1/(kT)$. The sum over configurations in this equation is rather daunting since, for a system of N spins, there are 2^N different configurations.

The thermodynamics of the model follows from the configurational sum. For example, the free energy is

$$F = -kT \ln Z \tag{2.5}$$

The mean magnetization per site, M, can be expressed as an expectation value with respect to the probability weight $P = \exp(-\beta S)$,

$$M = \left\langle \frac{1}{N} \sum_{n} s(n) \right\rangle = \sum_{\text{configs}} \left(\frac{1}{N} \sum_{n} s(n) \right) \exp(-\beta S)/Z \tag{2.6}$$

which can also be written in terms of the free energy, given above, as

$$M = \frac{1}{N} \frac{\partial F}{\partial B} \tag{2.7}$$

We will discuss the symmetry-breaking significance of M below when we consider the physics of the Ising model in some detail.

Another measure of the response of the spins in the model to an external magnetic field is given by the susceptibility per site,

$$\chi = \frac{\partial M}{\partial B} \tag{2.8}$$

This susceptibility is particularly illuminating for infinitesimal magnetic field $B \to 0$. Using the explicit form for the magnetization in terms of a configurational average, as stated above, we can derive

$$\chi = \frac{1}{NkT}\left[\langle s_{\text{tot}}^2\rangle - \langle s_{\text{tot}}\rangle^2\right] = \frac{1}{NkT}\left[\langle (s_{\text{tot}} - \langle s_{\text{tot}}\rangle)^2\rangle\right] \tag{2.9}$$

where $s_{\text{tot}} = \sum_n s(n)$. This equation shows that the zero-field susceptibility is a measure of the fluctuations in the spins comprising the model. This result is closely related to the "fluctuation-dissipation" theorem. χ can also be written in the form

$$\chi = \frac{1}{NkT}\left(\sum_{n,m}\langle s(n)s(m)\rangle - \left\langle \sum_n s(n)\right\rangle^2\right) \tag{2.10}$$

which shows that χ is a measure of the range of the correlations between the spins. For example, if we consider the system at a temperature at which it has no net magnetization, $\langle s(0)\rangle = 0$, then

$$\chi = \frac{1}{kT}\sum_n \Gamma(n) \tag{2.11}$$

where $\Gamma(n)$ is the spin–spin correlation function,

$$\Gamma(n) = \langle s(n)s(0)\rangle \tag{2.12}$$

and we used the translational invariance of the system to simplify the sums over the lattice sites.

This algebra teaches us a lesson that is of central importance in statistical physics and will play an important role in lattice-gauge theory: susceptibilities can become singular and can diverge in the thermodynamic limit of a macroscopic system, if the system develops sufficiently long-range correlations. Note that we are always considering the Ising model, whose energy is constructed from local interactions of just one lattice spacing. Nonetheless, the dynamics may generate a spin–spin correlation function that falls off so slowly across the lattice that the sum in the equation for χ diverges. This behavior lies at the heart of second-order phase transitions and continuum limits of lattice-gauge theory, so we will return to it several times more.

Consider the phenomenology of phase transitions, in particular, the behavior of the system for temperatures at and near a second-order phase transition. The magnetization M is a function of T and B. It is interesting to let B tend to zero for a fixed temperature T. If M remains nonzero, the system is said to experience "spontaneous magnetization." This occurs only for T below a critical value T_c, the Curie temperature in the Ising model. The existence of a nonzero magnetization indicates that the solution of the system does not share the global up–down symmetry of its energy expression. One says that the global Z_2 symmetry is "spontaneously broken" and that M serves as an "order parameter" to label the phases and symmetries of the system.

M is nonzero for $T < T_c$, but it vanishes continuously at a second-order phase transition, $T \to T_c$. In the Ising model, as in many physical systems, the order parameter vanishes as a power in the limit $T \to T_c$,

$$M \sim (T_c - T)^{\beta_{\text{mag}}} \tag{2.13}$$

where β_{mag} is the magnetization critical exponent, which has been calculated for many solvable models, including the Ising model in two dimensions, and has been measured in experiments and in computer simulations. It takes the value $\frac{1}{8}$ in the two-dimensional Ising model, and is $\frac{1}{2}$ in mean-field theory, a simple but extraordinarily useful approximation that will be discussed extensively below.

Another quantity that is particularly important and informative in the context of critical phenomena is the spin–spin correlation function $\Gamma(n)$. At sufficiently high temperatures, we expect thermal fluctuations to dominate the tendency of distant spins to align and correlate. Very general considerations lead to the expectation that $\Gamma(n)$ should fall off exponentially with the distance between the spins for any $T > T_c$,

$$\Gamma(n) \sim \exp(-|n|/\xi(T)), \qquad |n| \gg 1 \tag{2.14}$$

where $\xi(T)$ is the system's "correlation length." ξ gives a measure of the sizes of patches of correlated spins in the system. For high T, $\xi(T)$ will be comparable to a lattice spacing. As T approaches T_c, $\xi(T)$ should grow large and eventually diverge when $T = T_c$. In many models the singularity is power-behaved and we write

$$\xi(T) \sim (T - T_c)^{-\nu} \tag{2.15}$$

where ν is another critical index of fundamental significance in the theory of phase transitions. In the two-dimensional Ising model $\nu = 1$, whereas in mean-field theory $\nu = \frac{1}{2}$.

Since the correlation length diverges at the critical point T_c, the correlation function $\Gamma(n)$ is no longer an exponential there. Instead, it falls slowly, as a power of the distance n between the spins,

$$\Gamma(n) \sim |n|^{d-2+\eta}, \qquad T = T_c \tag{2.16}$$

where d is the dimensionality of the lattice and η is another critical index, which is $\frac{1}{4}$ in the two-dimensional Ising model.

At a temperature T below the critical point where the system develops a spontaneous magnetization, we must define the correlation fuction as

$$\Gamma(n) = \langle s(n)s(0)\rangle - \langle s(0)\rangle^2 \qquad (2.17)$$

in order that it fall to zero as $|n|$ grows large. With this definition $\Gamma(n)$ measures real correlations between individual spins and it falls off exponentially in $|n|$ for all $T < T_c$.

Finally, there is the specific heat,

$$C = -T\frac{\partial^2 F}{\partial T^2} \qquad (2.18)$$

which is also expected to be singular at the critical point. In fact, if the singular piece of the specific heat is power-behaved, we introduce the critical index α and write

$$C \sim (T - T_c)^\alpha \qquad (2.19)$$

It is amusing that $\alpha = 0$ in the two-dimensional Ising model and the actual critical singularity in the specific heat is a logarithm.

The final critical index we will discuss is another one specific to $T = T_c$ and is called δ. It measures the response of the system at the critical point to an infinitesimal external magnetic field,

$$M \sim B^{1/\delta}, \qquad T = T_c \qquad (2.20)$$

In the two-dimensional Ising model $\delta = 15$, so the response of the magnetization to the external field is very abrupt.

There is much more to be said about basic critical behavior of classical spin models, and we will introduce more ideas as we turn to applications below. However, the lesson we should take from this discussion is that, at the critical point, the correlation length diverges and this causes large fluctuations and singular behavior in various thermodynamic functions. This occurs even though the basic coupling in the system is short-ranged.

2.2 Correlation-length scaling and universality classes

One of the most striking features of the experimental study of physical systems at and/or near their critical points is that apparently different physical systems have identical critical singularities [2]. Various ferromagnets have the same critical indices as various liquid crystals, etc. Systematic experiments concerning a wide range of materials have led to the idea that physical systems can be arranged into "universality classes" and that within each class each substance has a common critical behavior. These classes are sensitive to just a few features of each physical system. In fact, the classes are labeled by the following parameters.

1. The dimension of the physical system.

2. The symmetry group of the local variables in the system's energy.

3. The symmetry of the coupling between the local variables.

4. The range of couplings in the energy.

For example, the two- and three-dimensional Ising models have different critical behaviors. In addition, the Z_2 Ising model has different critical indices than the three-state Pott model. Models with sufficiently long-range coupling in their energy expressions belong to the mean-field universality class whereas short-range-coupled models made from the same degrees of freedom formulated on identical lattices do not. One of the triumphs of K. G. Wilson's renormalization-group approach to critical phenomena was that it provided a natural framework for describing different universality classes through the notion of fixed points of renormalization-group transformations and scaling dimensions of operators [2].

At the descriptive, phenomenological level, one traces the singular behavior of diverse physical systems down to the divergent characters of their correlation lengths. The "correlation-length-scaling hypothesis" claims that the divergence of this single length is responsible for the singular behavior of each of the system's thermodynamic functions. This is a crucial hypothesis because it pinpoints the source of the phase transition in a single, singular quantity. Insofar as scaling laws and singular behavior of thermodynamic functions are concerned, the hypothesis states that only the single correlation length matters. In particular, the lattice spacing of the model, although it has the same dimensions as ξ, drops out of the picture. The long-range correlations between the microscopic degrees of freedom are the essential ingredients in understanding how the system can engage in a continuous phase transition.

The droplet picture of continuous phase transitions captures the essence of this phenomenon. It is a visualization of a scale-invariant field theory. As the critical temperature is approached, $T \rightarrow T_c$, the correlation length $\xi(T)$ becomes very large and long-range correlations begin to appear in the configurations. The droplet picture states that regions of "all up" and regions of "all down" spins, in the case of the Z_2 Ising model, begin dominating the configurations. The sizes of such "droplets" range from zero to $\xi(T)$ itself. It is important to appreciate that no particular size scale dominates: there are considerable numbers of droplets

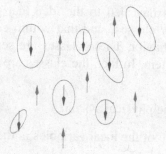

Figure 2.2. Droplets on all length scales at a second-order phase transition.

on all possible length scales, as shown in Fig. 2.2. This is the idea of "scale invariance" in the context of field theory or statistical mechanics. It is the physical phenomenon that underlies "critical opalescence," the fact that a liquid–gas system near a second-order phase transition appears "milky" when it is irradiated with ordinary light because droplets scatter the light on all possible length scales. Percolation phase transitions also constitute a precise realization of these ideas and show how singular thermodynamic functions with power-law scaling behaviors result. In the context of field theory, one is developing a physical picture that underlies conformal field theory with anomalous dimensions.

The reader is referred to the huge statistical mechanics literature for model building and examples of critical phenomena based on these ideas [3].

The scaling relations that follow from the correlation-length-scaling hypothesis are very important and very general. The idea is that, since there is but one relevant length in the problem, $\xi(T)$, which controls the long-range correlations and the singular behavior of thermodynamic functions, various of the singularities are related by dimensional-analysis arguments. This also implies that there are precise relations among the various critical indices characterizing the phase transition. These are called "hyperscaling relations" and read

$$\beta_{\mathrm{mag}} = \tfrac{1}{2}\nu(d - 2 + \eta) \tag{2.21}$$

$$\gamma = \nu(2 - \eta) \tag{2.22}$$

$$\alpha = 2 - \nu d \tag{2.23}$$

$$\delta = (d + 2 - \eta)/(d - 2 + \eta) \tag{2.24}$$

One can check that these relations are true for the two-dimensional Ising model we have reviewed, and they have been checked for a wealth of other two- and three-dimensional statistical models and materials. It is interesting that mean-field-theory critical exponents do not satisfy these scaling results except in four dimensions. Since mean-field theory replaces fluctuating quantites with their supposed averages, this failure of an otherwise extremely important and useful model should come as no surprise.

2.3 Properties of the Ising model

The two-dimensional Ising model played a most significant role in the development of the modern theory of critical phenomena. Its exact solution established the existence of a critical point with power-law singularities that were not given by Landau mean-field predictions. This result begged for understanding, both theoretical and intuitive, and it inspired several generations of physicists to make major contributions to the field. Many of these contributions, such as high-temperature and low-temperature expansions, duality, correlation-length scaling, finite-size scaling, the operator-product expansion, the renormalization group, and the ϵ expansion, had applications much broader than the specific Ising model and led to fundamental advances.

We wrote down the action for the model above. To study its thermodynamics, we consider its partition function,

$$Z(K) = \sum_{\sigma_N = \pm 1} \cdots \sum_{\sigma_1 = \pm 1} e^{K \sum_{\langle ij \rangle} \sigma_i \sigma_j} \tag{2.25}$$

where $K = J/(kT)$, the notation $\langle ij \rangle$ indicates a sum over nearest-neighbor sites i and j, and N is the total number of lattice sites in the square array.

Instead of doing sophisticated constructions and calculations, let's investigate the possibility that this system displays spontaneous magnetization at low temperatures, $\langle \sigma_i \rangle \neq 0$, by "following our noses" and expanding $Z(K)$ beginning with this "guess" for the ground state. In other words, we begin with all spins aligned, as would be the case at vanishing temperature, and expand $Z(K)$ in the number of flipped spins.

Choose the $T = 0$ ground state to be all spins "up." When one spin is flipped, four bonds are broken, since each spin has four nearest neighbors on a symmetric, two-dimensional lattice. If two spins are flipped, then six bonds are broken if those spins are nearest neighbors and eight bonds are broken otherwise. The first several terms in this "low-temperature expansion" are visualized in the Fig. 2.3 and the corresponding expression for $Z(K)$ is

$$Z(K)e^{-2NK} = 1 + Ne^{-8K} + 2Ne^{-12K} + \tfrac{1}{2}N(N-5)e^{-16K} + \cdots \tag{2.26}$$

where the factors of N record the number of ways the graph can occur on

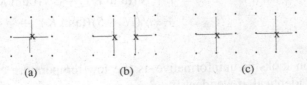

(a) (b) (c)

Figure 2.3. The first few terms in the low-temperature expansion of the two-dimensional Ising model. The crosses label flipped spins.

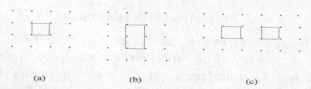

(a)   (b)   (c)

Figure 2.4. The first few terms in the high-temperature expansion of the two-dimensional Ising model.

the lattice. This expression is not very illuminating, so let's turn to the high-temperature properties of the model. In that case K is very small and we can write each bond term in the partition function in the convenient form

$$\exp(K\sigma) = \cosh K + \sigma \sinh K \qquad (2.27)$$

where σ is a generic Z_2 variable ($\sigma = \pm 1$). Applying this relation to the original expression for the partition function gives

$$Z(K) = \sum_{\sigma_N=\pm 1} \cdots \sum_{\sigma_1=\pm 1} \prod_{\langle ij \rangle} (\cosh K + \sigma_i \sigma_j \sinh K)$$

$$= (\cosh K)^{2N} \sum_{\sigma_N=\pm 1} \cdots \sum_{\sigma_1=\pm 1} \prod_{\langle ij \rangle} (1 + \sigma_i \sigma_j \tanh K) \qquad (2.28)$$

This form of the partition function is now a suitable basis from which to develop an expansion for $Z(K)(\cosh K)^{-2N}$ in the small variable $\tanh K$. To identify nonzero contributions to this expansion, we use the trivial identities

$$\sum_{\sigma=\pm 1} \sigma = 0, \qquad \sum_{\sigma=\pm 1} 1 = 2 \qquad (2.29)$$

Therefore, as we expand $\prod(1 + \sigma_i \sigma_j \tanh K)$, only those terms without a free spin variable will contribute. The product $\prod_{\langle ij \rangle}$ indicates that we place "bonds," $\sigma_i \sigma_j \tanh K$, on the links of the lattice and only if the bonds form closed paths will the term contribute. In addition, no link can be covered more than once. The several terms in the expansion are shown in Fig. 2.4 and the expression for $Z(K)$ is

$$Z(K)(\cosh K)^{-2N}2^{-N} = 1 + N(\tanh K)^4 + 2N(\tanh K)^6$$
$$+ \tfrac{1}{2}N(N-5)(\tanh K)^8 + \cdots \qquad (2.30)$$

This expression looks as uninformative as the low-temperature expansion. In fact, they look identical, if we identify

$$\tanh K = \exp(-2K^*) \qquad (2.31)$$

and compare the two series term by term,

$$\frac{Z(K^*)}{(e^{2K^*})^N} = \frac{Z(K)}{2^N(\cosh^2 K)^N} \qquad (2.32)$$

The last two expressions can be written in more symmetric fashion, as a short exercise in hyperbolic functions shows:

$$\sinh(2K)\sinh(2K^*) = 1 \qquad (2.33)$$

and

$$\frac{Z(K^*)}{\sinh^{N/2}(2K^*)} = \frac{Z(K)}{\sinh^{N/2}(2K)} \qquad (2.34)$$

In other words, we have discovered that the low-temperature properties of the model map onto its high-temperature properties: if we understand the magnetized phase, we understand the disordered phase equally well. We have discovered a "duality map" relating the two phases of the model. One says that the Ising model is self-dual and, if we assume that there is but one phase transition in the model, then it must occur at the self-dual point, that temperature at which $K = K^*$, or

$$\sinh^2(2K_c) = 1 \qquad \text{or} \qquad e^{2K_c} = \sqrt{2} + 1 \qquad (2.35)$$

This argument is correct and we have found the critical point *exactly*. It is interesting that this self-dual property of the model was known long before it was solved. In fact, the duality map can be derived elegantly without the use of the expansions used here, and more detail about the mapping between correlation functions in the low- and high-temperature phases can be extracted.

Duality maps are extremely important in lattice-gauge theory and field theory in general, and we will see more of them later.

2.4 The Kosterlitz–Thouless model

Up to this point we have used the Ising model as our means to illustrate notions of statistical mechanics that will play an important role in lattice-gauge theory. However, continuous variables will become more important, so we turn to the simplest case of them now. We replace the Z_2 Ising variables on the two-dimensional square lattice with two-dimensional, planar vectors of unit length,

$$s_1(n) = \cos\theta(n), \qquad s_2(n) = \sin\theta(n) \qquad (2.36)$$

where $\theta(n)$ is an angular variable that resides on each site n. The energy of the entire system becomes

$$S = -J\sum_{n,\mu} \vec{s}(n) \cdot \vec{s}(n+\mu) = -J\sum_{n,\mu}\cos(\theta(n) - \theta(n+\mu)) \qquad (2.37)$$

where μ labels the two directions between nearest-neighbor sites. As was the case for the Ising model, when J is chosen positive the system is ferromagnetic. We might expect this model to have an order–disorder transition at some sufficiently low T, but, although there is a phase transition in this model, it is particularly subtle and relevant to various lattice-gauge theories in higher dimensions. In fact, the planar model experiences a phase transition without a local order parameter, as we shall see shortly.

This model has a global continuous rotation symmetry: all the spins can be rotated through a common angle α, and the energy remains unchanged. The analogous global symmetry was broken at low temperatures in the Ising model. Will there be a similar ordering of spins in this model?

To gain an idea about the possible phases of the planar model, consider it first at high temperatures T. We will discuss low temperatures in due course. Let's use the language of two-dimensional field theory, as a foretaste of later discussions of lattice-gauge theory, and call the expression S in the previous equation the "action" of the Euclidean field theory. We write it in a slightly more convenient fashion,

$$S = -J \sum_{\langle n,m \rangle} \cos(\theta(n) - \theta(m)) = -\frac{1}{2} J \sum_{\langle n,m \rangle} \left(e^{i(\theta(n) - \theta(m))} + \text{h.c.} \right) \quad (2.38)$$

where $\langle n, m \rangle$ is a sum over nearest-neighbor sites of the two-dimensional lattice.

At high temperatures we expect that the system will be disordered and the spin–spin correlation function will be short-ranged. This is easy to verify. Consider the correlation function

$$\langle e^{i\theta(0)} e^{-i\theta(n)} \rangle = \int \prod_m d\theta(m) \, e^{i(\theta(0) - \theta(n))}$$

$$\times \exp\left(-\frac{J}{kT} \sum_{\langle n,m \rangle} \cos(\theta(n) - \theta(m)) \right) \bigg/ Z \quad (2.39)$$

The exponential of the action S can be expanded in powers of $J/(kT)$ for sufficiently high temperatures, which will allow us to obtain a useful upper bound on the correlation function. We use the elementary integrals

$$\int_0^{2\pi} d\theta(m) = 2\pi, \qquad \int_0^{2\pi} d\theta(m) \, e^{i\theta(m)} = 0 \quad (2.40)$$

So, the leading contribution to the correlation function will occur in the expansion of the exponential when a string of $e^{-i(\theta(k) - \theta(l))}$ factors extends from site 0 to site n, because otherwise there will be an uncancelled phase $e^{i\theta(j)}$ on a site, which, when it is integrated over, will average to zero. If the sites 0 and n are on an axis, the first nonzero contribution appears at order $[(J/(kT)]^{|n|}$, and the correlation function behaves as

$$\langle e^{i(\theta(0)-\theta(n))} \rangle \sim [J/(kT)]^{|n|} = \exp[-|n|\ln(kT/J)] \qquad (2.41)$$

which falls to zero exponentially as $|n|$ grows. In the language of scaling and the correlation length, we see that $\xi(T) \sim 1/\ln(kT/J)$.

Now we turn to the low-temperature behavior of the correlation function. In this limit $\theta(n)$ is expected to vary slowly over the lattice, so the argument of the cosine in the action should be very small. Then the cosine can be expanded and only the quadratic term need be accounted for,

$$S \approx \frac{1}{2} J \sum_{n,\mu} (\Delta_\mu \theta(n))^2 \qquad (2.42)$$

where Δ_μ denotes a discrete difference, $\Delta_\mu \theta(n) \equiv \theta(n+\mu) - \theta(n)$. We note further that the action in this approximation resembles that of a free, massless scalar field. The only subtlety is that θ is an angular variable, ranging from 0 to 2π. However, if it were pinned to zero at some site on the lattice, we would expect its values everywhere on a finite lattice to be near zero, rendering its periodic nature irrelevant. Certainly we must be careful in the thermodynamic infinite-volume limit, however.

If the action becomes a quadratic form and θ remains small over the entire lattice, then the model is integrable. The correlation function becomes

$$\langle e^{i(\theta(0)-\theta(n))} \rangle \sim \int \prod_m d\theta(m) \, e^{i(\theta(0)-\theta(n))} \exp\left(-\frac{J}{2kT} \sum_{n,\mu}(\Delta_\mu \theta(n))^2\right) \Big/ Z \qquad (2.43)$$

which can be evaluated explicitly because it involves only Gaussian integrals. If $\Delta(n)$ denotes the massless scalar propagator on the finite system, then

$$\langle e^{i(\theta(0)-\theta(n))} \rangle \sim \exp\left(\frac{kT}{J}\Delta(n)\right) \qquad (2.44)$$

where the propagator behaves as

$$\Delta(n) \underset{n \gg 1}{\sim} -\frac{1}{2\pi}\ln|n| \qquad (2.45)$$

on the finite $L \times L$, but large, $L \gg n \gg 1$, system. A physical interpretation follows when we substitute back into the expression for the correlation function,

$$\langle e^{i(\theta(0)-\theta(n))} \rangle \sim \left(\frac{1}{|n|}\right)^{kT/(2\pi J)} \qquad (2.46)$$

We see that the correlation fuction always falls to zero, even for infinitesimal temperature T. This means that the two-dimensional planar model never magnetizes. Apparently, the fluctuations in two dimensions are considerable enough at any nonzero T to average any particular spin to zero. This follows formally from this expression because the expectation value of the product of two spins should approach the product of their expectation values as $|n|$ grows large,

$$\left\langle e^{i(\theta(0)-\theta(n))}\right\rangle \sim \left\langle e^{i\theta(0)}\right\rangle \left\langle e^{-i\theta(n)}\right\rangle \tag{2.47}$$

Since the correlation function falls to zero in this limit, we must have

$$\left\langle e^{i\theta(0)}\right\rangle = 0 \tag{2.48}$$

precisely. This result, the absence of spontaneous magnetization at low temperatures, is in accord with general results in this field that state that continuous global symmetries cannot break down in two-dimensional systems with just local, short-range interactions [4]. In addition, the result indicates that the model is critical for any low temperature because the spin–spin correlation function is power-behaved, with a critical index $\eta = kT/(2\pi J)$ that varies continuously for low T. In other words, in contrast to the cases discussed above, this model has a line of critical points rather than an isolated T_c. We learn that, for any T that is sufficiently low, the correlation length ξ is divergent.

If we put our low- and high-T analyses together, we conclude that the model must have at least one phase transition at an intermediate temperature between a low-temperature region where the spin–spin correlation function is a power and a high-temperature region where the spin–spin correlation function is an exponential. The correlation length $\xi(T)$ is infinite in the low-temperature phase and finite otherwise. This behavior is distinct from the Ising model, for which the correlation length is finite both in the magnetized phase and in the unmagnetized phase and diverges only at T_c. The planar model has a phase transition, but no global order parameter because $\langle s(0)\rangle = 0$ in both phases.

What is the nature of the planar model's phase transition? An intriguing and very influential answer was suggested by Kosterlitz and Thouless [5], who pointed out that the periodic nature of the angular variables $\theta(n)$ allowed for singular spin configurations – vortices – which could appear in the system at sufficiently high temperatures and disorder it, changing the correlation function from a power to an exponential. They began their considerations with the Gaussian, low-temperature approximation to the action S, but they supplemented it with the periodicity of the angular variables $\theta(n)$,

$$\nabla^2\theta = 0, \quad 0 < \theta < 2\pi \tag{2.49}$$

The solutions to this periodic Gaussian system can be labeled by their winding number,

$$\oint \vec{\nabla}\theta \cdot d\vec{s} = 2\pi q, \qquad q = 0, \pm 1, \pm 2, \ldots \tag{2.50}$$

In the spin-wave analysis we presented above at low temperature, we ignored the periodicity of the $\theta(n)$ variables because they should vary very slowly across

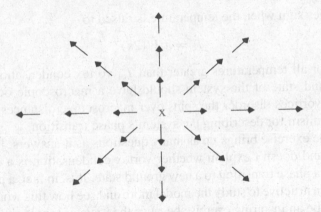

Figure 2.5. A vortex in the two-dimensional planar model.

the lattice. We were effectively dealing with just the $q = 0$ topological sector of the theory. A spin configuration with $q = 1$ is shown in the Fig. 2.5. It can be depicted as a "porcupine" or a "vortex." Clearly the configuration is singular, has a "center" about which the spins vary rapidly on the scale of the lattice spacing, and disorders the system over arbitrarily large distances.

Kosterlitz and Thouless presented a famous plausibility argument suggesting that vortex condensation drives the phase transition between the low- and high-temperature phases at a nonzero T_c. The argument begins by evaluating the action for the vortex in Fig. 2.5, using the Gaussian approximation

$$S \approx \pi J \ln(L/a) \qquad (2.51)$$

where L is the linear dimension of the lattice, and a is the short-distance cutoff, the lattice spacing. Since S diverges whenever $L/a \to \infty$, one might think that vortices are irrelevant in the continuum system. However, to estimate their importance to the thermodynamics, we must consider their contribution to the free energy, $F = S - TE$, where E is the entropy. We have used field-theory notation and have written S for "action" where one would have written "energy" in statistical mechanics. We see a competition between energy and entropy in the free-energy expression that decides whether the vortex is physically significant. To make this argument quantitative, we must estimate the entropy of a free vortex. This is a counting problem because the entropy is the logarithm of the multiplicity of the vortex configuration. Since the vortex is specified by its location, as is clear in Fig. 2.5, its multiplicity is just the number of sites of the lattice, entropy $= k \ln(L/a)^2$.

Therefore,

$$F = (\pi J - 2kT) \ln(L/a) \qquad (2.52)$$

which changes sign when the temperature is raised to

$$T_c = \pi J/(2k) \tag{2.53}$$

Therefore, for all temperatures greater than T_c, vortex condensation is favored and the ground state of the system should have a macroscopic occupation of them. Since vortices disorder the spins over macroscopic distances as well, we have a mechanism for describing the system's phase transition.

This simple exercise brings up as many questions as it answers. It ignores all interactions and doesn't explain whether vortex condensation is a catastrophic instability or a phase transition to a new ground state. It is, in fact, a phase transition and it is instructive to study the model more and see how this works out. This will prove to be an inspiring exercise because the issue of topological causes for disordering a spin system in two dimensions will prove to be closely related to discovering the mechanisms of confinement and chiral-symmetry breaking in four dimensional lattice gauge theory.

2.5 Coulomb gas, duality maps, and the phases of the planar model

Now let's take the ideas of the last section and make them quantitative. The motivation for this comes from lattice-gauge theory: many of the concepts that we can illustrate straightforwardly in two-dimensional spin models, like the planar model, generalize to four-dimensional lattice-gauge theory. The condensation of topological configurations in two-dimensional spin systems will find analogs in the condensation of various topological excitations in four-dimensional gauge theories. The analogs of disordered spin configurations will become quark confinement and chiral-symmetry breaking.

To carry through this program, we introduce a slightly simpler, for the purposes of analysis, formulation of the two-dimensional planar model, namely the periodic Gaussian model. To motivate the use of this model, consider a single link in the expression for the partition function of the planar model,

$$e^{-\beta[1-\cos(\theta(n)-\theta(m))]} = \sum_{l=-\infty}^{\infty} e^{il(\theta(n)-\theta(m))} I_l(\beta) \tag{2.54}$$

where we have written out the Fourier series of the periodic-link contribution to Z and $I_l(\beta)$ is the Bessel function of imaginary argument. Now specialize to low temperatures (large β) at which $I_l(\beta)$ is well approximated by a Gaussian,

$$e^{-\beta[1-\cos(\theta(n)-\theta(m))]} \approx \sum_{l=-\infty}^{\infty} e^{il(\theta(n)-\theta(m))} e^{-l^2/(2\beta)} \tag{2.55}$$

The right-hand side of this expression has the same "essential" ingredients as the original planar model: it has local, ferromagnetic couplings between spins that preserve the periodic nature of the action. It has the great advantage that it is a simple Gaussian and can be analyzed elegantly. The model was introduced by Villain and will be analyzed now for all temperatures β. Since it has the same essential features as the original planar model, we expect its critical behavior to be the same. However, because the nasty nonlinearities of the cosine interaction have been simplified, we will be able to rewrite the partition function into two factors: one describing the spin waves and another describing the vortices. The partition function for the vortices will be given by the two-dimensional Coulomb gas and will quantify the descriptive analysis of Kosterlitz and Thouless reviewed above.

The partition function of the periodic Gaussian model now reads, introducing discrete link variables $n_\mu(r)$,

$$Z = \int \prod_{r'} d\theta(r') \prod_{r,\mu} \sum_{n_\mu(r)=-\infty}^{\infty} e^{in_\mu \Delta_\mu \theta} e^{-n_\mu^2(r)/(2\beta)} \tag{2.56}$$

where Δ_μ is the discrete-difference operator, $\Delta_\mu f(n) = f(n+\mu) - f(n)$. We can do the integrals over $\theta(r')$ because θ enters only a phase factor. Each such integral generates a constraint, $\Delta_\mu n_\mu(r) = 0$, so the partition function becomes

$$Z = \prod_{r,\mu} \sum_{n_\mu(r)=-\infty}^{\infty} \delta_{\Delta_\mu n_\mu(r),0} e^{-n_\mu^2(r)/(2\beta)} \tag{2.57}$$

It is easy to solve the constraint and obtain a simpler form for Z. The constraint states in discrete form that $n_\mu(r)$ is a divergence-free integer-valued vector field. This implies that it is the discrete curl of an integer-valued scalar field,

$$n_\mu(r) = \epsilon_{\mu\nu} \Delta_\nu n(r) \tag{2.58}$$

Using the result $n_\mu^2(r) = (\Delta_\mu n(r))^2$, the partition function becomes

$$Z = \sum_{n_\mu(r)=-\infty}^{\infty} \exp\left(-\frac{1}{2\beta} \sum_{r,\mu} (\Delta_\mu n(r))^2\right) \tag{2.59}$$

It is interesting to step back for a moment and see what all this trickery has accomplished. We started with the planar model at temperature T and mapped it onto a new model, our last expression for Z, at temperature T^{-1}. This is an example of a duality transformation. The last expression for Z is called the "interface-roughening" model, and is of considerable fame in the statistical mechanics of crystal growth. Think of a crystal of steps, each of height $n(r)$, growing on a table top. At small values of β, the energetics of the system favors a flat crystal, all the $n(r)$ at a common value, with the finite temperature fluctuations inducing lumpy waves on it. However, as β decreases a point at which the localization of the surface is lost is reached and it becomes "rough," with $\langle n(r) \rangle$

diverging logarithmically as the linear dimensions L of the system are taken large, and the step–step correlation function behaving as $\langle n(r)n(0)\rangle \sim \ln|r|$. We will see shortly how to establish these results. One of the useful points about a duality mapping is that the high-temperature region of one model is mapped onto the low-temperature region of another. The relevant degrees of freedom are also mapped onto one another, so one can learn which variables expose the physics in the clearest way. In the case developed here, we also learn that two apparently different models, with totally different physical settings, one a conventional spin model and the other a crystal-growth model, have the same critical behavior.

We are not done with our maps in this case. It would be convenient to obtain a representation of the planar model in which the integer-valued field $n(r)$ were replaced by an ordinary scalar field $\phi(r)$. The trick to do this is the Poisson summation formula,

$$\sum_{n=-\infty}^{\infty} g(n) = \sum_{m=-\infty}^{\infty} \int_{-\infty}^{\infty} d\phi \, g(\phi) e^{2\pi i m \phi} \tag{2.60}$$

where g is an arbitrary but sensible function. The identity is rather trivial: to establish it, just do the sum over m before the integral over ϕ. Applying it to the case at hand,

$$Z = \int_{-\infty}^{\infty} \prod_{r} d\phi(r) \sum_{m(r)=-\infty}^{\infty} \exp\left(-\frac{1}{2\beta}\sum_{r,\mu}(\Delta_\mu \phi(r))^2 + 2\pi i \sum_r m(r)\phi(r)\right) \tag{2.61}$$

The symbols in this expression have simple interpretations in terms of the original model: the field $\phi(r)$ is the spin wave and the $m(r)$ are the vortices. The expression for Z states that the vortices act as sources of the spin waves, which are ordinary massless fields. The identification of the $m(r)$ with vortex variables becomes clear if we integrate out the spin waves. The integration over ϕ in the expression for Z is a simple Gaussian, so we obtain

$$Z = Z_{SW} \sum_{m(r)=-\infty}^{\infty} \exp\left(-2\pi^2\beta \sum_{r,r'} m(r)G(r-r')m(r')\right) \tag{2.62}$$

where Z_{SW} is the spin-wave contribution produced by the Gaussian integrations and $G(r-r')$ is the lattice propagator for a massless field.

This expression for the partition function is surprisingly simple, physical, and important. It shows that the spin waves are independent of the vortices in the periodic Gaussian model because Z has factored into two terms, a spin-wave piece Z_{SW} that has no phase-transition physics in it, and the vortex piece displayed above. The vortex piece shows that vortices interact through the "potential"

$G(r - r')$. The vortices are pointlike excitations with possible charges m ranging from $-\infty$ to $+\infty$. To extract the physics from this expression we must understand $G(r - r')$ better.

The propagator $G(r - r')$ appeared in the calculation when we did the Gaussian integral and had to invert the discrete, lattice Laplacian on a two-dimensional square array,

$$\Delta^2 G(r) = \delta_{r,0} \tag{2.63}$$

where $\Delta^2 = \Delta_x^2 + \Delta_y^2$ and Δ_x^2 is the simplest discrete form of the second derivative,

$$\Delta_x^2 G(r) = G(r + x) + G(r - x) - 2G(r) \tag{2.64}$$

We solve for the propagator G by passing to momentum space, as in the corresponding calculation in continuum field theory,

$$G(r) = \int_{-\pi}^{\pi} \frac{dk_x}{2\pi} \int_{-\pi}^{\pi} \frac{dk_y}{2\pi} \frac{e^{i\vec{k} \cdot \vec{r}}}{4 - 2\cos k_x - 2\cos k_y} \tag{2.65}$$

We will need some of the properties of $G(r)$. First, it is easy to check that, at large $|r| \gg 1$, $G(r)$ is well approximated by the continuum massless propagator,

$$G(r) \approx -\frac{1}{2\pi} \ln(r/a) - \frac{1}{4}, \qquad |r| \gg 1 \tag{2.66}$$

The behavior of $G(r)$ for small r can also be determined from the integral above,

$$G(0) \approx \frac{1}{2\pi} \ln(L/a) \tag{2.67}$$

where L is the linear dimension of the lattice. $G(0)$ is important because it contributes to the partition function through the $r = r'$ term in the sum, so it represents the "self-interaction mass" of each vortex. Its logarithmic dependence on the infrared cutoff L and the ultraviolet cutoff a was anticipated earlier in our review of the Kosterlitz–Thouless ideas. We also see that the $r \neq r'$ terms in the double sum of Z indicate that the vortices interact through spin-wave exchange, which generates the two-dimensional Coulomb potential $G(r - r')$. Since the possible charges m are positive or negative, we have a gas with opposite charges attracting and same-sign charges repelling: a Coulomb gas in which the charges are quantized and vary from $-\infty$ to $+\infty$.

It is instructive to write $G(r)$ as two terms, one of which is infrared-finite and the other of which is not,

$$G(r) = [G(r) - G(0)] + G(0) \equiv G'(r) + G(0) \tag{2.68}$$

The sum over the vortices in the partition function can then be simplified,

$$Z = Z_{SW} \sum_{m(r)=-\infty}^{\infty} \exp\left[-2\pi^2 \beta G(0) \left(\sum_r m(r) \right)^2 \right]$$

$$\times \exp\left(-2\pi^2 \beta \sum_{r,r'} m(r) G'(r - r') m(r') \right) \tag{2.69}$$

The first exponential eliminates all vortex configurations that are not neutral, $\sum_r m(r) \neq 0$, in the thermodynamic limit $L \to \infty$. Only configurations for which $\sum_r m(r) = 0$ contribute to Z. Therefore, we can write

$$Z = Z_{SW} \sum_{m(r)}' \exp\left(-2\pi^2 \beta \sum_{r,r'} m(r) G'(r - r') m(r') \right) \tag{2.70}$$

where the prime on the sum means "neutral configurations of vortices only." It is also useful to use the large-r expression for the propagator G, since it is actually quite accurate even for r near the ultraviolet cutoff,

$$Z = Z_{SW} \sum_{m(r)}' \exp\left(\frac{\pi^2 \beta}{2} \sum_{r,r'}' m(r) m(r') + \pi\beta \sum_{r,r'}' m(r) \ln(|r - r'|/a) m(r') \right)$$

$$\tag{2.71}$$

where the prime on the double sum over sites means that the $r = r'$ term is omitted. Using the neutrality condition, we can simplify the first double sum in the expression for Z,

$$0 = \sum_{r,r'} m(r) m(r') = \sum_{r,r'}' m(r) m(r') + \sum_r m(r)^2 \tag{2.72}$$

Now,

$$Z = Z_{SW} \sum_{m(r)}' \exp\left(-\frac{\pi^2 \beta}{2} \sum_r m^2(r) + \pi\beta \sum_{r,r'}' m(r) \ln(|r - r'|/a) m(r') \right)$$

$$\tag{2.73}$$

The first term in the exponential records the chemical potential of each vortex and the second gives the logarithmic interaction between different ones.

This Coulomb-gas representation of the planar model gives a precise underpinning to the ideas of Kosterlitz and Thouless about the model's phases and phase transition. The only possible configurations of vortices must be neutral. If β is very large, then the chemical-potential term suppresses the vortex contribution to the physics, leaving just the spin waves. In fact, the strong logarithmic

potential between vortices suggests that any vortex–anti-vortex pairs that populate the ground state are tightly bound together into neutral molecules. However,
as $\beta \equiv J/(kT)$ is reduced and becomes of the order of unity, one would expect
that the probability of finding vorticies would not be suppressed and they should
contribute to the properties of the ground state. Since they experience long-range
interactions, however, they cannot be treated as free excitations. In effect, as the
temperature T grows, the neutral molecules of vortex–anti-vortex pairs unbind
and the vortices form a plasma that screens external and internal charges and
dynamically generates a finite screening length. A renormalization-group analysis is needed in order to derive this fact, which also uncovers the character of
the phase transition: it is an infinite-order transition with the correlation length
growing as T approaches T_c from above as

$$\xi(T) \sim \exp[b\sqrt{1/(T/T_c - 1)}] \tag{2.74}$$

which is the famous essential singularity of the subtle planar-model transition
[6].

This result has been obtained by various methods in addition to using the
traditional langauge of the Coulomb gas of screening lengths and plasmas. A
field-theoretical approach follows from the expression for the partition function
in terms of ϕ and m,

$$Z = \int_{-\infty}^{\infty} \prod_r d\phi(r) \sum_{m(r)=-\infty}^{\infty} \exp\left(-\frac{1}{2\beta}\sum_{r,\mu}(\Delta_\mu\phi(r))^2 + 2\pi i\sum_r m(r)\phi(r)\right)$$

$$\tag{2.75}$$

We have already realized that the coupling of the vortices to the spin waves will
generate a mass, effectively a chemical potential, for the vortices. To accomodate
this effect we generalize the partition function to one with an additional variable
y,

$$Z(\beta, y) = \int_{-\infty}^{\infty} \prod_r d\phi(r)$$

$$\times \sum_{m(r)=-\infty}^{\infty} \exp\left(-\frac{1}{2\beta}\sum_{r,\mu}(\Delta_\mu\phi(r))^2 + \ln y\sum_r m^2(r)\right.$$

$$\left. + 2\pi i\sum_r m(r)\phi(r)\right) \tag{2.76}$$

If y is very small so that the vortices are heavy, then only the terms in the sum
with $m(r) = 0, \pm 1$ should be numerically significant, and the partition function

can be simplified,

$$\sum_{m(r)=0,\pm 1} e^{\ln y\, m^2(r) + 2\pi i m(r)\phi(r)} \approx 1 + e^{\ln y}(e^{2\pi i \phi(r)} + \text{h.c.})$$

$$\approx 1 + 2y\cos(2\pi\phi(r))$$

$$\approx e^{2y\cos(2\pi\phi(r))} \tag{2.77}$$

The partition function becomes

$$Z(\beta, y) \approx \int_{-\infty}^{\infty} \prod_r d\phi(r) \exp\left(-\frac{1}{2\beta}\sum_{r,\mu}(\Delta_\mu\phi(r))^2 + 2y\sum_r \cos(2\pi\phi(r))\right) \tag{2.78}$$

Finally, to restore conventional field-theoretical notation, we rescale $\phi \to \phi/\sqrt{\beta}$, so

$$Z(\beta, y) \approx \int_{-\infty}^{\infty} \prod_r d\phi(r)$$

$$\times \exp\left(-\frac{1}{2}\sum_{r,\mu}(\Delta_\mu\phi(r))^2 + 2y\sum_r \cos(2\pi\sqrt{\beta}\phi(r))\right) \tag{2.79}$$

which we recognize as the quantum sine-Gordon equation. The periodicity of the original planar model shows up in this expression in the cos interaction term. A renormalization-group analysis of this model in the β–y plane produces the phase diagram and scaling laws of the model in an elegant fashion, with all the universal features the same as those obtained from the Coulomb-gas representation.

We have dwelt on this model in considerable detail because it illustrates many physical and technical issues that arise in four-dimensional lattice-gauge theory. These include topological excitations as the source for disorder and a phase transition, duality maps to relate the high- and low-temperature features of the model in terms of weakly fluctuating variables, and correlation-length scaling.

2.6 Asymptotic freedom in two-dimensional spin systems

Another class of two-dimensional spin systems that will teach us what to expect of four-dimensional gauge theories consists of the nonlinear sigma models. We consider a square, two-dimensional lattice again with a unit vector spin at each site, coupled to its nearest neighbors through an inner product,

$$S = -\frac{2}{g}\sum_{\langle n,m\rangle} \vec{s}(n)\cdot\vec{s}(m), \qquad |\vec{s}| = 1 \tag{2.80}$$

The simplest case we will concentrate on is the O(3) model, in which we can visualize each spin as having its end on a site of the square lattice and its head on a unit sphere. We easily check that this action is the simplest lattice version of the continuum O(3) "sigma" model,

$$Z = \int \prod_n d\vec{s}(n) \exp\left(-\frac{1}{2g^2} \int d^2r \, \partial_\mu \vec{s} \cdot \partial_\mu \vec{s}\right) \qquad (2.81)$$

We will ultimately be interested in understanding this model's phase diagram as a function of the coupling g. Is there a region where there are long-range correlations, and another where there are only short-range correlations? What are the disordering mechanisms that control the behavior of the spin–spin correlation function at long distances?

We take a renormalization-group view to answer these questions. We are interested in extracting the long-distance physics from this model. Equivalently, we want the low-energy, low-momentum behavior. The theory has fluctuations and excitations on all length scales from the ultraviolet to the infrared cutoff. The long-distance/low-energy character of the model is not apparent from the short-distance, fundamental formulation of the model given by the partition function. The partition function gives us no clue about the theory's possible phases, the possibility of dynamical symmetry breaking, and generation of a correlation length. We need to integrate out the high-energy fluctuations of the theory, to find the relevant low-energy degrees of freedom and their interactions. For example, in the case of quantum chromodynamics, we would like to start with the theory of quarks and gluons and end up with an effective Lagrangian whose degrees of freedom are the known mesons and baryons, coupled together by the residual nuclear force. Physics is certainly far from this goal, but this program, the renormalization group, has had its share of successes in simpler models [2].

In the context of the O(3) nonlinear sigma model in two dimensions, let's integrate out high-frequency modes and find the effective degrees of freedom and interactions for which perturbation theory is adequate. This is an extremely modest calculation, on the scale of our discussion above, but we will find an interesting result, asymptotic freedom, along the way. We will see that the curvature of the sphere of the O(3) spins causes the interactions between them to be stronger at low energies than at high energies. This is the same trend as that found in QCD, which is responsible for its parton description at short distances and its confining features at long distances. QCD will be discussed below. Here we wish to see that the O(3) spin system in two dimensions is asymptotically free. This means, in our statistical-mechanics language, that the theory's only phase transition is at zero coupling, for which it becomes disordered. The theory is more strongly coupled at long distances than at short, as our perturbative asymptotic-freedom calculation will show, and it will dynamically generate a correlation length that will allow its spin–spin correlation function to be exponentially small at large distances. Actually, since our calculation will be perturbative, we won't be able to discuss very large scales in a controlled fashion, because the effective coupling will be too large there.

Since asymptotic freedom is a consequence of the geometry of this model, we will parametrize the spin variable $\vec{s}$ judiciously. $\vec{s}$ has low- and high-frequency fluctuations and we choose a parametrization in which two components, s_1 and s_2, are both slowly varying and large, $O(1)$, while the third component, s_3, is rapidly varying and small, $O(\sqrt{g})$. We will verify in the course of the calculation that such an arrangement is possible if we are content to consider only very small couplings g. Since each spin has unit length, we can use the parametrization

$$s_1 = \sqrt{1 - s_3^2} \sin\theta, \qquad s_2 = \sqrt{1 - s_3^2} \cos\theta \qquad (2.82)$$

Then the Lagrangian becomes

$$L = \frac{1}{2g}(\partial_\mu \vec{s})^2 = \frac{1}{2g}\left((1 - s_3^2)(\partial_\nu\theta)^2 + \frac{(\partial_\mu s_3)^2}{1 - s_3^2}\right) \qquad (2.83)$$

If we anticipate that, for small g, $s_3 \sim O(\sqrt{g})$, we can approximate L by

$$L = \frac{1}{2g}\left[(\partial_\nu s_3)^2 + (1 - s_3^2)(\partial_\nu\theta)^2 + s_3^2(\partial_\mu s_3)^2 + \cdots\right] \qquad (2.84)$$

Finally, it is convenient to rescale s_3 to eliminate the overall factor of g, $h(r) \equiv s_3(r)/\sqrt{g}$, so

$$L = \frac{1}{2}\left[(\partial_\nu h)^2 + \left(\frac{1}{g} - h^2\right)(\partial_\nu\theta)^2 + gh^2(\partial_\mu h)^2 + \cdots\right] \qquad (2.85)$$

This expression illustrates the good features of this parametrization of the sphere. As h fluctuates it affects the contribution of the θ field to the action. However, fluctuations in θ do not react back onto h.

Now consider the theory with a momentum-space cutoff, $|p| < \Lambda$. Introduce a high-momentum slice, $\Lambda' < |p| < \Lambda$, and a low-momentum slice, $0 < |p| < \Lambda'$, and organize the partition function accordingly,

$$Z = \int_{0<p<\Lambda} D\theta(p) \exp\left(-\frac{1}{2g}\int(\partial_\nu\theta)^2 \, d^2x\right)$$

$$\times \int_{\Lambda'<p<\Lambda} Dh(p) e^{-\frac{1}{2}\int(\partial_\nu h)^2 \, d^2x} e^{\frac{1}{2}\int h^2(\partial_\nu\theta)^2 \, d^2x - \frac{1}{2}\int gh^2(\partial_\nu h)^2 \, d^2x + \cdots} \qquad (2.86)$$

For small g the second term in the last exponential can be dropped. Then the integral over $h(p)$ is a Gaussian and can be computed exactly,

$$\int_{\Lambda'<p<\Lambda} Dh(p)e^{-\frac{1}{2}\int(\partial_\nu h)^2 d^2x} e^{\frac{1}{2}\int h^2(\partial_\nu\theta)^2 d^2x} = Ne^{\frac{1}{2}\int \langle h^2\rangle(\partial_\nu\theta)^2 d^2x} \quad (2.87)$$

$$= N\exp\left(\frac{1}{2}\int\int_{\Lambda'}^{\Lambda}\frac{d^2p}{(2\pi)^2 p^2}d^2x\int(\partial_\nu\theta)^2 d^2x\right)$$

$$= N\exp\left(\frac{1}{4\pi}\ln(\Lambda'/\Lambda)\int(\partial_\nu\theta)^2 d^2x\right) \quad (2.88)$$

where the overall integration constant N will not be important. On putting this result back into the partition function, we see that the θ term in the effective action is

$$L' = \frac{1}{2}\left(\frac{1}{g} - \frac{1}{2\pi}\ln(\Lambda'/\Lambda)\right)(\partial_\nu\theta)^2 + \cdots \quad (2.89)$$

On comparing this expression with the θ term in the original action, $[1/(2g)](\partial_\nu\theta)^2$, we find that integrating out high-frequency fluctuations has caused coupling-constant renormalization,

$$\frac{1}{g'} = \frac{1}{g} - \frac{1}{2\pi}\ln(\Lambda'/\Lambda) \quad (2.90)$$

Although this parametrization of the model does not leave its O(3) symmetry manifest, it is good enough for this leading-order calculation. In fact, it is better to convert this result into a traditional differential equation for the running coupling constant. Let the high-momentum slice be infinitesimal, $\Lambda' = \Lambda + \delta\Lambda$, so that

$$\ln(\Lambda/\Lambda') = -\ln(1 + \delta\Lambda/\Lambda) \approx -\delta\Lambda/\Lambda \quad \text{and} \quad \frac{1}{g'} - \frac{1}{g} = d\frac{1}{g} = -\frac{1}{g^2}dg \quad (2.91)$$

Therefore,

$$\Lambda\frac{dg}{d\Lambda} = -\frac{1}{2\pi}g^2 \quad (2.92)$$

Or, if we write this in terms of a real-space cutoff, like the lattice spacing a,

$$a\frac{dg}{da} = \frac{1}{2\pi}g^2 \quad (2.93)$$

These equations state that the theory is asymptotically free. We see that the effective coupling is a decreasing function of the momentum cutoff. Alternatively, in terms of lattice models, we learn that a smaller coupling must be used on a finer lattice to obtain the same low-energy, long-wavelength physics. The long-distance features of the theory would be described by a strongly coupled theory, if we can believe the indications of the weak-coupling calculation done here. The calculation suggests that the O(3) model should be disordered and have a finite correlation length ξ.

Let's develop this idea more thoroughly. Suppose that an asymptotically free theory has a finite correlation length ξ. In the language of high-energy physics, we would speak of the "mass gap" m, which is inversely related to the length ξ, $m = 1/\xi$. If the system is actually disordered and correlation-length scaling applies to it, then its correlation functions should be determined by ξ and behave as $\sim e^{-|r-r'|/\xi} = e^{-m|r-r'|}$. Since m characterizes the low-energy, long-distance features of the theory, it should be unaffected when we integrate out high-frequency modes in the lattice theory. In other words, the "bare" lattice theory with a momentum cutoff Λ and coupling g should give rise to the same m as the lattice theory with momentum cutoff Λ' and coupling g', as long as g' is related to g as stated above. It is more convenient to speak of the real-space cutoff a instead of Λ. So, we will consider the couplet (a, g) instead of (Λ, g). By dimensional analysis m can be written as

$$m = \frac{1}{a}F(g) \tag{2.94}$$

However, m is a physical quantity that does not depend on the cutoff scheme, so

$$\frac{d}{da}m = 0 \tag{2.95}$$

which implies that a and g must be related. We have already seen that the relation is given by asymptotic freedom in the case of the O(3) sigma model, and now let's see that this determines the functional dependence of $F(g)$. Substituting the dimensional-analysis statement for m into the last equation implies that

$$-\frac{1}{a^2}F(g) + F'(g)\frac{dg}{da} = 0 \tag{2.96}$$

but we have determined how g must depend on the ultraviolet cutoff a. Let's use standard and general notation for that result, the Callan–Symanzik relation,

$$a\frac{dg}{da} = \beta(g) \tag{2.97}$$

where $\beta(g)$ is the Callan–Symanzik function (not to be confused with $1/(kT)$ or the critical index for the magnetization, or the many other uses for the symbol β!). Now the differential equation for F becomes

$$\frac{d}{dg}F(g) = F(g)/\beta(g) \tag{2.98}$$

which can be integrated,

$$F(g) \sim \exp\left(\int \frac{dg'}{\beta(g')}\right) \tag{2.99}$$

If we choose $g^2 \ll 1$, we can evaluate this integral for the O(3) model,

$$F(g) \sim \exp(-2\pi/g) \tag{2.100}$$

There are several things to note about these curious results. First, the fixed points of the renormalization-group flows $g(a)$ are of central importance. We see that the field theory can be "renormalized" at the fixed point because here F diverges and $\xi = 1/m$ can be a macroscopic, physical length that is independent of the short-distance cutoff a. When we formulate other, more-challenging lattice-field theories we are particularly interested in understanding how their parameters depend on the lattice spacing and we are particularly interested in finding fixed points of the renormalization group because it is only at these points that the long-distance physics "decouples" from the cutoff and it is only at these points that a continuum limit can exist. Second, we need the renormalization-group trajectory $g(a)$ in the vicinity of the fixed point in order to determine the theory's scaling laws, to show how ξ depends on the coupling, and to understand its physical length scales. Recall that, in the case of the planar model, the scaling law was

$$\xi(T) \sim \exp[b\sqrt{1/(T/T_c - 1)}] \tag{2.101}$$

and the fixed point of the renormalization group was at a nonzero value $T_c \neq 0$. In the case of the planar model, the fixed point separated an ordered phase from a disordered phase and we needed to know the behavior of the model in the immediate vicinity of the transition in order to infer its long-distance physics. For the O(3) model, the phase transition occurs at zero coupling, a special feature of asymptotic freedom, and the model has only a disordered phase. We haven't really proven this point, of course, since we have done only weak-coupling analysis, but results from numerical studies back up this point of view. In fact, in the case of the O(3) model, numerical searches for other fixed points of the renormalization group have revealed none. Third, the dependence of the correlation length on the lattice coupling g is nonanalytic. The correlation length cannot be obtained by perturbing about weak, vanishing coupling $g = 0$. There is an instability at $g = 0$ that dynamically generates a finite correlation length. The weak-coupling calculation suggests that, even if the ultraviolet physics is charaterized by weak interactions, the infrared physics is strongly coupled. This point of view underlies modern thoughts on quark confinement in non-Abelian gauge theories. Those models are also asympotically free, so they are well approximated by free fields at short distances and the parton model works quite well, but at large distances they are strongly coupled, Coulomb's law is replaced by flux-tube formation and confinement results. There will be much more discussion on these points in the chapters that follow, needless to say!

Before turning to other topics, it is interesting to understand why the O(3) model was asymptotically free. This was a crucial result because it led to the scaling law for the correlation length and is an important part of the result that the model has just one phase, a disordered one with a dynamically generated correlation length. When we computed the effective action above by integrating out the high-frequency, small-amplitude fluctuations, the local curvature of the sphere produced an effective spin variable with $|\vec{s}'| < 1$. To restore the condition that the spin have unit length required a rescaling, $\vec{s}' \rightarrow \vec{s}'/|\vec{s}'|$, which was absorbed into a change of the coupling constant, $g' \rightarrow g'/|\vec{s}'|^2$. Therefore, the effective coupling increases as a result of integrating out the high-frequency fluctuations which feel the local curvature of the non-Abelian-group manifold.

When we turn to non-Abelian gauge theories, we will again find asymptotic freedom and it will be traced to the non-Abelian self-interactions among the vector gauge fields. Other interactions, such as fermion–gauge couplings, will show the more-familiar screening characteristics, known from quantum electrodynamics, and will oppose asympototic freedom. Recall that, in quantum electrodynamics, fermions with positive charges screen those with negative charges, and the effective coupling at large distances is reduced from one's classical-physics expectations. In fact, if there were "too many" colored quarks in QCD, screening would beat out asympototic freedom and, instead of the theory being weakly coupled at short distances and strongly coupled at long distances, it would behave in the opposite fashion, like quantum electrodynamics and Yukawa models.

Asymptotic freedom is paramount in understanding Nature.

2.7 Instantons in two-dimensional spin systems

Without topological excitations the planar model reduces to a free, infrared-sensitive scalar field. The topological excitations, vortices, provide a disordering mechanism that drives a phase transition to a high-temperature disordered phase in which the model has a finite correlation length. The phase transition is particularly interesting because it is driven by vortex condensation, has an essential singularity rather than the power-law singularity of the more-familiar Ising model, and has no global order parameter at any temperature. We will see that the planar model has much in common with phenomena in four-dimensional gauge theories.

The O(3) spin model is never well approximated by a free field. Asympototic freedom implies that it is strongly coupled and disordered for all couplings, if, as suspected, there are no additional phase transitions besides that at $g = 0$. Does this model have topological disordering mechanisms as well? Indeed it does. The model has instantons, named by G. 't Hooft and discovered by A. M.

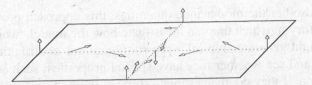

Figure 2.6. The instanton in the two-dimensional O(3) model.

Polyakov in this case [7]. These spin configurations, which constitute local minima of the action, carry a conserved topological charge, and disorder the spins over all distance scales from zero to infinity.

The instantons can be viewed as vacuum tunneling events in Minkowski space, or as stationary field configurations in the two-dimensional Euclidean formulation of the model. Take the $L \times L$ lattice and identify the points on the boundary to compactify the base space into a sphere. Then an instanton represents a mapping of that sphere onto the sphere of the O(3) target space. We know from elementary topology that such maps are labeled by their winding number, which ranges over the positive and negative integers. The winding number is given by

$$Q = \frac{1}{4\pi} \int \vec{s} \cdot (\partial_1 \vec{s} \times \partial_2 \vec{s}) \qquad (2.102)$$

A field configuration with $Q = 1$ and action $S = 2\pi/g$ is drawn in Fig. 2.6 and is given by the expression

$$\vec{s} = \left(\frac{2\rho x_1}{x^2 + \rho^2}, \frac{2\rho x_2}{x^2 + \rho^2}, \frac{x^2 - \rho^2}{x^2 + \rho^2} \right) \qquad (2.103)$$

As shown in Fig. 2.6, at large distances all spins point "up," in the center the spin points "down," and, at intermediate distances $x \sim \rho$, the spins are horizontal, pointing toward the center. The parameter ρ sets the size scale of a particular instanton and, in a partition-function formulation of the model, one would encounter an integral over all scale sizes.

Clearly the instantons of the O(3) model tend to disorder the spins. Since asymptotic freedom does the same, the instantons here do not bring in any qualitatively new physics. However, we shall see that instantons are very significant in QCD and appear to play a central role in its mass spectrum.

2.8 Computer experiments and simulation methods

Another tool for understanding simple statistical-mechanics models is provided by computer simulation. In addition to estimating order parameters, correlation

functions, and other thermodynamic quantities, this approach provides a theoretical laboratory in which one can investigate how the model "works." In other
words, we might generate "typical" spin configurations contributing to a partition function and see whether they have special properties, such as topological
excitations, and, if they do, then estimate some of their characteristics.

To carry out such a program we need an efficient method of generating typical
spin configurations. One very general method that applies to systems made of
classical variables is use of the Metropolis algorithm. The idea is to replace the
computation of the expectation value of a quantity A that depends on all the
spins of the system,

$$\langle A \rangle = \int \prod d\theta \, e^{-S} A \Big/\!\!\Big/ \int \prod d\theta \, e^{-S} \qquad (2.104)$$

which involves a very high dimensional integral, by an average over field configurations $\theta(r)$,

$$\langle A \rangle = \sum_i A(\theta(r)) \Big/ \sum_i 1 \qquad (2.105)$$

in which the configurations $\theta(r)$ are distributed according to the Boltzmann law
e^{-S}.

This approach to calculating expectation values is called "importance sampling" because it replaces the full multi-dimensional integral by a sample of
configurations that are highly likely. We will illustrate importance sampling by
discussing the Metroplis algorithm for generating an ensemble of configurations
that are distributed according to the Boltzmann weighting.

Suppose that we have a trial configuration on a computer. Then the Metropolis
algorithm gives a procedure for generating another configuration by replacing
one $\theta(r)$ variable at a time. The new configuration replaces the old one if the
rules of the algorithm are satisfied. After all the sites have been sampled, one
has "swept through" the lattice once. The aim of the algorithm is to generate,
after many sweeps, perhaps, a new, independent field configuration $\theta'(r)$, which
is a new member of the Boltzmann distribution. The value $A(\theta'(r))$ can be calculated for this configuration and a contribution to the average, $\langle A \rangle$, recorded.
Then the Metropolis algorithm is applied many more times to $\theta'(r)$ and another
configuration, a new, statistically independent member of the Boltzmann distribution, $\theta''(r)$, is generated. Then $A(\theta''(r))$ is computed and recorded. This
procedure is repeated again and again until $\langle A \rangle$ has been estimated to some sufficient, desired accuracy.

The Metropolis algorithm consists of the following procedure. We have a configuration in an Ising system, for example. This configuration may but need not
be an important member of the Boltzmann distribution for the parameter set of
interest. In any case, take the spin variable at a particular site n and flip it. Compute the change in the action, ΔS. If it is less than zero, the new configuration

is accepted and the old configuration is replaced. If ΔS is positive, the new configuration is accepted with the conditional probability $\exp(-\Delta S)$. This step can be done by picking a random number x between zero and unity and comparing it with $\exp(-\Delta S)$: if $\exp(-\Delta S) > x$, the configuration is accepted, otherwise it is not.

This Metropolis procedure is performed for a particular site and then another site is picked and the procedure continues until all the spins have been considered. This procedure is guaranteed to bring the system into thermal equilibrium and to produce a Boltzmann ensemble. The procedure is "local," that is, to update a spin at a particular site, one need know only those spins which are directly coupled to it. In the case of the nearest-neighbor coupled Ising model, this is just a group of four nearest neighbors. The procedure is very straightforward and easy to implement. One can even update several spins at a time, as long as they are outside each others' immediate spheres of influence. Of course, since the procedure is local, it may take many sweeps to generate a statistically independent member of the Boltzmann ensemble. Correlations between configurations in the Markov process must be monitored carefully in order to estimate the statistical errors in the procedure's estimate for $\langle A \rangle$.

Let's review the argument that states that the Metropolis algorithm works, in the sense that the probability of finding a configuration $\theta(r)$ after N sweeps is

$$P_N(\theta(r)) \underset{N \to \infty}{\sim} e^{-S(\theta(r))} \tag{2.106}$$

To do this, consider the relationship between the Nth and the $(N + 1)$th sweeps. Let $W(\theta \to \theta')$ be the probability of the transition $\theta \to \theta'$. Then the procedure generates a new distribution, $P_{N+1}(\theta)$, given by

$$P_{N+1}(\theta) = \sum_{\theta'} W(\theta' \to \theta) P_N(\theta') + \left(1 - \sum_{\theta'} W(\theta \to \theta')\right) P_N(\theta) \tag{2.107}$$

$$= P_N(\theta) + \sum_{\theta'} [P_N(\theta') W(\theta' \to \theta) - P_N(\theta) W(\theta \to \theta')] \tag{2.108}$$

A stationary distribution will satisfy

$$P_N^S(\theta') W(\theta' \to \theta) = P_N^S(\theta) W(\theta \to \theta') \tag{2.109}$$

which is the statement of detailed balance. An important property of the procedure is that it always makes configurations that are closer to satisfying detailed balance, so we are guaranteed that thermalization will occur eventually. To see this, suppose that

$$\frac{P_N(\theta_a)}{P_N(\theta_b)} > \frac{W(\theta_b \to \theta_a)}{W(\theta_a \to \theta_b)} \tag{2.110}$$

then, by substituting into the evolution equation for P_N, we find $P_{N+1}(\theta_a) < P_N(\theta_a)$ and $P_{N+1}(\theta_b) > P_N(\theta_b)$.

Finally, the Metropolis algorithm chooses the transition to be

$$W(\theta \to \theta') = 1, \qquad \text{if } S(\theta) > S(\theta') \tag{2.111}$$

$$W(\theta \to \theta') = e^{-(S(\theta')-S(\theta))}, \qquad \text{if } S(\theta) < S(\theta') \tag{2.112}$$

This implies that

$$\frac{W(\theta_b \to \theta_a)}{W(\theta_a \to \theta_b)} = e^{-(S(\theta_a)-S(\theta_b))} \tag{2.113}$$

and substituting back into the detailed-balance equation gives

$$P_N^S(\theta) = P^{eq}(\theta) \sim e^{-S(\theta)} \tag{2.114}$$

as desired. Note that we never need to calculate the entire action $S(\theta)$ in order to implement the algorithm – all we need are the nearest neighbors of the updated variable. Locality in the action guarantees that the computer time needed for a single sweep of the lattice grows in proportion to its size.

The final ingredient in the Metropolis scheme is the specification of the transition, $\theta \to \theta'$. The exact procedure depends on the character of the variables in the configuration, namely whether they are Ising spins, unbounded scalar fields, phase angles, vectors, or matrices. The essential ingredient is that one must choose a scheme that will cover the space of the local variable uniformly. For example, in the case of a model with SU(3) matrices at each site, one might set up a table of random SU(3) matrices and their Hermitian adjoints. Then, in the Metropolis algorithm one might specify a site, change its variable by multiplication by a randomly chosen matrix from the table and proceed with the rest of the algorithm as presented above. This procedure works because one can prove that repeated application of matrices from the table to any SU(3) matrix brings it arbitrarily close to any other member of the group. In fact, one could even bias the table of matrices toward the identity, so that the changes in the configuration are small at each step, thus improving the probability that the change in the matrix at each site is "accepted," and thus improving the efficiency of the algorithm for dealing with an ordered phase.

Classical statistical-mechanics models have been studied in great detail using the Metropolis algorithm and other, similar, algorithms. Many practical problems can arise in these studies.

1. There are considerable correlations between successive configurations in these local procedures, so the lattice must be swept through many times before an independent measurement of the observable A can be made. This problem is particularly severe near a critical point, where the correlation length becomes large. Typically, the number of sweeps needed to generate a new configuration under these conditions diverges as a power of

$T - T_c$. Statistical analyses of the set of configurations generated must be carried out in order to deal with this problem. In the vicinity of a first-order transition there are often metastable states that also slow the Metropolis algorithm's approach to an equilibrium configuration. Tunneling between metastable states is often seen in computer simulations. Nonlocal, global changes in the configuration can be substituted for the local changes discussed above in order to deal with these problems, but such algorithms must be carefully crafted for the particular model being studied, in order to keep the "acceptance probability" high enough to make the procedure successful. "Cluster algorithms" are of this class and have proven to be very worthwhile for particular, simple models [8].

2. The Monte Carlo procedure produces statistical estimates for observables and the uncertainties in the averages fall to zero as $1/\sqrt{N}$, where N is the number of independent configurations in one's equilibrated ensemble, according to the central limit theorem. This character of the procedure must be faced after problem 1 has been dealt with. This point emphasizes how computer-intensive accurate Monte Carlo studies must be.

3. Since the simulations are done on systems of finite size, one must monitor finite-size effects effects carefully, because they can interfere with one's real interest, the thermodynamic limit of the model. This limitation is particularly severe near a critical point, where the correlation length diverges. Since the critical points are especially important, simulations of a range of lattices are usually needed in order to extract universal features of the model such as critical indices and amplitude ratios. Finite-size scaling or renormalization-group methods may be needed in order to extract critical features of the model reliably. Again, vast computer resources are usually needed in quantitative studies.

One of the real challenges in lattice-gauge theory is the simulation of fermion systems. Since Fermi fields do not commute, they represent another hurdle for computer simulations. As we shall discuss later, lattice-gauge theory has excellent, general algorithms for simulating fermions in many, but, alas, not all, environments.

2.9 The transfer matrix in field theory and statistical physics

The purpose of this section is to discuss alternative formulations of lattice-gauge theory and statistical mechanics of spin systems that might help in their solutions. It should be clear to the reader that the partition-function formulation of four-dimensional statistical systems is analogous to the Euclidean path-integral formulation of field theories. We have seen already, and will see again as gauge theories are introduced, that the temperature T of the statistical system corresponds to the coupling g of the field-theory problem. High temperature corresponds to strong coupling and low temperature corresponds to weak coupling, all other things being fixed.

There are other equally popular and effective formulations of field theories besides the Euclidean path integral. The Hamiltonian formulation, in which one deals with time evolution, a Schrödinger-like equation, operator equations of motion and matrix elements, etc., is very important. How is this formulation related to those we have already seen?

If we begin with the Ising model in two dimensions, then it is traditional to formulate it on a symmetric lattice. In the light of universality, other cutoff procedures are equally valid if one's real interest is in the critical behavior of the model. In lattice-gauge theory we shall be interested in the continuum limits of cutoff systems where the correlation length grows large relative to the system's lattice spacing, so we must concern ourselves with the systems' critical points and scaling laws. Since the scaling laws are universal, they should be the same for any lattice formulation of a given system as long as we use the same degrees of freedom and local couplings in each case. This gives us tremendous freedom. The lattice could be very, very anisotropic and the critical behavior, or, equivalently, the continuum limit in the language of field theory, would be unchanged. In fact, the Hamiltonian version of a four-dimensional Euclidean lattice-gauge theory uses a spatial lattice as a continuum cutoff, but treats 'time' as a continuous variable, so that a Schrödinger equation and a system of equal time-commutation relations can be set up.

The goal of this section is to illustrate how this works in simple cases. Since the physics of interest is independent of the lattice, the ultraviolet cutoff, there is enormous freedom here that we should learn to use to our advantage.

2.9.1 *The simple harmonic oscillator*

A great deal of the mathematics and physics we wish to illustrate comes up in the context of nonrelativistic quantum mechanics, so we start here. We want to find the relation between the Feynman path integral and the Schrödinger equation for a one-dimensional-potential model problem.

Consider the Lagrangian for a one-dimensional simple harmonic oscillator,

$$L = \tfrac{1}{2}(\dot{x}^2 - \omega^2 x^2) \tag{2.115}$$

In the path-integral formulation of quantum mechanics, one considers the probability amplitude that the particle is initially at the spacetime point (x_a, t_a) and ends up at the spacetime point (x_b, t_b). The amplitude for this transition is postulated to be

$$Z = \sum_{\text{paths}} \exp\left(\frac{i}{\hbar} S_{\text{m}}\right) \tag{2.116}$$

where the sum is over all world lines between the initial and final points and S_{m} is the Minkowski-space action for a particular path,

$$S_{\text{m}} = \int_{t_a}^{t_b} L \, dt \tag{2.117}$$

Therefore, each path contributes to the transition amplitude through a weight $\exp[(i/h) \int L \, dt]$.

There are two challenges in this formulation of quantum mechanics. First, we must clarify what the "sum over all paths" means in quantitative terms. Next, we need to pinpoint the important paths in the sum and deal with the constructive and destructive interference expected among the complex contributions of individual paths, $\exp[(i/h) \int L \, dt]$.

To begin, introduce a temporal lattice and call the spacing between time slices ϵ,

$$t_{i+1} - t_i = \epsilon \tag{2.118}$$

Next, formulate the problem in imaginary time, so that the oscillating weights for individual paths become damped, real numbers:

$$t = -i\tau \tag{2.119}$$

Now the Minkowski-space action becomes, continued to imaginary time τ,

$$\frac{i}{\hbar} S_{\text{m}} = -\frac{1}{2\hbar} \int \left[\left(\frac{dx}{d\tau}\right)^2 + \omega^2 x^2\right] d\tau \tag{2.120}$$

To make contact with conventional statistical physics with its positive partition function, with positive weights, define the Euclidean action,

$$S = \frac{1}{2} \int \left[\left(\frac{dx}{d\tau}\right)^2 + \omega^2 x^2\right] d\tau \tag{2.121}$$

For discrete time slices, the action is approximated by

$$S = \frac{\epsilon}{2} \sum_i \left[\left(\frac{x_{i+1} - x_i}{\epsilon}\right)^2 + \omega^2 x_i^2\right] \tag{2.122}$$

and the path integral becomes

$$Z = \int_{-\infty}^{\infty} \prod_i dx_i \, \exp\left(-\frac{1}{\hbar} S\right) \tag{2.123}$$

It is interesting to interpret these formulas before proceeding to the transfer matrix. The path integral and the discrete action define a statistical mechanics problem in one dimension. The lattice sites are labeled with an index i and on each site there is a variable x_i that can vary from $-\infty$ to $+\infty$. The action couples nearest-neighbor sites and each configuration $\{x_i\}$ contributes to the sum over configurations, $\int \prod dx_i$, through a positive weight, $\exp[(-1/\hbar)S]$. We can interpret this weight as a Boltzmann factor if we identify $\hbar$ and temperature T. This is a useful correspondence because, in statistical physics, temperature is a measure of fluctuations through the heat bath and in quantum mechanics $\hbar$ is a measure of fluctuations through the uncertainty relation. In quantum mechanics the theoretical limit $\hbar \to 0$ picks out the unique classical-physics path, with fluctuations suppressed to zero. In statistical physics the $T = 0$ case produces a unique, completely "frozen" configuration.

Now let's return to the theme of this discussion, the development of the transfer matrix. This will replace the one-dimensional classical statistical-physics problem with a zero-dimensional quantum problem by viewing the sums over configurations on the temporal axis labeled by sites i as quantum evolution from one state at i to the next at $i + 1$. The nearest-neighbor coupling is important here because it allows us to write the partition function as

$$Z = \int \prod_i [dx_i \, T(x_{i+1}, x_i)] \tag{2.124}$$

where

$$T(x_{i+1}, x_i) = \exp\left[-\frac{1}{2\hbar}\left(\frac{1}{\epsilon}(x_{i+1} - x_i)^2 + \frac{1}{2}\omega^2 \epsilon x_{i+1}^2 + \frac{1}{2}\omega^2 \epsilon x_i^2\right)\right] \tag{2.125}$$

Since $T(x_{i+1}, x_i)$ has the same form for any two neighboring sites, we can write it as a matrix element of an operator, the transfer matrix. Set up a quantum basis at each site. Let there be position and momentum operators $\hat{x}$ and $\hat{p}$ with the commutator $[\hat{p}, \hat{x}] = -i\hbar$. Eigenstates of $\hat{x}$ satisfy

$$\hat{x}|x\rangle = x|x\rangle \tag{2.126}$$

are normalized,

$$\langle x'|x\rangle = \delta(x' - x) \tag{2.127}$$

and are complete,

$$1 = \int dx \, |x\rangle\langle x| \tag{2.128}$$

Now we can construct the operator $\hat{T}$ with the matrix elements

$$\langle x'|\hat{T}|x\rangle = T(x', x) \tag{2.129}$$

where $T(x', x)$ is given by Eq. (2.125). Identify $\hat{T}$ as the time-evolution operator of quantum mechanics (imaginary time) evaluated over the interval ϵ. Now Z can be written in the simpler form, using completeness,

$$
\begin{aligned}
Z &= \int \prod_i dx_i \, \langle x_{i+1}|\hat{T}|x_i\rangle \\
&= \int \langle x_{\mathrm{b}}|\hat{T}|x_{n-1}\rangle \, dx_{n-1} \, \langle x_{n-1}|\hat{T}|x_{n-2}\rangle \cdots dx_1 \, \langle x_1|\hat{T}|x_{\mathrm{a}}\rangle \\
&= \langle x_{\mathrm{b}}|\hat{T}^n|x_{\mathrm{a}}\rangle
\end{aligned}
\tag{2.130}
$$

where $n-1$ is the number of τ slices between the τ_{a} and τ_{b} initial and final conditions. If we impose periodic boundary conditions and sum over all possible initial conditions, we have

$$
Z = \operatorname{tr} \hat{T}^n
\tag{2.131}
$$

Finally, we need an explicit expression for the transfer matrix. Recalling the matrix element expressing the fact that the Fourier transform of a Gaussian is another Gaussian,

$$
\langle x'|e^{-\frac{1}{2}\epsilon \hat{p}^2}|x\rangle \sim \exp\left(-\frac{1}{2\epsilon\hbar^2}(x'-x)^2\right)
\tag{2.132}
$$

we can derive the exact result,

$$
\hat{T} \sim \exp\left(-\frac{1}{4\hbar^2}\epsilon\omega^2\hat{x}^2\right) \exp\left(-\frac{1}{2\hbar}\epsilon\hat{p}^2\right) \exp\left(-\frac{1}{4\hbar^2}\epsilon\omega^2\hat{x}^2\right)
\tag{2.133}
$$

Since $\hat{x}$ and $\hat{p}$ do not commute, the operator ordering here is important.

This expression for the evolution operator is probably not familiar. However, if we take the "time-continuum" limit and expand the exponentials, the operator ordering is unimportant to lowest order in ϵ, $\epsilon \to 0$, and we easily identify the Hamiltonian of the harmonic oscillator,

$$
\hat{T} \underset{\epsilon \to 0}{\to} 1 - \frac{\epsilon}{\hbar}\hat{H}
\tag{2.134}
$$

where

$$
\hat{H} = \tfrac{1}{2}(\hat{p}^2 + \omega^2\hat{x}^2)
\tag{2.135}
$$

Furthermore, the Hamiltonian $\hat{H}$ and the path integral produce the Schrödinger equation. The Feynman interpretation of the path integral Z means that, if the wavefunction of the particle is $\psi(x, \tau)$, then, at a later time,

$$
\psi(x', \tau') = \int Z(x', \tau'; x, \tau)\psi(x, \tau) \, dx
\tag{2.136}
$$

Choosing $\tau' - \tau = \epsilon$ to be infinitesimal, the path integral simplifies to

$$Z = \langle x'|\hat{T}|x\rangle = \langle x'|\exp\left(-\frac{\epsilon}{\hbar}\hat{H}\right)|x\rangle$$

$$\approx \langle x'|1 - \frac{\epsilon}{\hbar}\hat{H}|x\rangle = \delta(x'-x) - \frac{\epsilon}{\hbar}H\delta(x'-x) \qquad (2.137)$$

Then Eq. (2.136) becomes

$$\psi(x', \tau') = \psi(x, \tau) - \frac{\epsilon}{\hbar}H\psi(x', \tau) \qquad (2.138)$$

or,

$$\hbar\frac{\partial}{\partial\tau}\psi(x, \tau) = -H\psi(x, \tau) \qquad (2.139)$$

which is the Euclidean version of Schrödinger's equation. H in this equation is just the x-space realization of the Hamiltonian.

2.9.2 The transfer matrix for the Ising model

These ideas generalize to field theories in higher dimensions. One can establish correspondences between mass gaps and correlation lengths, free energies and vacuum energies, and correlation functions and composite propagators in statistical physics and field theory [9]. Here we will work through a specific example because it will bring up issues of immediate relevance to QCD in extreme environments.

Consider the two-dimensional Ising model again, but let the couplings in the two directions be independent,

$$S = -\sum_n [\beta_\tau \sigma(n + \hat{\tau})\sigma(n) + \beta\sigma(n + \hat{x})\sigma(n)] \qquad (2.140)$$

where we have chosen one direction $\hat{\tau}$ as the temporal direction and the other $\hat{x}$ as the spatial direction. The variables $\sigma(n)$ are classical Ising variables, with the two possible values of ± 1.

We are interested in constructing this model's transfer matrix, taking the τ-continuum limit and finding the Hamiltonian and then understanding its phases from this perspective. This will give us an interesting and illuminating series of insights into the model with very little effort. Since Hamiltonian lattice-gauge theory is an important formulation of QCD, especially in its physical content, this exercise will provide good background for later developments.

To begin, write the action in the form that makes the transfer matrix easier to identify,

$$S = \sum_n \{\beta_\tau[\sigma(n + \hat{\tau}) - \sigma(n)]^2 - \beta\sigma(n + \hat{x})\sigma(n)\} \qquad (2.141)$$

which differs from our original expression by an unimportant constant. Next consider two neighboring spatial rows and label the spins in one row $\sigma(m)$ and those in the next row $s(m)$. The action can now be written as a sum over these rows,

$$S = \sum_{n_0} L(n_0 + 1, n_0) \tag{2.142}$$

where

$$L = \frac{1}{2}\beta_\tau \sum_m [s(m) - \sigma(m)]^2 - \frac{1}{2}\beta \sum_m [\sigma(m+1)\sigma(m) + s(m+1)s(m)] \tag{2.143}$$

In setting up the transfer matrix note that, if there are M sites on each spatial row, then there are 2^M spin configurations. Therefore, the transfer matrix will be a $2^M \times 2^M$ matrix. Let's consider some of its elements, with an eye toward the τ-continuum limit. For a diagonal element, $\sigma(m) = s(m)$ and Eq. (2.143) reduces to,

$$L(0 \text{ flips}) = -\beta \sum_m \sigma(m+1)\sigma(m) \tag{2.144}$$

and, if there are n spin flips between the rows,

$$L(n \text{ flips}) = 2n\beta_\tau - \frac{1}{2}\beta \sum_m [\sigma(m+1)\sigma(m) + s(m+1)s(m)] \tag{2.145}$$

Next we must determine how the parameters β_τ and β must behave so that the transfer matrix has the form

$$\hat{T} \approx 1 - \tau\hat{H} \tag{2.146}$$

in the τ-continuum limit. To do this, consider several matrix elements,

$$\hat{T}(0 \text{ flips}) = \exp\left(\beta \sum_m \sigma(m+1)\sigma(m)\right) \approx 1 - \tau\hat{H}_{0 \text{ flips}}$$

$$\hat{T}(1 \text{ flip}) = \exp(-2\beta_\tau)\exp\left[\frac{1}{2}\beta\left(\sum_m \sigma(m+1)\sigma(m) + s(m+1)s(m)\right)\right]$$

$$\approx \tau\hat{H}_{1 \text{ flip}} \tag{2.147}$$

and we read off these equations that $\beta \sim \tau$ and $\exp(-2\beta_\tau) \sim \tau$. This means that β and $\exp(-2\beta_\tau)$ are proportional and we are free to define the temporal variable τ as

$$\tau = \exp(-2\beta_\tau) \qquad (2.148)$$

and introduce a coupling λ so that

$$\beta = \lambda\tau \qquad (2.149)$$

We learn from these details that, in order to define a smooth τ-continuum theory, the couplings must be adjusted so that the temporal coupling grows large while the spatial coupling becomes weak. If we rewrite Eq. (2.147) using the variable τ, we can identify the quantum Hamiltonian version of the transfer matrix,

$$\hat{T}(0 \text{ flips}) \approx 1 + \tau\left(\lambda \sum_m \sigma(m+1)\sigma(m)\right) \approx 1 - \tau\hat{H}_{0\ \text{flips}}$$

$$\hat{T}(1 \text{ flip}) \approx \tau \approx \hat{H}_{1\ \text{flip}} \qquad (2.150)$$

with multiple spin flips vanishing in the τ-continuum limit. This simplification of the model will prove to be very useful and insightful.

To write $\hat{H}$ in operator form, we need to set up a quantum basis of states at each site and we need the operators that implement transitions between those states. We represent spin "up" and spin "down" at site m as column vectors,

$$|\text{up}\rangle = \begin{pmatrix} 1 \\ 0 \end{pmatrix}, \qquad |\text{down}\rangle = \begin{pmatrix} 0 \\ 1 \end{pmatrix} \qquad (2.151)$$

Then we place a Pauli matrix $\hat{\sigma}_3$ on each site to measure whether a spin is up or down and we place a $\hat{\sigma}_1$ on each site to flip that spin,

$$\sigma_3 = \begin{pmatrix} 1 & 0 \\ 0 & -1 \end{pmatrix}, \qquad \sigma_1 = \begin{pmatrix} 0 & 1 \\ 1 & 0 \end{pmatrix} \qquad (2.152)$$

This notation has been chosen so that the $\hat{H}$ expression is particularly elegant,

$$\hat{H} = -\sum_m \hat{\sigma}_1(m) - \lambda \sum_m \hat{\sigma}_3(m+1)\hat{\sigma}_3(m) \qquad (2.153)$$

It cannot be stressed too much that we are interested in just the universal features of the Ising model in this case, and the universal features of lattice-field theories in general. So, the precise relations between the parameter λ and the couplings of the anisotropic Ising model, β_τ and β, are not primary. The qualitative character of λ is important, however. It is clear from its derivation that $1/\lambda$ is a measure of temperature. The τ-continuum Hamiltonian formulation of the Ising model is "hot" when λ is small and "cold" when λ is large. More detail on the connection between the isotropic and the anisotropic Ising models can be found in the literature [1].

2.9.3 Self-duality and kink condensation through the eyes of the transfer matrix

It is interesting and relevant to the search for new types of condensates in QCD in extreme environments to discuss the physics of the Ising model from the perspective of the τ-continuum limit. This search is aided by finding an operator interpretation of the self-duality property of the Ising model. In this discussion we will be using operators everywhere, but we will drop the "hats" from the symbols for notational simplicity.

To begin, construct the system dual to $\hat{H}$ given above. Sites of the original lattice will be associated with links of the dual lattice and vice versa. We can think of the dual lattice as another one-dimensional array of sites that is shifted by half a spatial lattice length relative to the original lattice.

Define operators μ_1 and μ_3 on the dual lattice,

$$\mu_1(n) = \sigma_3(n+1)\sigma_3(n), \qquad \mu_3(n) = \prod_{m<n} \sigma_1(m) \qquad (2.154)$$

So, $\mu_1(n)$ tells us whether spins on adjacent sites are aligned, and $\mu_3(n)$ flips all the spins to the left of n. These dual operators have several crucial properties.

- They satisfy the same Pauli spin algebra as σ_1 and σ_3.

- The Hamiltonian retains its form when it is written in terms of the dual variables, $H(\sigma; \lambda) = \lambda H(\mu; \lambda^{-1})$.

To check the first point, recall that the Pauli matrices σ_1 and σ_3 anti-commute on the same site and commute otherwise, and the square of any Pauli matrix is the identity. To check that

$$\mu_1(n)\mu_3(n) = -\mu_3(n)\mu_1(n) \qquad (2.155)$$

we note that, when it is written out in terms of the original Pauli matrices, only one factor of σ_1 and one factor of σ_3 are on the same site, and their anti-commutation produces the desired result. The fact that μ_1 and μ_3 commute on different sites follows from the fact that an even number of σ_1's and σ_3's is interchanged when μ_1 and μ_3 are interchanged.

To establish the second point, we note that

$$\sigma_1(m) = \mu_3(m+1)\mu_3(m) \qquad (2.156)$$

so the Hamiltonian can be written

$$H = -\sum_n \mu_3(n)\mu_3(n+1) - \lambda\sum_n \mu_1(n) \tag{2.157}$$

$$= \lambda\left(-\sum_n \mu_1(n) - \lambda^{-1}\sum_n \mu_3(n)\mu_3(n+1)\right) \tag{2.158}$$

which is the desired expression of the self-duality of the model. Note that this result implies that each eigenvalue of the Hamiltonian satisfies $E(\lambda) = \lambda E(\lambda^{-1})$, which relates high- and low-temperature properties of the model. In particular, apply it to the theory's mass gap, which vanishes at any critical point the theory might have. If we suppose that the theory has just one critical point, we learn that it must lie at

$$\lambda_c = 1 \tag{2.159}$$

Explicit calculations on the model confirm this result. In particular, perturbation-theory expansions of the mass gap $M(\lambda)$ in the coupling λ are easy to carry out to high orders [9] and produce the result

$$M(\lambda) = 2|1 - \lambda| \tag{2.160}$$

exactly. In other words, all the perturbation-theory graphs vanish beyond second order, and an exact result emerges. The result determines the correlation-length critical index ν exactly, $\nu = 1$. This success illustrates the power and use of universality. The mass gap determines the propagator in the model which corresponds to the correlation function in the isotropic formulation of the model. By dimensional analysis, the mass gap corresponds to the reciprocal of the correlation length of the isotropic model, so the linear vanishing of the mass gap in the vicinity of the critical point translates into the linear divergence of the correlation length as the critical point is approached [9]. In many field-theory applications, it is advantageous to use a Hamiltonian rather than a Euclidean path integral.

Further analysis of the quantum Hamiltonian gives the other critical indices of the traditional Ising model. In fact, it is elementary to solve the quantum Hamiltonian version of the Ising model explicitly [9].

We end this discussion of the Ising model with some observations about order–disorder transitions, kink condensation, and the magnetization. The magnetization $M = \langle\sigma_3\rangle$ serves as the global order parameter for the Ising model. It is nonzero at low temperature, vanishes continuously as the temperature passes through the critical point with a critical index of $\frac{1}{8}$, and remains identically zero for all higher temperatures. The magnetization indicates that the "up–down" global symmetry of the system's Hamiltonian is spontaneously broken in the low-temperature phase.

The self-duality of H allows us to view the magnetization and order in the low-temperature phase from a different, interesting perspective. Place the system into an external magnetic field h and describe it with the perturbed Hamiltonian

$$\frac{1}{\lambda}H + h\sum_n \sigma_3(n) \tag{2.161}$$

Apply the duality transformation to this expression and find

$$\frac{1}{\lambda}H + h\sum_n \sigma_3(n) = H(\mu; \lambda^{-1}) + h\sum_n \prod_{m<n} \mu_1(m) \tag{2.162}$$

where we observed that $\sigma_3(n) = \prod_{m<n} \mu_1(m)$, which follows from the definition of μ_1 in Eq. (2.154). However, now duality gives the matrix-element identity,

$$\langle 0|\sum_n \sigma_3(n)|0\rangle_\lambda = \langle 0|\sum_n \prod_{m<n} \mu_1(n)|0\rangle_{\lambda^{-1}} \tag{2.163}$$

where $|0\rangle_\lambda$ is the vacuum of $H(\sigma; \lambda)$ and $|0\rangle_{\lambda^{-1}}$ is the vacuum of $H(\mu; \lambda^{-1})$. Since σ_1 and σ_3 are equivalent to μ_1 and μ_3, this equation tells us that the operator $\prod_{m<n} \sigma_1(n)$ has a nonvanishing expectation value in the high-temperature phase of the original model. For this reason it is called a "disorder" operator. It has an interesting physical interpretation. Suppose that it acts on a completely ordered state having all spins "up" in the basis where σ_3 is diagonal. Then it flips all the spins to the left of site n. It creates a "kink" in the ordered low-temperature phase and the expectation value of the disorder operator vanishes identically in the low-temperature phase. The disorder operator has a nonzero expectation value in the high-temperature phase, which we can now interpret as a "kink condensate."

We can make a kink state having zero momentum in the low-temperature phase,

$$|1\text{ kink}\rangle \sim \sum_n \prod_{m<n} \sigma_1(n)|0\rangle_{\lambda^{-1}=0} \tag{2.164}$$

It follows from the self-duality of the model that the computation of the kink's mass in powers of λ^{-1} is identical to that of the flipped spin states' mass in powers of λ. Therefore, the mass gap formula, $2|1 - \lambda|$ applies to the kink in the low-temperature phase.

Since the high-temperature phase of the the model is a "kink condensate," it provides an illuminating interpretation of the short-range character of the the spin–spin correlation function in this phase. Since the disorder operator flips spins over an infinite region of space, a condensate of kinks randomizes the spin variables and leads to short-range correlations in this variable. The kinks have a topological significance. Applying a kink operator to the low-temperature ground state produces a state that interpolates between the two degenerate ground states of the low-temperature phase. Kinks alter the boundary conditions of the system. In fact, by inspecting the extremities of the spatial lattice one can determine whether there is an even or odd number of kinks in the system.

The concepts of topological disorder and operator condensation are very important in field theory, statistical physics, and applications to QCD in extreme environments.

3

Gauge fields on a four-dimensional euclidean lattice

3.1 Lattice formulation, local gauge invariance, and the continuum action

In its standard formulation, lattice-gauge theory replaces the continuum spacetime of Euclidean quantum-field theory with a four-dimensional lattice of spacetime points and links. If the lattice is hypercubic, then its lattice spacing a provides an ultraviolet cutoff. Although such a spacetime cutoff is not convenient in most analytic calculations, it has several extraordinary advantages. First, it allows strong-coupling calculations both through expansions and by simulation methods. These approaches provide a framework for formulating and developing physical pictures of confinement, chiral-symmetry breaking, and other nonperturbative phenomena that are so challenging in ordinary continuum formulations of quantum-field Theory.

The various fields of the lattice version of quantum chromodynamics exist on the sites and links of the regular lattice. In the standard formulation, the fermion fields reside on the sites and the gauge fields reside on the links [10]. The fermion fields, matter fields in general, carry the color quantum number of the gauge group SU(3). It is convenient, therefore, to imagine that there is a color frame of reference at each site so that the fermion field with color index α, ψ_α, can be visualized. (This geometrical point of view dates back to the original Yang–Mills paper which introduced non-Abelian gauge fields into high-energy physics when they generalized the notion of isospin symmetry from the familiar global symmetry of nuclear physics to a local symmetry of high-energy field theory [11].) The gauge fields are then the connections in this network of color coordinate systems and with them one can implement the notion of local color-gauge invariance in a precise and elegant fashion.

Let's illustrate the lattice approach with a construction of the pure Yang–Mills field and action. To keep the algebra simple we will specialize to SU(2) fields, but the generalization to SU(3) and, in fact, SU(N) will be evident.

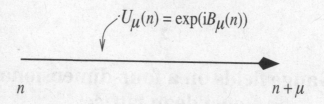

Figure 3.1. The link variable on a four-dimensional Euclidean lattice.

Consider a four-dimensional hypercubic Euclidean lattice with a spacing a. On each link place an SU(2) matrix, as shown in Fig. 3.1,

$$U_\mu(n) = \exp(iB_\mu(n)) \tag{3.1}$$

where $B_\mu(n) = \frac{1}{2}ag\tau_i A^i_\mu(n) = \frac{1}{2}ag\vec{\tau} \cdot \vec{A}_\mu(n)$ and $\vec{\tau}$ consists of the three Pauli matrices. Each link carries a direction, (n, μ) with $\mu = 1, 2, 3,$ or 4 and connects site n with site $n + \mu$. If we denote a link in the backward direction, we associate with it $U_\mu^{-1}(n)$,

$$U_{-\mu}(n + \mu) = U_\mu^{-1}(n) \tag{3.2}$$

We note that the $[U_\mu(n)]_{ij}$ are SU(2) rotation matrices. Following Yang and Mills again, we imagine a color frame of reference at each site of the lattice approximation to continuum spacetime. $U_\mu(n)$ is defined to be the local rotation that takes a color frame of reference at site n to that at site $n + \mu$. We want to write down a system of rules of dynamics, an action in particular, which will be invariant with respect to the orientations of these color frames. The symmetry should be a local one – if a color frame is rotated at one site, then the action and "physics" in general should be unaffected. The SU(2) color symmetry will become a local, gauge symmetry.

A rotation in color space can be done at each site with an SU(2) rotation matrix,

$$G(\chi(n)) = \exp\left(-i\frac{1}{2}\vec{\tau} \cdot \vec{\chi}(n)\right) \tag{3.3}$$

When G acts on the sites of the lattice, the link variables ("connections") should transform as

$$U_\mu(n) \to G(n)U_\mu(n)G^{-1}(n + \mu) \tag{3.4}$$

because they relate the color frames at sites n and $n + \mu$. Note the n and $n + \mu$ in this equation, and the local character of the symmetry group. Writing out the transformation law with indices gives

$$[U_\mu(n)]_{ij} \to \Sigma_{kl}\left[e^{-i\frac{1}{2}\vec{\tau} \cdot \vec{\chi}(\mathbf{n})}\right]_{ik}\left[e^{-i\frac{1}{2}\vec{\tau} \cdot \vec{\chi}(\mathbf{n}+\mu)}\right]_{jl}[U_\mu(n)]_{kl} \tag{3.5}$$

Our challenge is to construct a local action, made out of gauge variables $U_\mu(n)$, for which this local symmetry transformation is an exact symmetry. A little thought and experimentation convinces us that the action can be built only out of the products of U matrices taken around *closed* paths. These quantities are locally gauge-invariant because the color indices are all contracted into local group singlets. On a four-dimensional Euclidean lattice, the simplest and shortest closed paths are elementary squares ("plaquettes"). So a candidate action reads

$$S = -\frac{1}{2g^2} \Sigma_{n,\mu,\nu} \operatorname{tr} U_\mu(n) U_\nu(n+\mu) U_{-\mu}(n+\mu+\nu) U_{-\nu}(n+\nu) + \text{h.c.} \quad (3.6)$$

We shall check that the prefactor, $-1/(2g^2)$, is needed in order to establish conventional notation. We need to check the following attributes of this action.

1. Its classical continuum limit ($a \to 0$) reproduces ordinary Yang–Mills theory.

2. Its strong-coupling limit confines quarks.

Once these exercises have been done, we shall have a candidate formulation of gauge-invariant dynamics that can be used to study the self-interactions of non-Abelian gauge fields for all cutoffs a and all couplings g. There are many other hurdles that the lattice action must jump over before it becomes of real physical interest, but these two constitute a necessary and nontrivial beginning in those more-demanding discussions.

To take the classical continuum limit, we presume that the gauge fields are slowly varying on the length scale of the cutoff. Then we can Taylor expand the fields $B_\mu(n)$ in the action,

$$B_\nu(n+\mu) = B_\nu(n) + a\,\partial_\mu B_\nu(n) + O(a^2) \quad (3.7)$$
$$B_{-\mu}(n+\mu+\nu) = -B_\mu(n+\nu) = -[B_\mu(n) + a\,\partial_\nu B_\mu(n)] + O(a^2) \quad (3.8)$$
$$B_{-\nu}(n+\nu) = -B_\nu(n) \quad (3.9)$$

Now the expression for the action can be approximated,

$$UUUU \approx e^{iB_\mu} e^{i(B_\nu + a\,\partial_\mu B_\nu)} e^{-i(B_\mu + a\,\partial_\nu B_\mu)} e^{iB_\nu} \quad (3.10)$$

Finally, we use the operator identity

$$e^x e^y = e^{x+y+\frac{1}{2}[x,y]+\cdots} \quad (3.11)$$

and, after some easy algebra,

$$UUUU \approx e^{ia(\partial_\mu B_\nu - \partial_\nu B_\mu) - [B_\nu, B_\mu]} \tag{3.12}$$

However, using the definition

$$B_\mu(n) = \tfrac{1}{2}ag\tau_i A_\mu^i(n) \equiv agA_\mu(n) \tag{3.13}$$

we can identify the conventional Yang–Mills field strength $F_{\mu\nu} = (\partial_\mu A_\nu - \partial_\nu A_\mu) - ig[A_\nu, A_\mu]$ in the action,

$$UUUU \approx e^{ia^2 g F_{\mu\nu}} \tag{3.14}$$

where the corrections in the exponent are of higher order in a^2 and will not contribute to the classical continuum limit. In particular, we consider only smooth, classical fields that have long wavelengths on the scale of the lattice cutoff a,

$$a^2 g F_{\mu\nu} \ll 1 \tag{3.15}$$

so then we can simplify the action,

$$\operatorname{tr} UUUU \approx \operatorname{tr} e^{ia^2 g F_{\mu\nu}} = \operatorname{tr}\bigl(1 + ia^2 g F_{\mu\nu} - \tfrac{1}{2}a^4 g^2 F_{\mu\nu}^2 + \cdots\bigr) \tag{3.16}$$

$$\approx \operatorname{tr} 1 - \tfrac{1}{2}a^4 g^2 \operatorname{tr} F_{\mu\nu}^2 + \cdots \tag{3.17}$$

where $\operatorname{tr} F_{\mu\nu} = 0$ because the trace of a generator vanishes. Finally, we should write $\operatorname{tr} F_{\mu\nu}^2$ in terms of the vector potential A_μ^i,

$$\operatorname{tr} F_{\mu\nu}^2 = \operatorname{tr}\bigl[\tfrac{1}{2}\tau_i\bigl(\partial_\mu A_\nu^i - \partial_\nu A_\mu^i - g\epsilon_{kli} A_\mu^k A_\nu^l\bigr)\bigr]$$
$$\times \bigl[\tfrac{1}{2}\tau_{i'}\bigl(\partial_\mu A_\nu^{i'} - \partial_\nu A_\mu^{i'} - g\epsilon_{k'l'i'} A_\mu^{k'} A_\nu^{l'}\bigr)\bigr] \tag{3.18}$$

where we used $[\tau_i, \tau_j] = 2i\epsilon_{kij}\tau_k$. Finally, using $\operatorname{tr} \tau_i \tau_j = 2\delta_{ij}$, we have

$$\operatorname{tr} F_{\mu\nu}^2 = \tfrac{1}{2}\bigl(\partial_\mu A_\nu^k - \partial_\nu A_\mu^k - g\epsilon_{kij} A_\mu^i A_\nu^j\bigr)^2 \tag{3.19}$$

which is the square of the gauge-covariant field-strength tensor. Now the action can be written using continuum notation,

$$S = \frac{1}{2g^2} \int \tfrac{1}{2}\bigl(\partial_\mu A_\nu^k - \partial_\nu A_\mu^k - g\epsilon_{kij} A_\mu^i A_\nu^j\bigr)^2 + O(a^2) \tag{3.20}$$

where we used the familiar replacement,

$$\sum_{n,\mu\nu} \rightarrow \int \frac{d^4x}{a^4} \sum_{\mu\nu} \qquad (3.21)$$

which is adequate for smooth fields that vary little on the scale of the lattice spacing a, and a factor of 2 has appeared from the Hermitian conjugate.

Now, collecting everything, and naively taking the lattice spacing to zero,

$$S = \frac{1}{4} \int d^4x \left(F_{\mu\nu}^i \right)^2 \qquad (3.22)$$

which we identify as the usual Euclidean action of the classical Yang–Mills approach.

This result has several important properties.

1. The final result is Euclidean O(4) invariant. This is crucial for physical applications. The result is familiar and expected: the analysis was done only for fields that vary negligibly on the scale of the lattice spacing a, for which differences between the cubic and full rotational symmetry are significant. Therefore, classically one expects that "full rotational symmetry will be a property of the continuum limit of the lattice physics." If one pursued the $O(a^2)$ corrections to the lattice action, one would indeed find dependences on the difference between four-dimensional cubic and O(4) symmetry.

2. The final result has the standard gauge invariance of traditional Yang–Mills theory and is proportional to the square of the gauge-covariant field-strength tensor, $F_{\mu\nu}^i$. The local implementation of gauge invariance through elements of the gauge group in the lattice construction guaranteed that the classical continuum limit had the local invariance described by the algebra of the gauge group.

There is a great deal more to be said about the continuum limit of the lattice theory in the quantum case. We will return to this subject after we have discussed further properties of the cutoff lattice theory. The major issues center on the construction of operators, products of fields, that can be written down and could contribute to the the action and be compatible with the spacetime and other symmetries the theory possesses.

3.2 Confinement and the strong-coupling limit

One of the most interesting and important features of strong-coupling lattice-gauge theory is the property of confinement. We will learn that the pure gauge-field dynamics are highly nonlinear and the force between a widely separated quark and anti-quark rises linearly with the distance between them. The non-Abelian version of Gauss' law implies that one unit of flux emanates from the quark and is absorbed by the anti-quark. Unlike in weak-coupling quantum electrodynamics, the flux does not spread out in the typical dipole spatial pattern, but rather forms a flux tube that is long and thin.

How do we see this behavior in Euclidean lattice-gauge theory?

We begin with a *Gedankenexperiment*. Imagine creating a $Q\bar{Q}$ pair at space-time point $x_\mu = 0$ and adiabatically separating them to a relative distance R. We hold this configuration for a time $T \to \infty$. Finally, we bring the $Q\bar{Q}$ pair back together and let them annihilate. The Euclidean amplitude for this process is the matrix element of the Hamiltonian evolution operator e^{-HT} in the state $|Q\bar{Q}\rangle$ in which the members of the $Q\bar{Q}$ pair are a distance R apart. Note that we are ignoring the contribution to the amplitude coming from separating and rejoining the members of the $Q\bar{Q}$ pair because we are interested in the leading behavior of the amplitude as $T \to \infty$.

The answer can be expressed as a path integral,

$$\langle Q\bar{Q}|e^{-HT}|Q\bar{Q}\rangle = \int \left[DA_\mu^a\right] e^{-S + ig\int A_\mu^a J_\mu^a \, d^4x} \bigg/ \int\!\!\int \left[DA_\mu^a\right] e^{-S} \quad (3.23)$$

where we have written the functional integral without concerns about gauge fixing because they do not effect this generic argument and because the lattice form of this construction is manifestly gauge-invariant and finite, as we will discuss below. The current J_μ^a should describe the closed worldline C of the creation and annihilation of the $Q\bar{Q}$ pair. Then our expression for the amplitude simplifies to

$$\langle Q\bar{Q}|e^{-HT}|Q\bar{Q}\rangle = \int \left[DA_\mu^a\right] e^{-S + ig\int_C A_\mu^a \frac{1}{2}\lambda^a \, dx_\mu} \bigg/ \int\!\!\int \left[DA_\mu^a\right] e^{-S} \quad (3.24)$$

Now, use the fact that the state $|Q\bar{Q}\rangle$ represents a static $Q\bar{Q}$ pair whose members are a distance R apart, so H is purely potential and $\langle Q\bar{Q}|e^{-HT}|Q\bar{Q}\rangle = e^{-V(R)T}$, which produces the heavy quark potential $V(R)$. Taking the logarithm of our expression gives

$$V(R) = -\lim_{T\to\infty} \frac{1}{T} \ln \left\langle \text{trP} e^{ig\int_C A_\mu^a \frac{1}{2}\lambda^a \, dx_\mu} \right\rangle / \langle \text{tr} \, 1 \rangle \quad (3.25)$$

where P ("path-ordered") guarantees that the operator ordering along the closed path is maintained.

Our last expression contains the famous "loop-correlation" function which is so basic to this subject,

$$Z(J) = \left\langle \mathrm{tr} \, \mathrm{P} \, e^{\mathrm{i}g \int_C A_\mu^a \frac{1}{2}\lambda^a \, dx_\mu} \right\rangle \tag{3.26}$$

which serves as an order parameter of pure gauge-field dynamics, as we shall see shortly.

Let's calculate the loop-correlation function in the strongly coupled lattice version of pure gauge-field dynamics. Clearly such a calculation gives us the heavy-quark potential and provides important insights into strong-coupling gauge-field dynamics. The loop-correlation function is the path-ordered product of the exponential phase factors constructed out of the gauge-field vector potential. These exponential factors indicate the locally gauge-invariant coupling of the colored heavy quark with the colored gauge fields. Similar factors but without the non-Abelian color matrix should be familiar to the reader from quantum electrodynamics – such phase factors are needed in order to make point-split fermion bilinears gauge-invariant and were introduced by J. Schwinger in the earliest days of quantum-field theory. Let's briefly review this construction since it is so central to lattice gauge theory, confinement, and the axial anomaly.

Schwinger began his discussion with the current operator in quantum electrodynamics,

$$J_\mu(x) = \langle \bar{\psi}(x)\gamma_\mu \psi(x) \rangle \tag{3.27}$$

However, when quantum-field-theory calculations are done using this bilinear form, singularities that must be interpreted and handled with care can appear. To isolate these singularities, Schwinger introduced the point-split version of the quantum electromagnetic current operator,

$$J_\mu(x, \epsilon) = \langle \bar{\psi}(x + \epsilon)\gamma_\mu \psi(x) \rangle \tag{3.28}$$

where ϵ is a four vector that must vanish at the end of a physical calculation. The point-split character of this operator causes new problems in a gauge theory like quantum electrodynamics. Recall that, under a local gauge transformation in which the vector potential transforms as $A_\mu(x) \to A_\mu(x) + \partial_\mu \Lambda(x)$, a fermion field of charge e transforms through a phase

$$\psi(x) \to e^{ie\Lambda(x)}\psi(x) \tag{3.29}$$

Clearly our point-split form for the current $J_\mu(x)$ is not gauge-invariant. However, Schwinger realized that it could be modified with the addition of a gauge-field phase,

$$J_\mu(x, \epsilon) = \left\langle \bar{\psi}(x + \epsilon)\gamma_\mu e^{ieA(x)\cdot\epsilon}\psi(x) \right\rangle \tag{3.30}$$

which makes the operator locally gauge-invariant, as desired. It was with this operator that Schwinger originally discussed the axial anomaly of quantum electrodynamics.

Schwinger's discussion has a great deal in common with our earlier discussion of local gauge invariance in lattice-gauge theory. In the non-Abelian version of his discussion, the fermion fields carry a color index $\psi_a(x)$ and, when a local rotation in color space occurs at x, the Fermi field rotates accordingly,

$$\psi_a(x) \to \Sigma_b \left[e^{ig\vec{\Lambda}(x) \cdot \vec{\tau}} \right]_{ab} \psi_b(x) \tag{3.31}$$

We imagine a color frame of reference at each spacetime point, and local gauge transformations as implementing rotations on these coordinate systems. The relative orientation of these coordinate systems is provided by the vector gauge fields and two color coordinate systems separated by a four vector ϵ are related by the rotation $e^{ig\frac{1}{2}\tau_a A_a(x) \cdot \epsilon}$. The construction of a gauge-invariant color bilinear fermion current is now clear, generalizing Schwinger's work from quantum electrodynamics.

In all of this we see again how exponentials of the color vector potential become the building blocks of the gauge-invariant dynamics.

Now we can return to the loop-correlation function and write down its lattice realization. The lattice version of the loop-correlation function is clearly

$$Z(C) = \left\langle \prod_C U_\mu(n) \right\rangle \tag{3.32}$$

where C is some closed path of links on the four-dimensional spacetime lattice.

Now we can ask for the behavior of the expectation value of $Z(C)$ in the strongly coupled lattice theory. We need to calculate

$$Z(C) = \int \prod_{n,\mu} [dU_\mu(n)] \prod_C U_\mu(n) e^{-S} \bigg/ \int \prod_{n,\mu} [dU_\mu(n)] e^{-S} \tag{3.33}$$

We want to confirm that the strongly coupled theory confines quarks through a linear potential and that the gauge field forms a flux tube between the quark and its anti-quark partner.

To evaluate the expression for $Z(C)$ we need to evaluate the sum over all the gauge-field configurations, i.e. the integrals $\int \prod_{n,\mu} [dU_\mu(n)]$. We could parametrize the group with generalized angles, but all we will need are two properties of these group integrals, called Haar measures in real-analysis texts.

The first is that these integrals are finite,

$$\int [dU] = 1 \tag{3.34}$$

where we have dropped the index μ for convenience. The second is the group property,

$$\int [dU] f(U) = \int [dU] f(U_\circ U) \tag{3.35}$$

where $U_\circ$ is an arbitrary element of the group and f is a sensible function.

Two integrals that we need are

$$\int [dU] U_{ij} = 0 \tag{3.36}$$

which states that the average over the group of a group element vanishes – there is no group-singlet contribution in U_{ij}.

Next we need the integral over the product of two link variables,

$$\int [dU] U_{ij} U_{kl}^\dagger = \frac{1}{2} \delta_{il} \delta_{jk} \tag{3.37}$$

which one can easily check by using the group invariance of the Haar measure and the unitary character of each link variable (this determines the factor of $\frac{1}{2}$).

Finally, we are interested at this moment only in the strong-coupling case, $\beta = g^{-2} \ll 1$, so we can expand each factor of $e^{-S} = \prod_P e^{-\beta \, \mathrm{tr}\, UUUU} \approx 1 - \beta \, \mathrm{tr}\, UUUU$ in the matrix element of the loop-correlation function $\langle \prod_C U_\mu(n) \rangle$. Take a rectangular loop C for clarity and imagine how we could obtain a nonzero contribution to $Z(C)$. If a term in the expansion of $Z(C)$ in powers of β has a single U factor on a particular link, its contribution to the matrix element will be zero. It is clear that factors of $\beta \, \mathrm{tr}\, UUUU$ from the expansion of the action must include each link of the loop operator $\prod_C U_\mu(n)$ so that each link on the curve C has a U and a $U^\dagger$ that can average to a nonzero value. Considerations such as these lead us to the conclusion that the first nonzero contribution to $\langle Z(C) \rangle$ must have a factor of $\beta \, \mathrm{tr}\, UUUU$ for each plaquette making up the interior of the rectangle C. Some algebra and counting then gives the result

$$\left\langle \prod_C U_\mu(n) \right\rangle = \left(\frac{1}{g^2} \right)^{N_C} = \exp(- \ln g^2 \cdot \mathrm{Area}) \tag{3.38}$$

where N_C is the number of plaquettes making up the rectangle bounded by the contour C. Since the contour is a rectangle of base R and height T, we can read off the heavy-quark potential here,

$$V(R) = \sqrt{\sigma}|R| \tag{3.39}$$

and identify the string tension in strong coupling,

$$\sqrt{\sigma} = \ln g^2 + \cdots \tag{3.40}$$

This little calculation also illustrates that the excess energy density in the gauge-field configuration is in a flux tube between the quark and the anti-quark. A time slice of the RT contour shows that the extra gauge-field flux and energy are on the line between the sources and sinks of color flux.

This calculation can be extended to a power series in the coupling β and one can argue, on the basis of counting diagrams, that the radius of convergence of this strong-coupling calculation is nonzero. This means that there is a finite extent in the variable $1/g^2$ where confinement, namely a linearly rising heavy-quark potential, is a true property of the theory. Of course, this tells us little about the continuum limit of the theory. There will be more about that challenging topic in the chapters ahead.

This famous calculation produces the "area" law for the loop-correlation function. We can label the phase of the large-g^2 region of the lattice theory as "confining." In this case the loop-correlation function falls off rapidly with the dimensions of the loop C and the phase is referred to as "disordered."

What are other possible, physical behaviors for the expectation value of the loop-correlation function? In a free-field theory in which all the U matrices are unity, we have the "perimeter law"

$$\left\langle \prod_C U_\mu(n) \right\rangle = \exp(-2mT) \tag{3.41}$$

and we recognize the heavy-quark potential $V(R) = 2m$, which is just the sum of the masses of the quark and anti-quark.

Another possibility is given by the single-photon exchange expected in weak-coupling electrodynamics. In this case a perturbation-theory calculation gives

$$\left\langle \prod_C U_\mu(n) \right\rangle = \exp[-\alpha T/(4\pi R)] \tag{3.42}$$

where $\alpha = e^2/(4\pi)$ and we have Coulomb's law $V(R) = -\alpha/(4\pi R)$. Naturally, this possibility is called the "Coulomb phase."

3.3 Confinement mechanisms in two and four dimensions: vortex and monopole condensation

There are several lattice models and continuum field theories in which confinement can be illustrated in great detail. We will consider one such model because it is particularly simple, and yet brings up many important issues. We shall also be able to build on our work with the planar model to address confinement and screening. Although most researchers believe that QCD confines quarks, the mechanism for this "miracle" is a matter of debate. Is it due to monopole loops hidden in non-Abelian configurations, vortex sheets, etc?

The model we will consider is the electrodynamics of the planar model and the action reads

$$S = 2\kappa^2 \sum_{r,\mu} \cos(\Delta_\mu \theta(r) + B_\mu(r)) + \frac{1}{2g^2 a^2} \sum_{r,\mu,\nu} \cos(\Delta_\mu B_\nu(r) - \Delta_\nu B_\mu(r))$$

$$(3.43)$$

$B_\mu(r) = ag A_\mu(r)$, and $A_\mu(r)$ is the lattice version of the vector potential of QED, with its usual normalization conventions. Note that both terms in the action are periodic. The electrodynamics terms are locally gauge-invariant,

$$B_\mu(r) \to B_\mu(r) + \Delta_\mu \Lambda(r), \qquad \theta(r) \to \theta(r) - \Lambda(r) \qquad (3.44)$$

The idea behind the model is that there are matter fields on the sites carrying charge g and coupled to the gauge fields $B_\mu(r)$ which reside on links. The matter fields are represented by planar model phases.

It will be particularly interesting to see the new, modified role of the vortices of the planar model in this setting and see that their topological structure plays a central role in the confining features of this model. To study confinement, we set up the Wilson loop-correlation function,

$$\left\langle \exp\left(ie \int_C A_\mu \, dx^\mu\right) \right\rangle = \frac{\int \prod_r d\theta(r) \prod_{r',\mu} dA_\mu(r') \exp(-S + ie \int_C A_\mu \, dx^\mu)}{\int \prod_r d\theta(r) \prod_{r',\mu} dA_\mu(r') \exp(-S)}$$

$$(3.45)$$

Note that we are labeling the "internal," dynamical charge g and the "external," impurity charge e. We will see that the distinction will prove useful. The contour C in the loop-correlation function will be chosen to be a rectangle with temporal extent T much greater than its spatial extent R. As discussed earlier, we view the contour as the worldline of a heavy quark and the energy required to separate a heavy quark from a heavy anti-quark a distance $|R|$ away is obtained from the loop-correlation function through the relation

$$E(R) = -\lim_{T \to \infty} \frac{1}{T} \ln\left[\left\langle \exp\left(ie \int_C A_\mu \, dx^\mu\right)\right\rangle\right] \qquad (3.46)$$

We shall see that $E(R)$ will have qualitatively different behavior depending on the possible values of the ratio of the "external" and the "internal" charges, e/g.

Next we must write a lattice form of the loop integral, $\langle \int_C A_\mu \, dx^\mu \rangle$. Introduce the vector field $J_\mu(r)$ and let it equal $+1$ if the link $r \to r + \mu$ is on the contour C, -1 if the link $r + \mu \to r$ is on the the contour C, and zero otherwise. Then we can write the loop integral, using $B_\mu = agA_\mu$, as

$$e \int_C A_\mu \, dx^\mu \to \frac{e}{g} \sum_C B_\mu J_\mu \qquad (3.47)$$

Finally, if we define

$$Z(J) = \int \prod_r d\theta(r) \prod_{r',\mu} dA_\mu(r') \exp\left(-S + i\frac{e}{g} \sum_C B_\mu J_\mu\right) \qquad (3.48)$$

then the loop-correlation function is the ratio $Z(J)/Z(0)$.

In order to make analytic progress, we borrow from the tricks we assembled for the planar model. In particular, we make Villain-model replacements for both terms in the action,

$$e^{-2\kappa^2[1-\cos(\Delta_\mu\theta - B_\mu)]} \to \sum_{n_\mu=-\infty}^{\infty} e^{in_\mu(\Delta_\mu\theta - B_\mu)} e^{-n_\mu^2/(4\kappa^2)}$$

$$\exp\left(-\frac{1}{2g^2a^2}[1 - \cos(\Delta_\mu B_\nu - \Delta_\nu B_\mu)]\right) \to \sum_{l_{\mu\nu}=-\infty}^{\infty} e^{il_{\mu\nu}(\Delta_\mu B_\nu - \Delta_\nu B_\mu)}$$

$$\times e^{-g^2a^2l_{\mu\nu}^2/4} \qquad (3.49)$$

Now, following the manipulations introduced for the planar model, the integrals over $\theta(r)$ and $B_\mu(r)$ can be done and they generate constraints among the integer-valued auxiliary fields. The partition function $Z(J)$ reads

$$\prod_{r,\mu,\nu} \sum_{l_{\mu\nu}(r)} \prod_{r',\mu'} \delta_{\Delta_\nu l_{\mu\nu}(r') + (e/g)J_\mu(r') + n_\mu(r'),0}$$

$$\prod_{r''} \sum_{n_\nu(r'')} \prod_{r'''} \delta_{\Delta_\mu n_\mu(r'''),0} e^{-n_\mu^2(r'')/(4\kappa^2)} e^{-g^2a^2l_{\mu\nu}^2(r)/4} \qquad (3.50)$$

It is best to solve the constraints explicitly and reduce the number of auxiliary, integer-valued fields to a minimum. Since n_μ and J_μ are divergence-free, they can be written as the curls of other integer-valued fields,

$$n_\mu = \epsilon_{\mu\nu} \nabla_\nu n, \qquad J_\mu = \epsilon_{\mu\nu} \nabla_\nu J \tag{3.51}$$

Finally, the constraint

$$\nabla_\nu l_{\mu\nu} + \frac{e}{g} J_\mu + n_\mu = 0 \tag{3.52}$$

can be solved,

$$l_{\mu\nu} = m^\mu (m \cdot \nabla)^{-1} \left(\frac{e}{g} J^\nu + n^\nu \right) - m^\nu (m \cdot \nabla)^{-1} \left(\frac{e}{g} J^\mu + n^\mu \right) \tag{3.53}$$

where $(m \cdot \nabla)^{-1}$ indicates a discrete line integral along the direction m. On choosing m to be the 2 direction,

$$(m \cdot \nabla)^{-1} f(r_1, r_2) = \sum_{m'=-\infty}^{r_2} f(r_1, m') \tag{3.54}$$

These auxiliary sums will soon disappear from the physics, but these and their generalizations to higher dimensions are important technical details in lattice-gauge theory.

Finally, we write the exponentials in our last expression for the partition function in terms of n and J,

$$n_\mu n_\mu = (\nabla_\mu n)^2, \qquad l_{\mu\nu} l_{\mu\nu} = 2 \left(\frac{e}{g} J + n \right)^2 \tag{3.55}$$

so $Z(J)$ becomes

$$Z(J) = \sum_{n(r)=-\infty}^{\infty} \exp\left(-\frac{1}{4\kappa^2} \sum_r (\Delta_\mu n(r))^2 - \frac{g^2 a^2}{2} \sum_r \left(n(r) + \frac{e}{g} J(r) \right)^2 \right) \tag{3.56}$$

which we recognize as a generalization of the interface-roughening model we discussed earlier in the context of the planar model. We see that, if $g^2 a^2$ is appreciable, then the sum over $n(r)$ is quickly convergent and this is a useful form for the partition function. We are more interested in small $g^2 a^2$, near the continuum limit, $a \to 0$, so we borrow the Poisson summation formula from our earlier discussion of the planar model in order to obtain a useful expression for small

$g^2 a^2$. Recall the identity

$$\sum_{n=-\infty}^{\infty} g(n) = \sum_{m=-\infty}^{\infty} \int_{-\infty}^{\infty} d\phi \, g(\phi) e^{2\pi i m \phi} \tag{3.57}$$

and apply this identity to $Z(J)$ in order to write it in terms of an ordinary scalar field $\phi(r)$ and vortices $m(r)$,

$$Z(J) = \int_{-\infty}^{\infty} \prod_r d\phi(r) \sum_{m(r)=-\infty}^{\infty} \exp\left[-\frac{1}{4\kappa^2} \sum_r (\Delta_\mu \phi)^2 - \frac{g^2 a^2}{2} \right.$$
$$\left. \times \sum_r \left(\phi + \frac{e}{g} J\right)^2 + 2\pi i \sum_r m(r)\phi(r) \right] \tag{3.58}$$

where we recognize that ϕ is a massive scalar field with mass $m_s = \sqrt{2}\kappa g$.

Before continuing with our investigation of quark confinement in this model, we should interpret our findings at this point. The construction of the model started with the planar model, which was then coupled in a gauge-invariant fashion to an Abelian gauge field. The planar model had two phases, a low-temperature spin-wave phase in which the spin–spin correlation function is power-behaved and a high-temperature vortex-condensate phase in which the spin–spin correlation function falls exponentially over the distance between the spins. We learn from this expression for the partition function that, when the gauge field is turned on, the spin wave becomes a massive scalar field. This weak-coupling behavior is analogous to the Higgs mechanism of four-dimensional electroweak models. In those cases one has a Goldstone boson, which, when it is coupled to an otherwise-massless gauge boson, acquires a mass when a global part of the full gauge symmetry is spontaneously broken. This analogy is not perfect, of course, since infrared singularities forbid the existence of Goldstone bosons in two dimensions, but we can state with full accuracy that, for large κ, the planar model has a power-behaved spin–spin correlation function, while for large κ and small electrodynamic coupling ga, the theory's correlation functions would be short-ranged, with $m_s = \sqrt{2}\kappa g$ setting the scale in the exponentials. The periodicity of the original action would be irrelevant in this phase of the full model, the string tension would vanish, and all interactions would be short-ranged, characterized by the screening (correlation) length $\xi = 1/m_s$. By contrast, for small κ and nonzero ga, we shall see that the vortices in the lattice model are relevant and cause confinement through the formation of flux tubes.

To back up these claims, we take the last expression for the partition function, integrate out the scalar field ϕ, and obtain, up to an unimportant overall scale factor,

$$
\begin{aligned}
Z(J) = \sum_{m(r)=-\infty}^{\infty} \exp\Bigg(& -4\pi^2\kappa^2 \sum_{r,r'} m(r)G(r-r';m_s a)m(r') \\
& - 4\kappa^2\pi i \frac{e}{g}(g^2 a^2) \sum_{r,r'} m(r)G(r-r';m_s a)J(r') \\
& + 4\kappa^2(g^2 a^2)^2 \frac{e^2}{g^2} \sum_{r,r'} J(r)G(r-r';m_s a)J(r') \\
& - \frac{1}{2} g^2 a^2 \frac{e^2}{g^2} \sum_{r} J^2(r) \Bigg)
\end{aligned}
\tag{3.59}
$$

where $G(r;m_s a)$ is the lattice propagator for a massive scalar field,

$$
G(r;m_s a) = \int_{-\pi}^{\pi} \frac{dk_x}{2\pi} \int_{-\pi}^{\pi} \frac{dk_y}{2\pi} \frac{e^{i\vec{k}\cdot\vec{r}}}{4 - 2\cos k_x - 2\cos k_y + m_s^2 a^2} \tag{3.60}
$$

To expose the physics here, consider the case $J = 0$, the partition function itself,

$$
Z(J=0) = \sum_{m(r)=-\infty}^{\infty} \exp\left(-4\pi^2\kappa^2 \sum_{r,r'} m(r)G(r-r';m_s a)m(r') \right) \tag{3.61}
$$

which expresses the partition function in terms of integer-valued fields $m(r)$ interacting through a short-ranged potential $G(r;m_s a)$. In contrast to the case in the planar model, in which the bare interaction between the vortices was long-ranged and singular in the infrared, here the mass m_s renders the bare interaction short-ranged and well-posed. The term $r = r'$ in the sum produces the vortex self-energy,

$$
G(0;m_s a) \sim -\frac{1}{2\pi} \ln(m_s a) \tag{3.62}
$$

so, if the vortices are dilute so that their short-ranged interactions can be neglected, the partition function becomes simply

$$
Z(J=0) \approx \sum_{m(r)} \exp\left(2\pi\kappa^2 \ln(\sqrt{2}\kappa g a) \sum_{r} m^2(r) \right) \tag{3.63}
$$

Now return to the loop-correlation function $Z(J)$ and estimate it under these conditions for which the vortex interactions can be ignored. Note that the function $J(r)$ is $+1$ inside the loop and zero otherwise. This behavior guarantees that $J_\mu = \epsilon_{\mu\nu}\Delta_\nu J$ is $+1$ on the left-hand vertical part of the contour C and is -1 on the right-hand vertical part, representing the charge flow of the heavy quark–anti-quark pair. Aside from an overall factor of $g^2 a^2 (e^2/g^2)$, the last two terms in the expression for $Z(J)$ become

$$(m_s a)^2 \sum_{r,r' \text{ inside } C} G(r - r'; m_s a) - RT \tag{3.64}$$

Let's show that this expression vanishes to leading order in RT. Consider both R and T large compared with $(m_s a)^{-1}$. Consider r' inside and not near the contour. Then the sum over r can be done because the extremities in the sum can be ignored. Using

$$\sum_r G(r; m_s a) = \frac{1}{(m_s a)^2} \tag{3.65}$$

we see that

$$(m_s a)^2 \sum_{r' \text{ inside } C} \left(\sum_r G(r; m_s a) \right) - RT = (m_s a)^2 \frac{1}{(m_s a)^2} RT - RT = 0 \tag{3.66}$$

to leading order in RT. Similar approximations give

$$\sum_{r,r'} m(r) G(r - r'; m_s a) J(r') = \frac{1}{m_s^2} \sum_{r \text{ inside } C} m(r) \tag{3.67}$$

So, the final expression for $Z(J)$ becomes

$$Z(J) \approx \sum_{m(r)} \exp\left(2\pi\kappa^2 \ln(\sqrt{2}\kappa g a) \sum_r m^2(r) - 2\pi i \frac{e}{g} \sum_{r \text{ inside } C} m(r) \right) \tag{3.68}$$

We are particularly interested in a parameter set for which the self-energy of each vortex is large, so that the sum over vortices can be truncated after $m(r) = 0, \pm 1$. Then, if r is inside the contour C,

$$\sum_{m(r)} \exp\left(2\pi\kappa^2 \ln(m_s a) m^2(r) - 2\pi i \frac{e}{g} m(r) \right)$$

$$\approx 1 + 2 \exp[2\pi\kappa^2 \ln(m_s a)] \cos\left(2\pi \frac{e}{g} \right)$$

$$\approx \exp\left[2 e^{2\pi\kappa^2 \ln(m_s a)} \cos\left(2\pi \frac{e}{g} \right) \right] \tag{3.69}$$

and we have for the loop-correlation function

$$\frac{Z(J)}{Z(0)} \approx \exp\left[2(m_s a)^{2\pi\kappa^2} \sum_{r \text{ inside } C} \cos\left(2\pi\frac{e}{g}\right) - 1\right] \qquad (3.70)$$

So

$$\frac{Z(J)}{Z(0)} \approx \exp\left\{-2(m_s a)^{2\pi\kappa^2}\left[1 - \cos\left(2\pi\frac{e}{g}\right)\right]RT\right\} \qquad (3.71)$$

from which we read off

$$E(R) = 2m_s^2(m_s a)^{2\pi\kappa^2-2}\left[1 - \cos\left(2\pi\frac{e}{g}\right)\right]R \qquad (3.72)$$

where we have restored physical units (i.e. R could be in GeV^{-1}).

This is the major result of this discussion. On choosing parameters so that the dilute-vortex condition holds, we see that the vortices disorder the system and lead to confinement through a linear potential. The dependence of the coefficient of the confining potential is significant: it is a periodic function of e/g, the ratio of the external and the internal charges. If we take the external charge to match the internal, dynamical charge, we see that the $|R|$ piece of the heavy-quark potential vanishes. We interpret this as screening: as the heavy quark and anti-quark are adiabatically separated on the contour C, it is energetically favorable for a dynamical pair to cancel out the charges of the "impurities" locally, screening the long-range confining electric flux entirely. If, however, we choose the dynamical charge to be twice the charge of the impurity, $g/e = 2$, then the screening cannot occur and there is a confining $|R|$ potential. This is significant because of the absence of the sort of screening that accompanies a plasma (Higgs) system. There, the charge is itself indefinite because of unbounded vacuum fluctuations and any charge impurity would be screened dynamically: an infinitesimal charge, or a large external charge, would all be screened to zero and a long-range confining potential would never occur. In a truly confining system, one expects screening of a different sort. The confinement charge remains an exact, quantized quantum number and the vacuum respects the associated symmetry.

We will not address the continuum limit of this model. We are content to demonstrate its confining and nonconfining features as a function of κ for fixed lattice spacing a. There are discussions in the literature about the character of its continuum limit.

It is natural to ask whether any of these particular methods of analysis cast light on four-dimensional theories. It is interesting that the disordering mechanisms of the planar model and the electrodynamics of the planar model apply to the U(1) gauge theory in four dimensions. Since the U(1) gauge group provides the dynamical basis for QED, our exercise might have more significance than planned. In addition, there are speculative discussions of the confinement mechanism in non-Abelian gauge theories that focus on "maximal Abelian subgroups" and the analysis of confinement here is much like the ideas of confinement in the U(1) theory we will now consider briefly.

The Abelian U(1) lattice-gauge theory in four dimensions follows the construction of the non-Abelian theories above, but it places phases on each link instead of a non-Abelian group element. The action reads

$$S = \beta \sum_{r,\mu,\nu} (1 - \cos \theta_{\mu\nu}(r)) \tag{3.73}$$

where $\theta_\mu(r)$ is an angular variable on the link $r \to r + \mu$, and $\theta_{\mu\nu}(r) \equiv \nabla_\mu \theta_\nu - \nabla_\nu \theta_\mu$ is the discrete form of the field strength.

In order to discuss confinement and the phases of the model we consider the loop-correlation function, $\langle \exp(ie \oint_C A_\mu \, dx^\mu) \rangle$. The analysis of this quantity uses the same tricks as those we presented for the planar model and its electrodynamics above and we shall summarize the resulting formulas below. However, before looking at those technicalities, it is interesting to discuss the physics issues surrounding U(1) gauge theory, which is often referred to as PQED ("periodic" QED). Since the theory contains the photon as its "spin wave," we would expect the theory to have a weakly coupled phase that would reproduce the ordinary electromagnetic field which could be coupled to electrons. This sector of the theory should provide a lattice-regulated version of textbook QED. Results of various studies indicate that this is the case and that the periodicity of the U(1) gauge group is indeed irrelevant with weak coupling. Certainly our work on the planar model prepares us for this result. However, with strong coupling, under which gauge-field configurations should explore the U(1) group thoroughly, the periodic nature of the group should become relevant. This is indeed the case and it shows up by the appearance of "monopole loops" in the partition function. The strong-coupling phase of PQED is a version of "quantum electromagneto dynamics" (QEMD), in which there are conserved monopole currents and invisible Dirac strings. The theory could be coupled to electrons and, at least when the lattice spacing is nonzero, it would confine those electrons through a linear potential. The electric and magnetic charges satisfy a Dirac quantization condition that might guarantee that the theory would be interacting and sensible in a continuum limit for which electrons would be confined in a monopole condensate. Since the theory has a separate weak-coupling phase in which monopoles are irrelevant and electrons propagate freely with interactions with the photons well described by perturbation theory, there must a phase

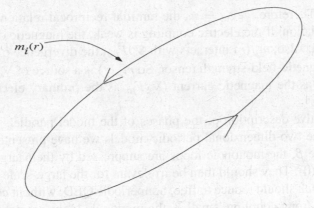

$m_i(r)$

Figure 3.2. A closed monopole loop in four-dimensional compact QED.

transition in the full PQED theory at some intermediate gauge coupling. If this phase transition were of second order, it might have the properties described above and be very intriguing.

Unfortunately, only parts of this scenario have been established. In particular, when the gauge coupling is large and the lattice cutoff is held fixed, the theory has the properties stated above. This can be shown by analyzing the loop-correlation function by the same tricks as those that were introduced for the planar model. The four-dimensional character of the model and the periodic nature of its interactions produce four-dimensional versions of the constraints and auxiliary fields which lead to an expression for the partition function,

$$Z(J) = \exp\left(-\frac{1}{2\beta}\sum_{r,r'}J_\mu(r)V(r-r')J_\mu(r')\right)\sum_{m_\mu(r)=-\infty}^{\infty}\prod_r \delta_{\nabla_\mu m_\mu(r),0}$$

$$\times \exp\left(-2\pi^2\beta\sum_{r,r'}m_\mu(r)V(r-r')m_\mu(r')\right.$$

$$\left. +2\pi i\sum_{r,r'}m_\mu(r)\epsilon_{\mu\nu\lambda\kappa}n_\nu V(r-r')\nabla_\lambda(n\cdot\nabla)^{-1}J_\kappa(r')\right) \qquad (3.74)$$

where $V(r)$ is the four-dimensional massless propagator. Let's interpret the various pieces in this formula. Since $\nabla_\mu m_\mu(r) = 0$, the vector field $m_\mu(r)$ can be visualized as traveling on closed loops as shown in Fig. 3.2. These are the worldlines of magnetic monopoles. This identification shows up in two aspects in the partition function. First we see that the heavy electron, described by the vector field $J_\mu(r)$, couples to the massless propagator $V(r)$ of the photon field, with the strength $g_E^2 = 1/(2\beta)$, and the magnetic loops interact with the strength

$g_M^2 = 2\pi^2\beta$. Therefore, $g_E g_M = \pi$, the familiar reciprocal relation of Dirac. It is a duality relation: if the electric coupling is weak, the magnetic one is strong and vice versa. Also, $m_\mu(r)$ interacts with $\nabla_\kappa \tilde{F}_{\mu\kappa}$, the divergence of the dual of the electromagnetic-field-strength tensor. So $m_\mu(r)$ is a source of $\nabla_\kappa \tilde{F}_{\mu\kappa}$, which we recognize as the magnetic current ($\nabla_\kappa F_{\mu\kappa}$ is the ordinary electromagnetic current).

The qualitative description of the phases of the model parallels our considerations for the two-dimensional periodic models we have presented. At weak coupling, large β, the monopole loops are suppressed by their large activation energy, $2\pi^2 V(0)$. They should then be irrelevant for the large-scale behavior of the theory, which should reduce to free, nonperiodic QED, without confinement. However, for strong coupling, small β, the monopole loops are not suppressed; they can grow to become macroscopic in size and there should be a finite probablity of finding a monopole in any 3 cube. These monopoles disorder the system, suppressing $Z(J)$ and leading to the area law and confinement through a linear potential. One would call this an "electric meissner effect."

One can make a crude estimate for the critical coupling in PQED using an energy–entropy argument. Consider a loop of L steps. Its activation energy is $2\pi^2\beta V(0)L$. The number of loops of length L passing through a given point on the lattice behaves roughly as 7^L. The free energy, action minus entropy times temperature, of such a loop is

$$F = 2\pi^2 V(0)L - T\ln(7^L) = (2\pi^2 V(0) - T\ln 7)L \qquad (3.75)$$

which becomes negative at

$$T_c = \frac{2\pi^2 V(0)}{\ln 7} \qquad (3.76)$$

and a condensate of monopole loops occurs for all couplings $T > T_c = g_c^2 \approx 1.57$, using the numerical result $V(0) \approx 0.155$.

In summary, according to these estimates, PQED has a monopole condensate and confinement for all $g^2 > 1.57$. This result compares favorably with Monte Carlo results even though it has neglected the $1/R$ potentials in the partition function and has overcounted loops, etc. These features appear to be subdominant effects. Computer simulations indicate that the strong-coupling phase is a plasma of condensed monopole loops that generates a finite correlation (screening) length ξ.

One would like to know the nature of the critical point and whether it can be used to define a renormalized, interacting field theory of QEMD. The answer to this question is not known. One would also like to know whether this confinement mechanism is related to that governing QCD. As mentioned above, there are speculations, based on gauge fixing and projecting to Abelian subgroups, that this is true. However, there are other proposals for topologically driven confinement mechanisms in QCD as well. At this time it is not known whether a simple, single mechanism controls confinement in QCD, let alone its physical and mathematical nature. One of the goals of studying QCD in extreme environments is the elucidation of its underlying dynamics, namely the sources of confinement on the one hand and chiral-symmetry breaking (dynamical mass generation for the hadronic mass spectrum) on the other.

We return to non-Abelian gauge theories that enjoy asymptotic freedom.

4

Fermions and nonperturbative dynamics in QCD

4.1 Asymptotic freedom and the continuum limit

The lattice formulation of pure $SU(N)$ gauge fields can be analyzed at strong coupling to show confinement, as we have discussed above. It can also be analyzed at weak coupling, even though a spacetime lattice provides an awkward framework for standard perturbation theory. Asymptotic freedom [12] can be derived in various ways. A relatively simple derivation uses the loop-correlation function and the heavy-quark potential as we have discussed here for strong coupling.

The heavy-quark potential can also be calculated in ordinary perturbation theory, by an expansion in powers of the gauge coupling g^2. To lowest order, the static quark and anti-quark exchange a colored gluon and there results an attractive force given by Coulomb's law,

$$V(R) = -\frac{4g^2/3}{4\pi R} \tag{4.1}$$

Next, quantum fluctuations dress this result and modify its R dependence with the logarithms of perturbation theory. Fermion vacuum-polarization effects modify the single-gluon propagator and cause partial screening of the color charge, reducing the potential from the classical result. On the other hand, color interactions between the gluons lead to partial anti-screening of the color charges of the quarks. The final, gauge-invariant result for the heavy-quark potential is

$$V(R) = -\frac{4/3}{4\pi R}\left(g^2 + \frac{22g^4}{16\pi^4}\ln(R/a)\right) \tag{4.2}$$

and we see a logarithmic correction to Coulomb's law that makes the attraction between the quark and the anti-quark stronger than would be predicted by a classical calculation. The interesting effect is in the logarithm which involves the

74

ratio of the physical distance R and the ultraviolet cutoff a, the lattice spacing. It is instructive and conventional to write this result in terms of a running coupling,

$$g^2(R) = g^2(a) + \frac{22}{16\pi^2}g^4(a)\ln(R/a) \tag{4.3}$$

The left-hand side of this equation should be a physical quantity that is independent of the cutoff a. This means that $g^2(a)$ must be tuned to the cutoff a in just the correct manner so that $g^2(R)$ and the heavy-quark potential do not vary with a. This requires that the lattice coupling $g^2(a)$ must vanish logarithmically in the continuum limit $a \to 0$. Alternatively, one can view the R dependence of $g^2(R)$ and consider the Callan–Symanzik function

$$\beta(R) = -R\frac{\partial}{\partial R}g^2(R) = -\frac{11}{16\pi^2}g^3(R) + \cdots \tag{4.4}$$

The minus sign on the right-hand side implies that the running coupling grows as the distance between the quarks increases and their attraction is more than one would have estimated classically.

Underlying this discussion is the full modern theory of the renormalization group. This includes the study of how the ultraviolet-sensitive parameters of a field theory must vary with the cutoff procedure such that physical phenomena remain independent of the regularization method. We cannot introduce this subject here, but rather refer the reader to basic texts on the subject [2]. We will be using many concepts of renormalization as we develop the physics of dense matter.

To take the continuum limit of the lattice theory we must understand lattice perturbation theory and respect asymptotic freedom, the fact that non-Abelian SU(3) gauge theory with a limited number of quarks in the fundamental representation of the gauge group becomes weakly coupled at short distance at a logarithmic rate. For N_f flavors of quarks, the asymptotic running of the coupling constant reads

$$\frac{dg(R)}{d\ln R} = \beta_0 g^3 + \beta_1 g^5 + \cdots \tag{4.5}$$

where

$$\beta_0 = \frac{1}{16\pi^2}\left(11 - \frac{2}{3}N_f\right), \qquad \beta_1 = \left(\frac{1}{16\pi^2}\right)^2\left(102 - \frac{38}{3}N_f\right) \tag{4.6}$$

and the coefficient β_1 is obtained in a two-loop perturbation-theory calculation. We see that the continuum limit lies far from the strong-coupling, coarse-lattice limit in which we have demonstrated confinement and will later demonstrate chiral-symmetry breaking, the Goldstone mechanism, etc. How then can we be assured that the continuum limit of the lattice theory has these physical properties? Using the full apparatus of the renormalization group, we must show that the confining, chiral-symmetry-breaking features of the theory are analytically connected to the short-distance asymptotic-freedom features of the running coupling constant. Numerical evidence indicating the veracity of this result is accumulating. We shall be assuming throughout this text that all will work out on this front.

4.2 Axial symmetries and the vacuum of QCD

The basic features of QCD consist in its phases and symmetries and their modes of realization. The most fundamental aspects of QCD include its asymptotic freedom, which teaches us that high-resolution experiments such as deep inelastic scattering can detect the pointlike charged constituents of the theory, namely spin-$\frac{1}{2}$ quarks in the fundamental representation of the color gauge group. Its presumed infrared-slavery property posits that the theory is strongly coupled at large distances and that it confines quarks via a linear potential that is generated by electric-flux-tube formation. Confinement is a property of the QCD vacuum, a relativistic electric Meissner effect. In addition, the fundamental spin-1/2 quarks have small bare masses, but the strong dynamics of QCD breaks the approximate chiral symmetry of these fermions and generates considerable constituent-quark masses dynamically. Dynamical chiral-symmetry breaking has the byproduct of producing an SU(3)$_{\text{flavor}}$ multiplet of almost-Goldstone bosons, with the three pions the clearest of the eight mesons filling out the octet, which plays a particularly important role in the low-energy features of hadronic physics. In fact, many of the properties of low-energy hadronic physics are simply consequences of the Goldstone nature of the theory's pions.

The QCD Lagrangian reads, in a standard but abbreviated notation,

$$L = \int d^3r \left(\bar{\psi}(i\,\slashed{\nabla} + gA_\alpha \lambda_C^\alpha)\psi - \frac{1}{4}F_{\mu\nu}^\alpha F_\alpha^{\mu\nu} + \bar{\psi}M\psi \right) \tag{4.7}$$

where λ_C^α are the eight 3×3 color matrices, $F_{\mu\nu}^\alpha$ is the field strength constructed from the non-Abelian potential A_α, and M is the quark-mass matrix. In the case that M vanishes, the eight flavor-bearing, chiral currents,

$$j_{5,i}^\mu = \bar{\psi}\gamma^\mu \gamma^5 F_i \psi \tag{4.8}$$

are conserved. In this expression the F_i are the eight SU(3) flavor matrices and the flavor, color, and spin indices on the fermion fields have been suppressed. If the QCD interactions generate a quark mass dynamically, then these global symmetries are spontaneously broken. Eight massless Goldstone bosons must then appear in the theory's spectrum and their properties are constrained by various low-energy theorems, as will be discussed throughout this book.

One can also check that the ninth axial-vector current associated with flavor-singlet chiral rotations,

$$j_5^\mu = \bar{\psi}\gamma^\mu\gamma^5\psi \tag{4.9}$$

is also conserved formally and should be associated with a ninth, flavor-singlet Goldstone boson. In fact, this "prediction" is a disaster. There is no such light state in the spectrum of QCD...the candidate η' is much too massive and does not have the expected properties of an "approximate" Goldstone particle.

In the early days of current algebra, before the development of QCD, it was realized both in model field theories such as the linear sigma model of the baryons and in QED that the flavor-singlet current's conservation is broken by quantum fluctuations, the famous triangle anomaly. It was hoped that this fact, the need to regulate the underlying field theory, would save the day and explain the heaviness of the would-be Goldstone particle, the η'. These hopes were never realized and the puzzle of the flavor-singlet current remained in the field for many years. In fact, other low-energy theorems associated with j_5^μ, such as the suppression of the low-energy process $\eta \rightarrow 3\pi$, were derived and were also at odds with experiment.

4.3 Two-dimensional fermionic models of confinement, axial symmetries, and θ vacua

Dynamical symmetry breaking is a major theme of this book and we will dissect, model and analyze it repeatedly. However, here we want to discuss the axial flavor-singlet symmetry because, although it is a perfect classical symmetry of massless quarks, it is broken by quantum effects, is sensitive to the strong-coupling, topological excitations in QCD, and also plays an important role in the properties of the theory's vacuum. We will illustrate the issues at play by considering two-dimensional models, the massless and massive Schwinger models, and will discuss the role of instantons in QCD in setting the scale of this symmetry breaking. The reason for looking at solvable models is their clarity in the face of some subtle issues involving boundary conditions, long-range forces, cluster properties of matrix elements, and gauge invariance...all those topics which bedevil the field and are often swept "under the rug."

Some progress in unraveling the puzzle of the flavor-singlet axial symmetry was made by considering the Schwinger model, garden-variety QED in $1 + 1$ dimensions [13]. In this case there is only the flavor-singlet symmetry if there is just one electron. Therefore, the formal chiral flavor-singlet symmetry would predict a massless Goldstone boson, which would be an infrared catastrophe in $1 + 1$ dimensions. In fact, the Schwinger model is solvable and consists of just a free massive neutral boson of mass $m = e/\sqrt{\pi}$, where e is the electric charge. The reader should recall that QED in $1 + 1$ dimensions is a super-renormalizable

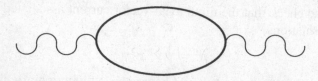

Figure 4.1. A vacuum-polarization loop in the two-dimensional Schwinger model.

theory and its coupling carries the dimensions of mass. The fact that it is super-renormalizable means that interactions can be ignored for distances small compared with $1/e$, the reciprocal of the theory's charge. So, the theory is simple and does not need an elborate renormalization scheme to define it, and we should be able to understand how it evades the disaster of the extra symmetry.

If one considers the Feynman diagrams of the model, one finds that the masslessness of the fermions and conservation of current conspire to render most of them zero. The one-loop correction, shown in Fig. 4.1, to the photon propagator is nonzero; however, it is easy to calculate and is physically relevant. As in four-dimensional QED, write the exact photon propagator in the form dictated by vector-current conservation and local gauge invariance,

$$D_{\mu\nu}(k^2) = \frac{g_{\mu\nu} - k_\mu k_\nu / k^2}{k^2(1 - \Pi(k^2))} \tag{4.10}$$

The one-loop contribution to the propagator is computed following the methods familiar from four-dimensional calculations in which gauge invariance is used to simplify its kinematic structure. The resulting expression is finite and yields

$$\Pi(k^2) = \frac{e^2/\pi}{k^2} \tag{4.11}$$

Note that the form of this answer is dictated by dimensional analysis. The contribution is of second order in e, the charge, so $\Pi(k^2)$ is proportional to e^2, which carries the dimensions mass^2. However, $\Pi(k^2)$ is dimensionless, so a factor of k^2 must appear in its denominator. The factor of π results from the Feynman rules. On substituting back into the expression for the photon propagator, we find

$$D_{\mu\nu}(k^2) = \frac{g_{\mu\nu} - k_\mu k_\nu / k^2}{k^2 - e^2/\pi} \tag{4.12}$$

and the photon has picked up a mass, $m = e/\sqrt{\pi}$, which is Schwinger's famous result that showed that gauge invariance alone does not guarantee a massless photon. We recognize that the model displays a very simple example of the dynamical Higgs mechanism. Now that the photon is massive, the potential between charged static impurites is short-ranged, of order $\epsilon^2 \exp(-m|x|)$, where ϵ

is the charge of the impurity and is arbitrary. We learn that an arbitrary charge is screened to zero in this model. This is a characteristic of the Higgs mechanism; the vacuum is a relativistic plasma. It is not true confinement as we have discussed above because there is no remnant of a long-range potential for any ϵ. We will see, however, that, if the fermion had a nonzero bare mass, confinement rather than the Higgs mechanism would return and there would be a linear potential between impurities separated by a distance $|x|$ if ϵ did not match a multiple of e.

The one-loop graph exhausts the dynamics of the massless Schwinger model. Further analysis shows that the result $m = e/\sqrt{\pi}$ is exact, and the spectrum of the model consists of a neutral, massive boson. Before we turn to a discussion of the vacuum of the Schwinger model, let's consider the conservation laws in the theory in the light of its simple solution.

Conservation of current gives

$$\partial_\mu j^\mu = 0, \qquad j^\mu = \bar{\psi}\gamma^\mu\psi \tag{4.13}$$

However, a conserved vector field can be written as the curl of a scalar,

$$j^\mu = \epsilon^{\mu\nu}\partial_\nu\phi/\sqrt{\pi} \tag{4.14}$$

In $1 + 1$ dimensions, the γ matrices satisfy

$$\gamma^\mu\gamma_5 = \epsilon^{\mu\nu}\gamma_\nu \tag{4.15}$$

so the axial-vector current, $j_\mu^5 = \bar{\psi}\gamma_\mu\gamma_5\psi$, is simply related to the vector current,

$$j_5^\mu = \epsilon^{\mu\nu} j_\nu \tag{4.16}$$

This means that j_μ^5 is related to ϕ,

$$j_5^\mu = \partial^\mu\phi/\sqrt{\pi} \tag{4.17}$$

and its divergence is

$$\partial_\mu j_5^\mu = \Box\phi/\sqrt{\pi} \tag{4.18}$$

However, if we return to Maxwell's equation,

$$ej^\nu = \partial_\mu F^{\mu\nu} \tag{4.19}$$

and realize that, in $1 + 1$ dimensions, where there is no "space" for magnetic effects, the field strength can be written as $F_{\mu\nu} = -\epsilon_{\mu\nu}E$, where E is the electric field, and we can identify the auxiliary field ϕ as

$$\phi = \frac{\sqrt{\pi}}{e}E \tag{4.20}$$

However, our brief analysis of the graphs of the Schwinger model showed us that the electromagnetic field develops a mass of $m = e/\sqrt{\pi}$ due to vacuum polarization, so E satisfies the massive Klein–Gordon equation,

$$(\Box - e^2/\pi)E = 0 \tag{4.21}$$

and our divergence equation for the axial-vector current becomes

$$\partial_\mu j_5^\mu = \frac{e}{\pi} E \tag{4.22}$$

which can be written in the explicitly covariant form

$$\partial_\mu j_5^\mu = \frac{e}{2\pi}\epsilon_{\mu\nu} F^{\mu\nu} \tag{4.23}$$

So, the fermion loop graph breaks the conservation of the flavor-singlet axial current. Recognize this as the $(1 + 1)$-dimensional version of the triangle anomaly – in $1 + 1$ dimensions the lack of conservation of the axial current comes from vacuum polarization and the divergence of the axial current is linear rather than quadratic in the field-strength tensor [14].

We learn that quantum fluctuations, which are not accounted for in classical field-theory discussions of conservation laws, save the day and remove the unwanted symmetry, and its unwanted consequences, from the theory. The theory has no axial-vector symmetry and, therefore, no massless (Goldstone or even light) mesonic state. The ϕ is the Schwinger model's analog of QCD's η' state.

The absence of the flavor-singlet axial symmetry changes the vacuum structure of the theory profoundly. The chiral charge is the zero-frequency mode of the momentum canonically conjugate to ϕ,

$$Q_5 = \int j_5^0 \, \mathrm{d}x = \int \psi^\dagger \gamma_5 \psi \, \mathrm{d}x = \frac{1}{\sqrt{\pi}} \int \partial_0 \phi \, \mathrm{d}x = \frac{1}{\sqrt{\pi}} \int \Pi_\phi \, \mathrm{d}x \tag{4.24}$$

Canonical quantization for ϕ then implies

$$[\phi(x), Q_5] = \mathrm{i}/\sqrt{\pi} \tag{4.25}$$

so the axial charge generates translations in the field ϕ. If α is a constant,

$$e^{\mathrm{i}\alpha\sqrt{\pi}Q_5} \phi e^{-\mathrm{i}\alpha\sqrt{\pi}Q_5} = \phi + \alpha \tag{4.26}$$

Since the chiral group is compact and $\exp(\mathrm{i}n\pi Q_5)$ leaves a Fermi field unchanged, up to an unphysical sign, for any integer n, the equivalent boson wavefunctions of the model should be unchanged under the translation

$$\phi \to \phi + n\sqrt{\pi} \tag{4.27}$$

So, the eigenvalue spectrum of Q_5 consists of the even integers, when one is acting on admissable Bose states. ϕ is effectively an angle variable and Q_5 is its canonically conjugate "angular momentum."

Since the chiral charge changes the ϕ field's boundary values at spatial infinity, one should write ϕ as the sum of a massive ($m = e/\sqrt{\pi}$) field $\hat{\phi}$ and a constant field θ [15]. The chiral transformation acts only on θ,

$$e^{i\alpha\sqrt{\pi}Q_5}\hat{\phi}e^{-i\alpha\sqrt{\pi}Q_5} = \hat{\phi} \tag{4.28}$$

$$e^{i\alpha\sqrt{\pi}Q_5}\theta e^{-i\alpha\sqrt{\pi}Q_5} = \theta + \alpha \tag{4.29}$$

and the chiral charge Q_5 is the momentum conjugate to θ, $Q_5 = (1/\sqrt{\pi})\Pi_\theta$. The full space of states of the model, $|\rangle$, can be spanned by product states: one factor $|\Psi\rangle$ describing the state of the massive field $\hat{\phi}$, and a second factor $|n\}$, where n is an integer, which is an eigenfunction of $\frac{1}{2}Q_5$. We can make raising and lowering operators for the states $|n\}$, $a^\pm = \exp(\pm 2i\sqrt{\pi}\theta)$, which change the chirality by two units,

$$a^\pm|n\} = |n \pm 1\} \tag{4.30}$$

We can also make the eigenfunctions of the raising and lowering operators $a^\pm$. They are coherent states,

$$|\theta_0\} = \frac{1}{\pi^{1/4}} \sum_n e^{-2i\sqrt{\pi}n\theta_0}|n\} \tag{4.31}$$

Note that these states $|\theta_0\}$ have a simple physical interpretation: since ϕ is proportional to the electric field E, the state $|\theta_0\}$ has a background, constant electric field of value $e\theta_0/\sqrt{\pi}$. In the Schwinger model with massless electrons, it costs no energy to screen any external electric field to zero, so only $\theta_0 = 0$ is a physical possibility. This is a consequence of chiral symmetry. θ_0 does not enter the theory's energetics.

In any case, what should the vacuum state be? The perturbative vacuum is $|0\rangle = |0\rangle|n = 0\}$. This is not a good candidate for the exact vacuum of the theory because there might be physical processes that change the chirality of the system and lift the system from $|n = 0\}$. These processes cause mixing between the different $|n\}$, so any particular $|n\}$ is not a stationary state. The "θ vacua" are eigenstates of these processes, so a candidate for the exact vacuum would be $|\theta_0\rangle = |0\rangle|\theta_0\}$ [16]. The complete screening in the model singles out $\theta_0 = 0$.

These considerations become more thought-provoking if we consider the massive Schwinger model, i.e. let the electron have a nonzero bare mass m_0 so we have an additional source of explicit chiral-symmetry breaking. In the massless Schwinger model just discussed the background electric field could always be removed through a chiral rotation, but that will not be true in the massive Schwinger model and we will see that the energetics and confinement mechanism will be closer to the electrodynamics of the planar model than to the massless Schwinger model. The θ vacua of the massive Schwinger model will teach us some subtleties we will see again in four-dimensional QCD [17].

The massive Schwinger model is conveniently analyzed using the equivalent-boson method ("bosonization") introduced for the massless model. The scalar field ϕ is used again and a thorough analysis of Feynman diagrams allows us to establish the equivalences between the model formulated in terms of fermions and one formulated just in terms of ϕ,

$$
\begin{aligned}
j^\mu &= \epsilon^{\mu\nu}\partial_\nu\phi/\sqrt{\pi} \\
j_5^\mu &= \partial^\mu\phi/\sqrt{\pi} \\
\bar{\psi}(1 \pm \gamma_5)\psi &= ce : \exp(\pm 2i\sqrt{\pi}\phi) : \\
\bar{\psi}\gamma^\mu\partial_\mu\psi &= \tfrac{1}{2}[(\partial_0\phi)^2 - (\partial_x\phi)^2]
\end{aligned}
\tag{4.32}
$$

where c is a known constant,[1] and the normal ordering is done with respect to the mass $e/\sqrt{\pi}$. The equivalence of the chiral bilinears $\bar{\psi}(1 \pm \gamma_5)\psi$ to the exponentials of the scalar field ϕ can be demonstrated by considering some mass-insertion graphs in the massless Schwinger model in which ϕ is a free field. The contractions of the exponentials can be done by applying Wick's theorem and the equivalence can be established directly in real x space. For short distances compared with the scale of the Schwinger model mass $e/\sqrt{\pi}$, the boson propagator depends on x logarithmically. Exponentials of the propagator then produce the power-law behavior of fermion propagators and the equivalence is established after some careful algebra and counting of graphs.

If we apply this set of equivalences to the Lagrangian written in the Coulomb gauge and use Maxwell's equations to eliminate the electric field in favor of the long-range confining Coulomb potential of $1 + 1$ dimensions, then

$$
\begin{aligned}
L = {} & \frac{1}{2}\int dx\,[(\partial_0\phi)^2 - (\partial_x\phi)^2] + \frac{mce}{2}\int dx : \cos(2\sqrt{\pi}\phi) : \\
& - \frac{e^2}{4\pi}\int dx\,dx'\,(\partial_x\phi)|x - x'|(\partial_{x'}\phi)
\end{aligned}
\tag{4.33}
$$

We know from our discussion of the massless Schwinger model that we must exercise care in dealing with the last term of L: a careless application of integration by parts to render the term local will fail. The correct procedure is again to write ϕ in terms of a piece $\hat{\phi}$ that falls off quickly at spatial ∞ plus a constant piece, the θ term,

$$
\phi = \hat{\phi} + \theta
\tag{4.34}
$$

----

[1] The value of the dimensionless constant $c = \pi^{-3/2}e^\gamma$ (see e.g. [18]), where $\gamma = 0.577\ldots$ is the Euler constant, is not important for our discussion.

Now,

$$L = \frac{1}{2} \int dx \left((\partial_0 \phi)^2 - (\partial_x \phi)^2 - \frac{e^2}{\pi} \hat{\phi}^2 \right) + \frac{mce}{2} \int dx : \cos[2\sqrt{\pi}(\hat{\phi} + \theta)] : \tag{4.35}$$

We learn that, as long as the fermion mass m is different from zero, the model is sensitive to θ because of the long-range confining forces. This is different from the massless Schwinger model, as we saw above, and suggests that this model confines rather than screens. To check this point, let's calculate the energy between heavy, static arbitrary charges in the model. We need to place equal and opposite external charges a distance $|R|$ apart. In the equivalent-boson language, this is done using

$$j_{ext}^0 = \frac{\epsilon}{\sqrt{\pi}} \partial_x \phi_{ext} \tag{4.36}$$

where ϕ_{ext} is chosen to be nonzero between the external charges,

$$\phi_{ext}^0 = \sqrt{\pi} \, \Theta\left(\frac{1}{2} |R| + x \right) \Theta\left(\frac{1}{2} |R| - x \right) \tag{4.37}$$

On taking the derivatives of the step function Θ to form j_{ext}^0, we see that there are pointlike charges $\pm \epsilon$ at $\pm \frac{1}{2} |R|$. In this environment the potential-energy term in the Lagrangian (4.33) reads

$$-\frac{e^2}{2} \int dx \, dx' \left(j^0(x) + \frac{\epsilon}{e} j_{ext}^0(x) \right) |x - x'| \left(j^0(x') + \frac{\epsilon}{e} j_{ext}^0(x') \right) \tag{4.38}$$

On introducing boson fields as before, and doing the integration by parts, we obtain

$$L = \frac{1}{2} \int dx \left[(\partial_0 \phi)^2 - (\partial_x \phi)^2 - \frac{e^2}{\pi} \left(\hat{\phi} + \frac{\epsilon}{e} \phi_{ext} \right)^2 \right]$$
$$+ \frac{mce}{2} \int dx : \cos[2\sqrt{\pi}(\hat{\phi} + \theta)] : \tag{4.39}$$

Finally, shift $(\epsilon/e)\phi_{ext}$ into the cosine term,

$$L = \frac{1}{2} \int dx \left((\partial_0 \phi')^2 - (\partial_x \phi')^2 - \frac{e^2}{\pi} (\hat{\phi}')^2 \right)$$
$$+ \frac{mce}{2} \int dx : \cos\left[2\sqrt{\pi} \left(\hat{\phi}' + \theta - \frac{\epsilon}{e} \phi_{ext} \right) \right] : \tag{4.40}$$

and, in the region between the external charges $-\frac{1}{2}|R| < x < \frac{1}{2}|R|$, L has the extra energy, proportional to $|R|$,

$$V(R) = \frac{mce}{2}\left\{\cos(2\sqrt{\pi}\theta) - \cos\left[2\sqrt{\pi}\left(\theta - \frac{\epsilon}{e}\sqrt{\pi}\right)\right]\right\}|R| \quad (4.41)$$

where we have set the fluctuating, dynamical field ϕ' to zero.

We learn from this expression that there is a linear confining energy, an unscreened flux tube, for any external charge that is not a multiple of the internal charge e. However, if ϵ is a multiple of e, then the long-range confining force is eliminated. The strength of the confining force is proportional to m. This result is interpreted in the same way as in the electrodynamics of the planar model. The nonzero value of m saves the model from experiencing the Higgs mechanism. The impossibility of screening for a general ϵ means that the theory is sensitive to the value of θ for the vacuum. The long-range forces in the model and its singlet chiral properties are intimately related. θ becomes a physically meaningful, measurable parameter in the solution of the model. Only for $\theta = 0$ will the theory's solution have the discrete symmetries that we expect of the strong interactions.

How much of this physics generalizes to four dimensions? This important, difficult question will be discussed below. The instanton solutions of QCD play an important role.

4.4 Instantons and the scales of QCD

We have discussed models of order–disorder and confinement. When fermions were introduced, axial-singlet symmetries and the vacuum acquired some unexpected, subtle features. Low-dimensional models allowed us to discuss flux-tube formation and symmetry-breaking phenomena that are hard to deal with analytically in four dimensions. Before we turn to a systematic look at lattice-gauge theory, which will require a long detour from the physics of QCD, let's discuss some of the theoretical and phenomenological features of QCD that lattice-gauge theory addresses in a fundamental, unbiased fashion.

In particular let's discuss the instantons of QCD, because they are believed to play an important role in the vacuum structure, chiral-symmetry breaking, and new phases of QCD at high temperature and chemical potentials which are the ultimate goal of this branch of research.

The reader should be aware that the role of instantons in QCD is not settled. However, a great deal of instanton physics will be addressed and their ultimate importance in QCD will be answered in the context of lattice-gauge theory. One of the finest attributes of lattice-gauge theory is that it promises to teach us how QCD works by providing a testbed for theoretical ideas.

To discuss the instantons of QCD, we begin with the theory's partition function after integrating out the fermions,

$$Z = \int DA_\mu \, e^{-S} \prod_f^{N_f} \det(\slashed{D} + m_f), \qquad S = \frac{1}{4g^2} \int d^4x \, F_{\mu\nu}^a F_{\mu\nu}^a \quad (4.42)$$

where the determinant of the Dirac operator $\gamma_\mu(\partial_\mu - iA_\mu)$ accounts for the anti-commuting fermions. When we come to lattice-gauge theory we will also integrate out the fermions in our first step of analyzing the system's partition function because a classical computer cannot handle nonclassical variables, non-commuting fermions, in any other way. This expression for the partition function is not easy to deal with because the fermion determinant hides nonlocal interactions induced by exchanges of light fermions. Nonetheless, in four dimensions it will prove to be a useful starting point in many, but certainly not all, cases.

In a semi-classical analysis of Z, we seek configurations that minimize the classical action S. We reviewed this exercise above in the case of the O(3) model and plotted its instantons. These solutions to its classical equations of motion tend to disorder its spin configurations and play a role in calculations of its correlation function. In the case of QCD, instantons contribute to the disorder in the system, and contribute to chiral-symmetry breaking, the resolution of the U(1) problem, and the inter-quark forces within hadronic bound states. We will touch on these points as we develop the subject [19].

An instructive way to write the pure gauge action for these purposes is

$$S = \frac{1}{4g^2} \int d^4x \left(\pm F_{\mu\nu}^a \tilde{F}_{\mu\nu}^a + \frac{1}{2}\left(F_{\mu\nu}^a \mp \tilde{F}_{\mu\nu}^a\right)^2 \right) \quad (4.43)$$

where $\tilde{F}_{\mu\nu}^a = \epsilon_{\mu\nu\rho\sigma} F_{\rho\sigma}^a$ is the dual field-strength tensor. Since the second term in this expression for S is always positive, the action is a minimum as long as the field is self-dual or anti-self-dual, $F_{\mu\nu}^a = \pm\tilde{F}_{\mu\nu}^a$. The action of such a field configuration is then the first term, which is

$$S = \frac{8\pi^2}{g^2}|Q|, \qquad Q = \frac{1}{32\pi^2} \int d^4x \, F_{\mu\nu}^a \tilde{F}_{\mu\nu}^a \quad (4.44)$$

Field-theory textbooks show that Q is a topological invariant with integral eigenvalues for finite-action field configurations and can be written as an integral over the divergence of a conserved four vector. The instanton field configurations with $Q = 1$ are

$$A_\mu^a(x) = \frac{2\eta_{a\mu\nu}x_\nu}{x^2 + \rho^2}, \qquad \left(F_{\mu\nu}^a\right)^2 = \frac{192\rho^4}{(x^2 + \rho^2)^4} \quad (4.45)$$

where the symbol $\eta_{a\mu\nu}$ was introduced by 't Hooft [20] and reduces to the completely anti-symmetric symbol $\epsilon_{a\mu\nu}$ as μ and ν range from 1 to 3, is $\delta_{a\mu}$ for

$\nu = 4$, and is $-\delta_{av}$ for $\mu = 4$. The parameter ρ characterizes the size of the instanton. The instanton's topological charge is not localized for this particular gauge-dependent configuration $A_\mu^a(x)$. One can choose a different gauge in which the topological charge is at a point singularity,

$$A_\mu^a(x) = \frac{2\eta_{a\mu\nu}x_\nu\rho^2}{(x^2 + \rho^2)x^2} \tag{4.46}$$

In order to specify the instanton completely, one must give its position x_μ, its size ρ, and its color orientation. In addition to instantons, there are also anti-instantons, whose topological charges range over the negative integers.

Instantons can be interpreted as tunneling events between vacua whose topological winding numbers differ by a unit. A pure gauge configuration can be written as $A_k = iU(\vec{x})\partial_k U^\dagger(\vec{x})$, in the temporal gauge $A_0 = 0$. In order to enumerate the classical vacua we must classify the gauge transformations U. U provides a mapping from 3-space onto the gauge group SU(N). Consider gauge transformations U that approach the identity as $x \to \infty$. Then, in the case of the gauge group SU(2), U provides a mapping of R^3 onto R^3 and the equivalence classes of these mappings are well understood. They are labeled by the Pontryagin number, which counts how many times the group manifold is covered and is given by

$$n_P = \frac{1}{24\pi^2}\int d^3x\,\epsilon_{ijk}\,\text{tr}[(U^\dagger\,\partial_i U)(U^\dagger\,\partial_j U)(U^\dagger\,\partial_k U)] \tag{4.47}$$

This expression can also be written in terms of the gauge potentials, in which case it is called the Chern–Simons characteristic,

$$n_{CS} = \frac{1}{16\pi^2}\int d^3x\,\epsilon_{ijk}\left(A_i^a\,\partial_j A_k^a + \frac{1}{3}f^{abc}A_i^a A_j^b A_k^c\right) \tag{4.48}$$

The important feature about these quantities is that their values are unaffected by small, perturbative fluctuations. Their values can be changed only by transformations that can unwrap their mappings. Instantons are precisely this sort of configuration. The topological charge, which was introduced above, can be written as a surface integral,

$$Q = \frac{1}{32\pi^2}\int d^4x\,F_{\mu\nu}^a\tilde{F}_{\mu\nu}^a = \int d^4x\,\partial_\mu K_\mu = \int d\sigma_\mu K_\mu \tag{4.49}$$

where the current K_μ carries the Chern–Simons index,

$$K_\mu = \frac{1}{16\pi^2}\epsilon_{\mu\alpha\beta\gamma}\left(A_\alpha^a\partial_\beta A_\gamma^a + \frac{1}{3}f^{abc}A_\alpha^a A_\beta^b A_\gamma^c\right) \tag{4.50}$$

So, we learn that the topological quantum number n is the difference between the Chern–Simons characteristic at $t \to \infty$ and that at $t \to -\infty$,

$$Q = n = \int d^3x\,K_0|_{-\infty}^{+\infty} = n_{CS}(t = \infty) - n_{CS}(t = -\infty) \tag{4.51}$$

This displays the tunneling feature of the instanton. It provides the mechanism for mixing between the different topologically distinct vacua. We learn that the true vacuum in the theory must be a θ vacuum, as introduced in our discussion of the Schwinger model. Recall that the θ vacua are coherent states, $|\theta\rangle = \sum_n e^{in\theta} |n\rangle$. The rate of tunneling between different components $|n\rangle$ of the θ vacuum behaves as

$$\left(\frac{8\pi^2}{g^2}\right)^{2N_c} \exp\left(-\frac{8\pi^2}{g^2(\rho)}\right) \frac{d^4z \, d\rho}{\rho^5} \tag{4.52}$$

where $g^2(\rho)$ is the running coupling constant evaluated at the size scale of the instanton. This factor suppresses small instantons. The regime of large ρ cannot be analyzed easily because the coupling grows large in that case and our perturbative formulas become unreliable. The realm of large ρ is not understood. Perhaps study of the lattice will eventually clarify this part of the configuration space. Our understanding of instantons will remain incomplete until the large ones are under control, either numerically or, better, analytically.

The role of instantons in establishing the physical vacuum when there are light fermions in the theory is similar to the phenomena we studied in the $(1 + 1)$-dimensional Schwinger model. The axial anomaly in $3 + 1$ dimensions originates from the famous triangle diagram coupling the divergence of the flavor-singlet axial current to two gluons,

$$\partial_\mu j_\mu^5 = \frac{N_f}{16\pi^2} F_{\mu\nu}^a \tilde{F}_{\mu\nu}^a \tag{4.53}$$

The right-hand side of this equation is proportional to the topological charge density Q/V, so in the background field of an instanton conservation of axial charge will be violated by $2N_f$ units. This breaking occurs through localized fermion states that are zero modes of the Dirac operator, $i\not{D}\psi_0(x) = 0$, in the background field of an instanton [20]. The zero mode is left-handed, $\gamma_5\psi_0 = -\psi_0$. During a tunneling event the instanton pushes a left-handed fermion state from negative to positive energy, changing the topological charge of the theory's vacuum, $|n\rangle \rightarrow |n+1\rangle$, and changing its axial charge as well. The Dirac sea acts as an infinite source of axial charge and the instanton's left-handed zero mode taps into this source and gently breaks the global conservation law. The θ vacua are, therefore, states of indefinite axial charge, not dissimilar to the physics of the Schwinger model. The lightest flavor-singlet pseudo-scalar meson, the η', is not a candidate for being a light Goldstone boson because the axial current is not conserved, and the vacuum does not support long-wavelength, low-energy fluctuations in this quantum number. The $(3 + 1)$-dimensional situation is similar, again, to the Schwinger model in which the long-range Coulomb potential gave the ϕ a large mass because of the anomaly. Here, however, it is the nonperturbative interactions induced by instantons that play the crucial role.

The importance of the fermion zero modes is hard to over-emphasize. The nonconservation of the axial charge is given by

$$\Delta Q_5 = Q_5(t = +\infty) - Q_5(t = -\infty) = \int d^4x \, \partial_\mu j_\mu^5 \tag{4.54}$$

In terms of the fermion propagator,

$$\Delta Q_5 = \int d^4x \, N_f \, \partial_\mu \text{tr}(S(x, x)\gamma_\mu \gamma_5) \tag{4.55}$$

However, the fermion propagator can be written as an eigenfunction expansion. Let ψ_λ be an eigenfunction of the Dirac operator, $\not{D}\psi_\lambda = \lambda\psi_\lambda$, then

$$S(x, y) = \sum_\lambda \frac{\psi_\lambda(x)\psi_\lambda^\dagger(y)}{\lambda} \tag{4.56}$$

which gives

$$\Delta Q_5 = N_f \int d^4x \, \text{tr}\left(\sum_\lambda \frac{\psi_\lambda(x)\psi_\lambda^\dagger(x)}{\lambda} 2\lambda\gamma_5\right) \tag{4.57}$$

However, for every nonzero λ, $\gamma_5\psi_\lambda = \psi_{-\lambda}$, so the $\lambda \neq 0$ terms in this expression give zero upon integration. Only the normalized zero modes contribute to ΔQ_5, and we have

$$\Delta Q_5 = 2N_f(n_L - n_R) \tag{4.58}$$

where n_L (n_R) is the number of left (right)-handed zero modes. However, this is exactly how instantons effect the axial charge: every instanton contributes one unit to the topological charge and has a left-handed zero mode, while every anti-instanton has topological charge minus one and a right-handed zero mode.

The fermion zero modes are essential in order that the instantons have physical effects in QCD with light quarks. Note that quarks contribute determinants to the QCD partition function,

$$\prod_f^{N_f} \det(\not{D} + m_f) \tag{4.59}$$

Since nonzero eigenvalues occur in pairs,

$$\det(\not{D} + m_f) = m^\nu \prod_{\lambda > 0}(\lambda^2 + m^2) \tag{4.60}$$

where ν is the number of zero modes. So we see again that individual instanton tunneling events cannot occur in the limit of massless fermions. We know the "escape" from this dilemma: during a tunneling event the axial charge of the vacuum must change by two units, so instantons must be accompanied by fermions. If one calculates the tunneling amplitude in the presence of external quark sources, the zero modes in the denominators of the quark propagators cancel out against zero modes in the determinant and leave a nonzero result, in appropriate cases.

Below we shall review how instantons cause chiral-symmetry breaking, giving the light quarks substantial dynamical masses. Then, in a self-consistent, but mean-field, approach to instanton and light-quark dynamics, we find a solution in which the instantons have a nonzero density and the quarks have a nonzero dynamical mass. Each nontrivial effect sustains the other.

Instantons are believed to have other physical effects in light-quark hadronic physics. In particular, although their role in confinement is unknown because they are not understood in the strong-coupling, long-distance regime, they appear to play an important role in the energetics of the theory at the intermediate distances which are important for much of hadronic spectroscopy. As stated above, it is believed that they are responsible for the spontaneous breaking of chiral symmetry and the associated phenomena of constituent-quark masses and the existence of the Goldstone pions. This occurs when one considers the quark–quark interactions induced by instantons. In fact, in the case of two flavors, four Fermi terms are induced into the low-energy effective Hamiltonian and the breaking of chiral symmetry and the appearance of Goldstone bosons occurs as in the Nambu–Jona Lasinio model.

Each tunneling event is accompanied by a change in the chirality of each of the N_f fermion species through the zero modes surrounding each instanton. This induces a $2N_f$-fermion interaction. For momenta small compared with the reciprocal of the size of the instanton, ρ^{-1}, this interaction can be written in a local form. After the instantons' color orientations and positions have been integrated over, the effective quark interaction becomes

$$L^{N_f=2} = \int d\rho \, n_0(\rho) \left[\prod_f \left(m\rho - \frac{4}{3}\pi^2\rho^3 \bar{q}_{f,\mathrm{R}} q_{f,\mathrm{L}} \right) + \frac{3}{32} \left(\frac{4}{3}\pi^2\rho^3 \right)^2 \right.$$

$$\left. \times \left(\bar{u}_\mathrm{R}\lambda^a u_\mathrm{L}\bar{d}_\mathrm{R}\lambda^a d_\mathrm{L} - \bar{u}_\mathrm{R}\sigma_{\mu\nu}\lambda^a u_\mathrm{L}\bar{d}_\mathrm{R}\sigma_{\mu\nu}\lambda^a d_\mathrm{L} \right) \right] \quad (4.61)$$

When this interaction is studied in a mean-field approximation, one finds that it induces chiral-symmetry breaking, a nonzero chiral condensate that co-exists with a nonzero instanton density. The effective quark mass is given by

$$m_f^* = m_f - \tfrac{2}{3}\pi^2\rho^2 \langle \bar{q}_f q_f \rangle \quad (4.62)$$

and the chiral condensate is estimated through the gap equation,

$$\langle \bar{q}q \rangle = -4N_c \int \frac{d^4k}{(2\pi)^4} \frac{M(k)}{M^2(k) + k^2} \tag{4.63}$$

where $M(0)$ is the constituent-quark mass whose k dependence is determined by the shape of the fermion zero modes. From the known values for the chiral condensate and the constituent-quark mass, $\langle \bar{q}q \rangle \approx -(255)^3 \, \text{MeV}^3$ and $M(0) \approx 320 \, \text{MeV}$, we obtain the estimates for the size and density of instantons,

$$\rho \approx \frac{1}{3} \, \text{fm}, \qquad N/V = 1 \, \text{fm}^{-4} \tag{4.64}$$

These quantities play an important role in instanton phenomenology, which is studied in this mean-field scenario. Since this work is a mix of theory and experiment, it may change considerably as more ambitious instanton programs are undertaken or as more is learned about instantons in lattice-gauge-theory simulations, as we will discuss below.

Hadronic phenomenology, namely results from spectroscopy and correlation functions, supports these values of the basic features of the instanton ensemble. Unfortunately, the issues of large instantons and confinement are not resolved, so there could be flaws hidden in this ambitious approach. However, accepting these caveats, a qualitative picture of the QCD vacuum emerges from these studies [19].

1. The instanton size appears to be a well-defined concept. The size is considerably smaller than the typical inter-instanton separation R, and the instanton ensemble is fairly dilute, $\rho/R \sim \frac{1}{3}$. The fraction of spacetime where there are strong fields is only a few percent.

2. The fields inside an instanton are very strong, $F_{\mu\nu}^a \gg \Lambda_{\text{QCD}}^2$. The typical instanton action is large, $S_0 = 8\pi^2/g^2(\rho) \sim 10\text{–}15$. The use of the semiclassical approach to the instanton vacuum appears to be self-consistent.

3. The interactions between instantons are small compared with S_0, but are not negligible, $|\delta S_{\text{int}}| \sim 2\text{–}3 \ll S_0$. The dilute-gas picture for the instanton ensemble is not good, and the ensemble is better thought of as a "liquid."

Tests of these ideas through lattice simulations are possible and should be decisive. Such studies are crucial because the instanton correlations in the instanton ensemble are not tractable analytically and probably must be determined numerically. Comparison with hadronic phenomenology, light-quark spectroscopy as well as light-heavy spectroscopy, is also informative and insightful. For example, within this picture, the strongly attractive interaction in the pseudo-scalar quark–anti-quark channel needed to explain the Goldstone nature of the pion also implies that there is an attractive interaction in the scalar quark–quark channel. Therefore, the instanton picture of the vacuum appears to support the old diquark model of the baryons. It can easily accomodate the splitting between the nucleon and the delta in this fashion. Unfortunately, decisive evidence for diquarks in nucleons is not apparent in lattice simulations and in purely phenomenological works. Since the instanton calculations cannot accomodate confinement, which disfavors the colored-diquark idea, it may be missing some important energetics and be misleading in some cases. Time, aided by lattice-gauge theory, will tell.

Instantons are believed to play an important role in determining the behavior of QCD in extreme environments, high temperature and/or high baryon density. These issues will be discussed at length later in this book, but a few words here are appropriate just to set the stage.

How does the instanton ensemble change as QCD is heated through the transition to the quark–gluon plasma? Since chiral symmetry is restored at the transition, the influence of the instantons on the quarks must be diminished dramatically. This could occur in many different ways: the density of instantons could fall dramatically and not be sufficient to drive chiral-symmetry breaking, or the correlations between instantons could become relatively more significant, diminishing their long-range effects. In fact, results from lattice studies indicate that the instanton density does not change dramatically at T_c, but there is some indication that instantons and anti-instantons pair up above T_c and form "molecules." This is analogous to the Coulomb-gas phase transition in the planar model, whereby, below the transition, the vortices are paired up into neutral molecules and the Kosterlitz–Thouless transition occurs when these molecules ionize. The temperature of QCD would correspond to the reciprocal of the temperature in the planar model. Such ideas are being developed by workers active in the field.

It is believed that the role of instantons at large chemical potentials is even more central. As the baryon chemical potential increases one expects a transition to a state of matter in which baryon matter condenses in the ground state. Ordinary nuclear matter with a density of $\approx 0.17\,\mathrm{fm}^{-3}$ is of this sort. However, if one continues to increase the chemical potential, the nucleons in the ground state should begin to overlap and it is expected that abrupt transitions to new states of matter can occur. Intermediate values for the chemical potential might produce a host of new phases in which the quark mobility is qualitatively enhanced over that in confining QCD. Only at very, very large chemical potentials do we know what to expect. There the Fermi surface of the quarks should be so high that asymptotic freedom should apply and admit a weak-coupling analysis. However, such a system should be unstable with respect to a BCS instability and Cooper pairs, scalar diquarks in the $\bar{3}$ representation, should form and make a color superconductor. This scenario was pointed out many years ago [21, 22] and has been considered by astrophysicists because the interiors of neutron stars are expected to be dense enough to be quark superconductors. However, for no good reason, this effect (the size of the pairing gap) was assumed to be tiny. This phenomenon has been rediscovered and reexamined recently [23, 24], and an attempt to estimate the size of the gap at moderate densities was made. In fact, as we reviewed briefly above, instantons have particularly large attractive forces in the $\bar{3}$ scalar-diquark channel. The strength of this attraction is such that it can generate BCS gaps of the order of 100 MeV [24] at moderate densities just a few times that of nuclear matter.

The study of QCD at chemical potentials and temperatures relevant to the real world is a Holy Grail of lattice-gauge theory. As we learn more about lattice-gauge theory and effective Lagrangians we will focus on this application, develop some useful methods, discuss problems and limitations, and, it is hoped, point out directions for progress.

5

Lattice fermions and chiral symmetry

5.1 Free fermions on the lattice in one and two dimensions

Lattice fermions, their symmetries, and their continuum properties are issues central to lattice studies of dense and hot matter. It is a subtle subject because there are fundamental restrictions on the number of fermion species, their handedness, and gauge invariance. To see what the challenges are, we will illustrate lattice forms of the Dirac equation in various settings.

First consider the free Klein–Gordon equation and free Dirac equation on a spatial lattice with continuum time variable. In a Hamiltonian lattice-gauge theory, one would have a three-dimensional lattice and a continuum of time. Excitations in the system could hop from site to site given the rules of the Hamiltonian, the discrete version of the spatial derivatives in the energy, etc. Although Hamiltonian lattice-gauge theory is an important subject, our emphasis here continues to be on the Euclidean version of the theory. However, a short look at Hamiltonian methods is very elementary and enlightening. The details of the problems of lattice versions of the Dirac equation are different depending on whether time is treated as a continuum variable or a discrete one. Euclidean lattice fermions will be discussed in detail below after our introduction.

Let there be a free boson field $\phi(x, t)$ in $1+1$ dimensions, namely one discrete spatial axis and one continuum temporal axis. The continuum Klein–Gordon equation reads

$$\ddot{\phi} = \vec{\nabla}^2 \psi - m^2 \phi \qquad (5.1)$$

which implies the energy–momentum relation for plane waves,

$$E^2 = \vec{p}^2 + m^2 \qquad (5.2)$$

We want to know how well the solutions to the lattice form of these equations approximate the continuum situation. Our first task is to place the Laplacian on

93

the spatial lattice. The standard discretization reads

$$a^2 \vec{\nabla}^2 \phi \to \phi(n+1) + \phi(n-1) - 2\phi(n) = (d^+ + d^- - 2)\phi(n) \quad (5.3)$$

where $d^{\pm}\phi(n) = \phi(n \pm 1)$ are shift operators. The lattice Klein–Gordon equation now reads

$$a^2 \ddot{\phi} = (d^+ + d^- - 2)\phi(n) - a^2 m^2 \phi \quad (5.4)$$

This discrete equation has plane-wave solutions,

$$\phi = \exp(ikna + iEt) \quad (5.5)$$

Here the wave vector k takes values within the Brillouin zone, $-\pi < ka < \pi$, which exhausts all the independent solutions. The lattice energy–momentum relation follows on substituting the plane wave back into the discrete equation of motion,

$$-E^2 \phi = \frac{e^{ika} + e^{-ika} - 2}{a^2}\phi - m^2 \phi \quad (5.6)$$

so

$$E^2 = m^2 - 2[\cos(ka) - 1]/a^2 \quad (5.7)$$

For small $ka \ll 1$, the continuum energy–momentum relation is reproduced,

$$E^2 = m^2 + k^2 + O(k^4 a^2) \quad (5.8)$$

and relativistic propagation is restored for wavelengths that are long on the scale of the cutoff. The energies of the states near the edges of the Brillouin zone behave as $O(1/a)$ and are classically irrelevant as $a \to 0$ in the continuum limit where we are interested only in fixed, finite energy–momentum states.

In conclusion, there are no surprises with the lattice form of the Klein–Gordon equation. This result will be in sharp contrast to that for the Dirac equation.

Consider the $(1+1)$-dimensional free Dirac equation. For the most part we shall set the Dirac mass m_0 to 0, and concentrate on chiral symmetry on the lattice, which will prove to be a challenging subject. The two-component spinor and the 2×2 Dirac matrices will be written

$$\psi = \begin{pmatrix} \psi_1 \\ \psi_2 \end{pmatrix}, \qquad \gamma_0 = \begin{pmatrix} 1 & 0 \\ 0 & -1 \end{pmatrix}, \qquad \alpha = \gamma_5 = \begin{pmatrix} 0 & 1 \\ 1 & 0 \end{pmatrix} \quad (5.9)$$

The continuum Dirac equation reads

$$i\dot{\psi} = -i\alpha\, \partial_z \psi = -i\gamma_5\, \partial_z \psi \quad (5.10)$$

The equation is simplified if we consider chiral eigenstates,

$$\gamma_5 \chi_{\pm} = \pm \chi_{\pm} \quad (5.11)$$

Consider plane-wave solutions of the Dirac equation,

$$\psi_\pm = e^{-ikz+iEt}\chi_\pm \qquad (5.12)$$

Substituting into the Dirac equation gives the energy–momentum relation,

$$E = \pm k, \qquad -\infty < k < \infty \qquad (5.13)$$

We identify massless left- and right-moving fermions and anti-fermions.

Now, to begin understanding the subtleties of the lattice Dirac equation, place

$$\psi = \begin{pmatrix} \psi_1 \\ \psi_2 \end{pmatrix} \qquad (5.14)$$

on a spatial lattice and replace ∂_z by a simple discrete difference,

$$\dot\psi = -\frac{i}{2a}\gamma_5[\psi(n+1) - \psi(n-1)] \qquad (5.15)$$

Again, look for plane-wave solutions with definite chirality,

$$\psi_\pm = e^{-i(-kna+Et)}\chi_\pm, \qquad \gamma_5\chi_\pm = \pm\chi_\pm \qquad (5.16)$$

Substituting back into the equation of motion gives the energy–momentum relation

$$-E = \pm\frac{i}{2a}(e^{ika} - e^{-ika}) \qquad (5.17)$$

which gives

$$E = \pm\sin(ka)/a \qquad (5.18)$$

There is something seriously "wrong" with this result! Note that, for small $ka \ll 1$, it reduces to

$$E \approx \pm k + O(k^3 a^2) \qquad (5.19)$$

as expected. However, for ka close to the edge of the Brillouin zone, $ka = \pi - k'a$, $k'a \ll 1$, we have additional low-energy excitations,

$$E \approx \mp k' + O(k'^3 a^2) \qquad (5.20)$$

The full energy–momentum dispersion relation is shown in Fig. 5.1. We identify two two-component Dirac particles in the continuum limit. In addition, the total chiral charge of the resulting multiplet is zero.

This catastrophe is even better seen by considering the field ψ_+ on the lattice. It describes the branch of the dispersion relation labeled $\gamma_5 = +1$ in Fig. 5.1. There is a right-mover ($k \approx 0$) and a left-mover ($k \approx \pi$). Since ordinary chirality

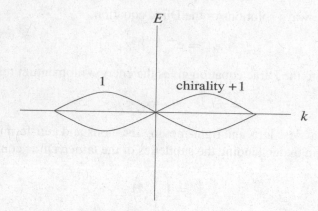

Figure 5.1. The energy–momentum relation for the naive Dirac equation on a one-dimensional spatial lattice.

is just velocity for massless particles in $1+1$ dimensions, the low-energy content of the lattice field ψ_+ is a pair of fermions whose chiralities add up to zero. This is the phenomenon of "species doubling" which is well known in condensed-matter theory.

H. B. Nielsen and M. Ninomiya [25] pointed out that this result is much more general than this specific exercise. They presented a series of "no-go" theorems concerning exact chiral symmetry on the lattice and species doubling. The idea is that species doubling occurs because of the continuity of the excitation branches in the Brillouin zone, independently of the particular lattice approximations as long as the lattice formulation of the theory has exact chiral symmetry associated with a compact, continuous group and the hopping and interaction terms in the lattice action are sufficiently local to insure continuity of the energy–momentum relations for the lattice fermions.

For example, one of their theorems considers the action

$$S = \int d^4x\, [-i\bar{\psi}_L\dot{\psi}_L + \bar{\psi}_L\vec{\sigma}\cdot(-i\,\vec{\partial} - \vec{A}(x))\psi_L] \qquad (5.21)$$

Recognize this as a simple theory of a left-handed neutrino coupled to a gauge field. It could be a piece of the standard-model action, for example. Nielsen and Ninomiya's theorem states that, when this theory is put onto the lattice with exact, continuous chiral symmetry and locality in its interactions and hopping (kinetic) terms, then the low-energy excitations of the theory will include an equal number of left- and right-handed Weyl fermions. The continuum limit of the lattice theory will be left–right symmetric, contrary to plan.

There are several methods to deal with the limitations placed upon us by these observations. The two most useful and straightforward sacrifice chiral symmetry to different degrees to avoid the full consequences of species doubling.

1. Thin the degrees of freedom.

2. Lift the energy at the edges of the Brillouin zone.

Let's discuss each approach in turn by illustration in $1 + 1$ dimensions.
Approach 1: "staggered fermions"
Begin by placing a single-component Fermi field $\phi(n)$ on each site. The equal-time anti-comutation relations of $\phi(n)$ read

$$[\phi(n), \phi(m)]_+ = 0, \qquad [\phi^\dagger(n), \phi(m)]_+ = \delta_{nm} \tag{5.22}$$

and let its dynamics be described with the Hamiltonian

$$H = -\frac{i}{2a} \Sigma [\phi^\dagger(n)\phi(n+1) - \phi^\dagger(n+1)\phi(n)] \tag{5.23}$$

The equation of motion for the Fermi field reads, $i\,\partial\phi/\partial t = [H, \phi]$,

$$\dot\phi(n) = -\frac{1}{2a}[\phi(n+1) - \phi(n-1)] \tag{5.24}$$

This doesn't resemble the two-component Dirac equation we expect in $1 + 1$ dimensions until we identify the upper and lower components of the spinor. To do this, consider the spatial lattice to consist of two, interleaved, "even" and "odd" lattices,

$$\begin{aligned} \psi_1(n) &= \phi(n), & n \text{ even} \\ \psi_2(n) &= \phi(n), & n \text{ odd} \end{aligned} \tag{5.25}$$

Now the equation of motion can be written in two-component form,

$$\dot\psi_1(n) = -\frac{1}{2a}[\psi_2(n+1) - \psi_2(n-1)]$$

$$\dot\psi_2(n) = -\frac{1}{2a}[\psi_1(n+1) - \psi_1(n-1)] \tag{5.26}$$

which gives a discretization of the continuum Dirac equation,

$$\dot\psi = -\gamma_5 \partial_z \psi = -\begin{pmatrix} 0 & 1 \\ 1 & 0 \end{pmatrix} \partial_z \psi \tag{5.27}$$

which reads, in component form,

$$\dot\psi_1 = -\partial_z \psi_2 \quad \text{and} \quad \dot\psi_2 = -\partial_z \psi_1 \tag{5.28}$$

We learn from this exercise that the staggered-fermion method avoids the species-doubling problem by doubling the size of the unit cube in real space and effectively halving the size of the Brillouin zone.

We must clarify the symmetries of the lattice Hamiltonian and equation of motion. This is necessary because, when we add interactions into the lattice model, the continuum limit its solutions will have only those symmetries built into the system. For example, if there is no remnant of chiral symmetry in the discrete equation of motion, then we cannot expect massless fermions to appear in the continuum propagators of the theory. In fact, the mass of the fermion will be of the order of the reciprocal of the lattice spacing and the fermion will be unphysical.

Consider some bilinears of the lattice system:

$$\int \psi^\dagger \psi \, dz = \int (\psi_1^\dagger \psi_1 + \psi_2^\dagger \psi_2) \, dz \rightarrow \Sigma_n \phi^\dagger(n)\phi(n)$$

$$\int \bar{\psi} \psi \, dz = \int (\psi_1^\dagger \psi_1 - \psi_2^\dagger \psi_2) \, dz \rightarrow \Sigma_n (-1)^n \phi^\dagger(n)\phi(n)$$

$$\int \bar{\psi} \gamma_5 \psi \, dz = \int (\psi_1^\dagger \psi_2 - \psi_2^\dagger \psi_1) \, dz$$

$$\rightarrow \Sigma_n (-1)^n [\phi^\dagger(n)\phi(n+1) - \phi^\dagger(n+1)\phi(n)]$$

$$\int \psi^\dagger \gamma_5 \psi \, dz = \int (\psi_1^\dagger \psi_2 + \psi_2^\dagger \psi_1) \, dz$$

$$\rightarrow \Sigma_n [\phi^\dagger(n)\phi(n+1) + \phi^\dagger(n+1)\phi(n)] \qquad (5.29)$$

Now we can classify the symmetries of the lattice Hamiltonian,

$$H = -\frac{i}{2a} \Sigma [\phi^\dagger(n)\phi(n+1) - \phi^\dagger(n+1)\phi(n)]$$

1. Translation of the spatial lattice by an even number of sites.
   The generator of ordinary translations in the continuum theory is the momentum operator,

$$p_z = -i \int \psi^\dagger \partial_z \psi \, dz = -i \int (\psi_1^\dagger \partial_z \psi_1 + \psi_2^\dagger \partial_z \psi_2) \, dz \qquad (5.30)$$

which, of course, does not mix the components, ψ_1 and ψ_2. So, its lattice version should not mix the two sublattices,

$$p_z \rightarrow \Sigma_n [\phi^\dagger(n+2)\phi(n) + \phi^\dagger(n)\phi(n+2)] \qquad (5.31)$$

is the corresponding lattice generator. Translations by two lattice spacings correspond to continuum translations. The lattice Hamiltonian is, of course, unaffected by such operations.

2. Tranlations by an odd number of sites.

Here we have a surprise. Looking at the bilinears above, we see that translations by one site are generated by the chiral charge, $\int \psi^\dagger \gamma_5 \psi \, dz$. Clearly the lattice Hamiltonian is exactly preserved by this operation, regardless of the physical dimensions of the lattice spacing. This lattice-fermion method preserves discrete γ_5 symmetry for any lattice spacing a. Since γ_5 simply interchanges the components of ψ,

$$\gamma_5 \begin{pmatrix} \psi_1 \\ \psi_2 \end{pmatrix} = \begin{pmatrix} \psi_2 \\ \psi_1 \end{pmatrix} \tag{5.32}$$

these are just chiral rotations through $\pi/2$ radians, $\exp(i\gamma_5\pi/2) = i\gamma_5$.

So, translation through an even number of sites is a pure translation whereas translation through an odd number of sites is a pure discrete chiral transformation. This curious mixing of spacetime and chiral symmetries occurs because of the extended nature of the Brillouin zone. The staggered fermion method in $(3 + 1)$-dimensional Hamiltonian systems (discrete spatial but continuum time lattices) and in four-dimensional Euclidean systems (symmetric four-dimensional lattices) has similar but even more intricate symmetry considerations because flavor symmetries (multiple species) will also play a role there.

Note that the standard mass term, $m_0 \Sigma_n (-1)^n \phi^\dagger(n)\phi(n)$, breaks the discrete chiral symmetry, but not the translational symmetry, of the staggered-fermion method. It is particularly important that the Hamiltonian with $m_0 = 0$ has an exact, discrete chiral symmetry no matter what the lattice spacing. This means that interactions that also preserve chiral symmetry on the lattice cannot generate mass terms in the theory's effective Hamiltonian. In other words, the symmetry is "protected." This is the reason why staggered fermions are so important in applications to QCD.

Now consider another lattice-fermion method that has poorer chiral properties but more convenient spinor properties.

Approach 2: Wilson fermions ("brute force")

Place a two-component Dirac spinor at every site of the spatial lattice, but add terms to H so that the energy–momentum dispersion relation does not have secondary minima at $ka = \pm\pi$,

$$\begin{aligned} H = &-\frac{i}{2a} \sum_n \psi^\dagger(n)\alpha[\psi(n+1) - \psi(n-1)] \\ &+ \frac{B}{2a} \sum_n \bar\psi(n)[2\psi(n) - \psi(n+1) - \psi(n-1)] \\ &+ \frac{m}{2a} g^2 \sum_n \bar\psi(n)\psi(n) \end{aligned} \tag{5.33}$$

Note that the second term in H is a "boson" kinetic energy multiplied by the lattice spacing a. This term is designed to lift the energy–momentum dispersion relation at the edge of the Brillouin zone, just like a bosonic term, for nonzero lattice spacings, but should become irrelevant in the continuum limit, $a \to 0$,

$$\frac{1}{a} \sum_n \bar{\psi}(n)[2\psi(n) - \psi(n+1) - \psi(n-1)] \to \frac{1}{a} \int \frac{dz}{a} a\bar{\psi} a^2 \nabla^2 \psi$$

$$= a \int \bar{\psi}\, \nabla^2 \psi\, dz$$

$$\to 0, \qquad \text{as} \qquad a \to 0 \quad (5.34)$$

for sufficiently smooth fields.

Omitting the last term in H for the moment, the equation of motion for ψ is

$$\dot{\psi}(n) = -\frac{i}{2a}\gamma_5[\psi(n+1) - \psi(n-1)]$$

$$+ \frac{B}{2a}\gamma_0[2\psi(n) - \psi(n+1) - \psi(n-1)] \quad (5.35)$$

For a plane-wave *Ansatz*, $\psi = e^{i(-kna+Et)}\chi$, the equation of motion becomes

$$-E\chi = -\frac{i}{2a}\gamma_5(e^{-ika} - e^{ika})\chi + \frac{B}{2a}\gamma_0(2 - e^{ika} - e^{-ika})\chi$$

$$-E\chi = -\gamma_5 \frac{\sin(ka)}{a}\chi + 2B\gamma_0 \frac{\sin^2(ka/2)}{a}\chi \quad (5.36)$$

So

$$E^2 = \frac{\sin^2(ka)}{a^2} + 4B^2 \frac{\sin^4(ka/2)}{a^2} \quad (5.37)$$

Letting $ka \to 0$, we find that $E^2 \approx k^2 + B^2 k^4 a^2 + \cdots$, which is the desired result. In addition, near the edge of the Brillouin zone, $ka \approx \pm\pi$,

$$E^2 \approx 4B^2/a^2 \quad (5.38)$$

and we see that the bosonic term in the Hamiltonian has opened a gap in the fermion's spectrum which diverges in the naive continuum limit. However, we have paid a heavy cost for this success.

1. H explicitly breaks both continuous and discrete γ_5 symmetry. There is no remnant of chiral symmetry left.

2. Since there is no symmetry prohibiting mass counterterms in the theory when additional interactions are put into H, delicate fine tuning of parameters will have to be done in the lattice model in order to have finite-mass fermions in the propagators of the continuum model.

5.2 Fermions and bosons on Euclidean lattices

Now let's turn to the propagation of free bosons and fermions in two and four Euclidean dimensions. Again we shall find that bosons are easy to deal with on the lattice, but fermions are not. There will be species doubling again, but the details will be lattice-dependent and the remnants of chiral symmetry different from the Hamiltonian case.

Begin with the boson propagator Δ in four-dimensional Euclidean space, described by the Klein–Gordon equation,

$$(m^2 - \Box_x)\,\Delta\,(x - x') = \delta^4(x - x') \tag{5.39}$$

Now place this equation on a four dimensional Euclidean lattice,

$$x_\mu = n_\mu a, \qquad n_\mu = 0, \pm 1, \pm 2, \ldots, \qquad \mu = 1, 2, 3, 4 \tag{5.40}$$

Replace the derivatives with finite differences,

$$\Box_x = \sum_\mu \frac{\partial^2}{\partial x_\mu^2} \to \frac{1}{a^2} \sum_\mu (d_\mu^+ + d_\mu^- - 2) \tag{5.41}$$

where $d_\mu^\pm$ is a shift operator carrying out shifts by one lattice spacing in the direction $\pm\mu$.

It is convenient to define a dimensionless boson propagator, $\Delta_n = a^2\,\Delta\,(n)$. Then the Klein–Gordon equation for the propagator becomes

$$a^{-2} m^2 \Delta_n = \sum_\mu (\Delta_{n+\mu} + \Delta_{n-\mu} - 2\,\Delta_n)/a^4 = \delta_{n0}/a^4 \tag{5.42}$$

or, in dimensionless form,

$$(8 + m^2 a^2)\Delta_n - \sum_\mu (\Delta_{n+\mu} + \Delta_{n-\mu}) = \delta_{n0} \tag{5.43}$$

Transforming to momentum space allows us to solve for the free propagator,

$$\Delta(p) = \frac{1}{1 - K \sum_\mu (e^{ip_\mu a} + e^{-ip_\mu a})}, \qquad K \equiv \frac{1}{8 + m^2 a^2} \tag{5.44}$$

which can be written

$$\Delta(p) = \frac{1}{1 - 2K \sum_\mu \cos(p_\mu a)} \tag{5.45}$$

The Brillouin zone covers the range $-\pi/a \le p_\mu \le \pi/a$ for each $\mu = 1, 2, 3$, or 4. If we continue back to Minkowski space, $p_0 \to iE$, we can identify the particle spectrum with the poles of $\Delta(E, \vec{p})$,

$$\Delta(E, \vec{p}) = \frac{1}{1 - 2K \cosh(Ea) - 2K \sum_i \cos(p_i a)} \tag{5.46}$$

The only poles in $\triangle(E, \vec{p})$ occur near the origin of the Brillouin zone, $Ea \ll 1$, $p_i a \ll 1$. Under these conditions we have

$$\triangle(E, \vec{p}) \approx \frac{1}{1 - 8K - K(E^2 a^2 - \vec{p}^2 a^2)} \tag{5.47}$$

which has a pole at

$$E^2 = \vec{p}^2 + (1 - 8K)/(Ka^2) \tag{5.48}$$

We recognize this result as the relativistic energy–momentum relation for a particle of mass $m^2 = (1-8K)/(Ka^2)$. So we could obtain sensible propagation in the continuum limit if we take the limit $K \to \frac{1}{8}$ from below, $1 - 8K > 0$, and take $a \to 0$ so that m^2 is held fixed in physical units.

Now consider the free fermion propagator in four dimensional Euclidean space. The Dirac equation with unit source reads

$$(\gamma^\mu \partial'_\mu - m)G(x' - x) = \delta^4(x' - x), \qquad [\gamma^\mu, \gamma^\nu]_+ = 2\delta^{\mu\nu} \tag{5.49}$$

The continuum action is

$$A = \int L \, d^4 x = \frac{1}{2} \int \bar{\psi}(\gamma^\mu \partial_\mu - m)\psi \, d^4 x \tag{5.50}$$

so a "naive" lattice action would read

$$A = \sum_{n,\mu} -\frac{1}{2a}[\bar{\psi}(n)\gamma_\mu \psi(n + a_\mu) - \bar{\psi}(n + a_\mu)\gamma_\mu \psi(n)]a^4$$

$$- m \sum_n \bar{\psi}(n)\psi(n)a^4 \tag{5.51}$$

We can expect from our previous discussion of the lattice Dirac equation that this lattice action will describe not one species, but $16 = 2^4$, because of the four-dimensionality of the Brillouin zone. Actually this lattice action is very useful in formal developments of the subject, so we will use it many times below. We will also elucidate its relation to other lattice fermion forms that have fewer low-energy species.

To reduce the number of species in the low-energy spectroscopy of the lattice fermion action, we can follow the first strategy used with Hamiltonian lattice actions and add a boson-like term to this expression, $r/(2a)\sum_{n,\mu}[\bar{\psi}(n)\psi(n + a_\mu) + \bar{\psi}(n + a_\mu)\psi(n) - 2\bar{\psi}(n)\psi(n)]a^4$, so the action becomes

$$A = \sum_{n,\nu} \frac{1}{2a}[\bar{\psi}(n)(r - \gamma_\mu)\psi(n + a_\mu) + \bar{\psi}(n + a_\mu)(r + \gamma_\mu)\psi(n)]a^4$$

$$+ \left(m - \frac{8r}{a}\right) \sum_n \bar{\psi}(n)\psi(n)a^4 \tag{5.52}$$

It is more convenient to introduce dimensionless fields, $\psi(n) \rightarrow (1/\sqrt{a^3})\psi(n)$, so the action becomes

$$A = \sum_{n,\nu} \bar{\psi}(n)(r - \gamma_\mu)\psi(n + \mu) + \bar{\psi}(n + \mu)(r + \gamma_\mu)\psi(n)$$

$$+ (ma - 8r) \sum_n \bar{\psi}(n)\psi(n) \tag{5.53}$$

The fermion propagator then satisfies

$$[(r - \gamma_\mu)d_\mu^- + (r + \gamma_\mu)d_\mu^+ + (ma - 8r)]G(n) = \delta_{n0} \tag{5.54}$$

Passing to momentum space, we obtain the propagator

$$G(p) = \frac{K}{1 - K\sum_\mu[(r + \gamma_\mu)e^{ip_\mu} + (r - \gamma_\mu)e^{-ip_\mu}]} \tag{5.55}$$

where the parameter K is $-1/(ma - 8r)$.

The propagator simplifies for some special choices of its parameters.

1. $m = 0, r = 0$. Then,

$$G(p) = \frac{1}{\sum_\mu \gamma_\mu \sin p_\mu} = \frac{\sum_\mu \gamma_\mu \sin p_\mu}{\sum_\mu \sin^2 p_\mu} \tag{5.56}$$

Generalizing our earlier discussion, we see that $G(p)$ describes 2^4 low-energy fermion excitations because it has zeros at the origin in momentum space as well as zeros at the edges of the four dimensional Brillouin zone. This fermion method is called "naive fermions."

2. $m = 0, r \neq 0$. Then,

$$G(p) = \frac{1}{2\sum_\mu \gamma_\mu \sin p_\mu - 2r\sum_\mu \cos p_\mu + 8r} \tag{5.57}$$

For nonvanishing r, we see that the fermion propagator does not have additional zeros at the edges of the Brillouin zone. It is also clear from the action, however, that, even though $m = 0$, the lattice system has neither continuous nor discrete chiral symmetry: the fact that $r \neq 0$ removes the unwanted species at the cost of removing all remnants of chiral symmetry. This "catastrophe" is a consequence of the general principles underlying the Nielsen–Ninomiya theorems.

In applications the choice $r = 1$ is quite handy for this class of fermion methods. In this case the propagator equation becomes

$$[(1 - \gamma_\mu)d_\mu^- + (1 + \gamma_\mu)d_\mu^+ - 8]G(n) = \delta_{n0} \tag{5.58}$$

It is easy to see that the operators $P_\mu^\pm = \frac{1}{2}(1 \pm \gamma_\mu)$ are projection operators, $(P_\mu^\pm)^2 = P_\mu^\pm$ and $P_\mu^+ P_\mu^- = P_\mu^- P_\mu^+ = 0$. In addition, with $r = 1$ and letting $K \to \frac{1}{8}$, we obtain the correct continuum form of the lattice propagator, $G \approx 1/p$, for low-momentum modes.

Clearly this fermion method has some useful properties, but the absence of chiral symmetry and the protection that lattice chiral symmetry provides to unwanted, unbounded bare-mass generation limits its usefulness.

5.3 Staggered Euclidean fermions

Now consider the Euclidean version of staggered fermions. This fermion method is particularly important in applications because it retains a continuous remnant of chiral symmetry and has a precisely massless Goldstone boson, a pion, whenever chiral symmetry is spontaneously broken by the lattice theory's dynamics. Although the Hamiltonian version of staggered fermions had only a discrete piece of chiral symmetry when the lattice spacing a was nonzero, the Euclidean version has more chiral symmetry insuring the existence of a massless Goldstone boson. Since these aspects of the theory dictate many of its low-energy features, the theoretical and practical importance of these features have proved very important in applications ranging from spectroscopy to finite-temperature transitions and low-energy theorems.

We will illustrate Euclidean staggered fermions in two dimensions so everything will be explicit and simple. The generalizations to higher dimensions can then be carried through more formally. We shall see that one of the drawbacks of the staggered fermion methodology is the identification of ordinary flavor symmetries and species (the method is challenging... "staggering," as its critics have complained).

To begin, we will choose a standard representation of the Dirac matrices in two dimensions. The Pauli matrices suffice,

$$\gamma_0 = \begin{pmatrix} 0 & 1 \\ 1 & 0 \end{pmatrix} = \sigma_1, \quad \gamma_1 = \begin{pmatrix} 0 & -i \\ i & 0 \end{pmatrix} = \sigma_2, \quad \gamma_5 = \begin{pmatrix} 1 & 0 \\ 0 & -1 \end{pmatrix} = \sigma_3 \quad (5.59)$$

We will begin with the "naive" fermion action and then "thin" its degrees of freedom and construct the "staggered" action which will describe a single relativistic iso-doublet. The naive action for massless fermions is, in dimensionless notation,

$$S = \sum_{i,\mu} \bar{\psi}(i)\gamma_\mu[\psi(i+\mu) - \psi(i-\mu)], \qquad i = (n_0, n_1) \quad (5.60)$$

To thin the degrees of freedom, we "spin diagonalize" this expression by introducing a Fermi field χ,

$$\psi = \gamma_0^{|n_0|} \gamma_1^{|n_1|} \chi \quad (5.61)$$

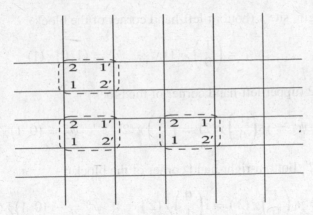

Figure 5.2. The basic Brillouin zone for the two-dimensional staggered-fermion method.

and, independently,

$$\bar{\psi} = \bar{\chi}\gamma_1^{|n_1|}\gamma_0^{|n_0|} \tag{5.62}$$

Writing the action in terms of χ, we have

$$S = \sum_{i,\mu} \bar{\chi}(i)(-1)^{\phi(i,\mu)}[\chi(i+\mu) - \chi(i-\mu)] \tag{5.63}$$

where the phases are

$$(-1)^{\phi(i,0)} = \gamma_1^{|n_1|}\gamma_0^{|n_0|}\gamma_0\gamma_0^{|n_0+1|}\gamma_1^{|n_1|} = +1 \tag{5.64}$$

$$(-1)^{\phi(i,1)} = \gamma_1^{|n_1|}\gamma_0^{|n_0|}\gamma_1\gamma_0^{|n_0|}\gamma_1^{|n_1+1|} = (-1)^{|n_0|} \tag{5.65}$$

So the matrix structure has collapsed to simple signs... S is "spin-diagonal," and one need keep only a single complex field $\chi(i)$ at each site.

We can expect from our earlier analysis of the Hamiltonian lattice (discrete space, continuum time), that this lattice system will describe two spin-$\frac{1}{2}$ fermion species. The basic Brillouin zone will consist of four lattice sites that are conveniently labeled as "even" and "odd," as shown in Fig. 5.2. Consider the four sites of each block shown in Fig. 5.2 and label them with γ matrices as suggested by the defining equation for χ above.

1. The origin, site 1 (bottom-left-hand corner of the block):

$$\psi_1^a = \begin{pmatrix} 1 \\ 0 \end{pmatrix} \chi(1), \qquad \bar{\psi}_1^a = (1 \ 0)\bar{\chi}(1) \tag{5.66}$$

2. γ_0, site 2 (upper-left-hand corner of the block):

$$\psi_2^a = \gamma_0 \begin{pmatrix} 1 \\ 0 \end{pmatrix} \chi(2) = \begin{pmatrix} 0 \\ 1 \end{pmatrix} \chi(2), \qquad \bar{\psi}_2^a = (0 \ 1)\bar{\chi}(2) \tag{5.67}$$

3. γ_1, site $2'$ (bottom-right-hand corner of the block):

$$\psi_2^b = \gamma_1 \begin{pmatrix} 1 \\ 0 \end{pmatrix} \chi(2') = i \begin{pmatrix} 0 \\ 1 \end{pmatrix} \chi(2'), \qquad \bar{\psi}_2^b = -i(0 \ 1)\bar{\chi}(2') \tag{5.68}$$

4. $\gamma_0\gamma_1$, site $1'$ (top-right-hand corner of the block):

$$\psi_1^b = \gamma_0\gamma_1 \begin{pmatrix} 1 \\ 0 \end{pmatrix} \chi(1') = i \begin{pmatrix} 1 \\ 0 \end{pmatrix} \chi(1'), \qquad \bar{\psi}_1^b = -i(1 \ 0)\bar{\chi}(1') \tag{5.69}$$

where the labels a and b will specify the two species of an iso-doublet.

Now, we can write the action in terms of the ψ fields,

$$\begin{aligned}
S &= \bar{\chi}(1)(\nabla_0\chi(2) + \nabla_1\chi(2')) + \bar{\chi}(2)(\nabla_0\chi(1) - \nabla_1\chi(1')) \\
&\quad + \bar{\chi}(1')(\nabla_0\chi(2') - \nabla_1\chi(2)) \\
&\quad + \bar{\chi}(2')(\nabla_0\chi(1') - \nabla_1\chi(1)) + \text{other blocks} \\
&= \bar{\psi}_1^a(\nabla_0\psi_2^a - i\nabla_1\psi_2^b) + \bar{\psi}_2^a(\nabla_0\psi_1^a + i\nabla_1\psi_1^b) + \bar{\psi}_1^b(\nabla_0\psi_2^b - i\nabla_1\psi_2^a) \\
&\quad + \bar{\psi}_2^b(\nabla_0\psi_1^b + i\nabla_1\psi_1^a) + \text{other blocks}
\end{aligned} \tag{5.70}$$

where $\bar{\chi}(1)(\nabla_0\chi(2))$ means $\bar{\chi}(1)(d_\mu^+ - d_\mu^-)\chi(2)$ with $\mu = 0$, etc. The site labeling in the equation follows the conventions used in Fig. 5.2.

We can diagonalize the action in the flavor indices and identify up and down quark fields by making appropriate linear combinations. Let

$$u_i = (\psi_i^a + \psi_i^b)/\sqrt{2}, \qquad \tilde{d}_i = (\psi_i^a - \psi_i^b)/\sqrt{2} \tag{5.71}$$

so that

$$\psi_i^a = (u_i + \tilde{d}_i)/\sqrt{2}, \qquad \psi_i^b = (u_i - \tilde{d}_i)/\sqrt{2} \tag{5.72}$$

On substituting into the action, we find

$$\begin{aligned}
S &= \bar{u}_1(\nabla_0 u_2 - i\nabla_1 u_2) + \bar{u}_2(\nabla_0 u_1 + i\nabla_1 u_1) \\
&\quad + \bar{\tilde{d}}_1(\nabla_0\tilde{d}_2 + i\nabla_1\tilde{d}_2) + \bar{\tilde{d}}_2(\nabla_0\tilde{d}_1 - i\nabla_1\tilde{d}_1)
\end{aligned} \tag{5.73}$$

and, on collecting terms,

$$S = \bar{u}Du + \bar{\tilde{d}}\tilde{D}\tilde{d} \qquad (5.74)$$

where the "Dirac" operators are

$$D = \gamma_0 \nabla_0 + \gamma_1 \nabla_1, \qquad \tilde{D} = \gamma_0 \nabla_0 - \gamma_1 \nabla_1 \qquad (5.75)$$

Finally, we can perform a linear transformation on $\tilde{d}$ to obtain a standard Dirac operator,

$$\tilde{d} = i\gamma_5\gamma_1 d = i\sigma_3\sigma_2 d = \sigma_1 d = \begin{pmatrix} 0 & 1 \\ 1 & 0 \end{pmatrix} d \qquad (5.76)$$

Now the second term in the action can be written

$$\bar{\tilde{d}}\tilde{D}\tilde{d} = \bar{d}\sigma_1 \tilde{D}\sigma_1 d = \bar{d}(\gamma_0 \nabla_0 - \gamma_0\gamma_1 \nabla_1\gamma_0)d = \bar{d}Dd \qquad (5.77)$$

So the action becomes

$$S = \bar{u}Du + \bar{d}Dd \qquad (5.78)$$

and we identify the physical content of the lattice theory: it consists of two fermions, an iso-doublet.

5.4 Block derivatives and axial symmetries

Since the fermion degrees of freedom of the iso-doublet are spread over four lattice sites, each comprising a "block," it is necessary to distinguish between couplings between blocks, which have a clear continuum meaning, and those within blocks. With this in mind, we introduce "block" derivatives and rewrite the lattice action in a fashion that displays the continuum symmetries and their breakings by finite lattice effects more explicitly.

To begin, go back to the lattice action written in terms of the ψ fields,

$$S = \bar{\psi}_1^a(\nabla_0\psi_2^a - i\nabla_1\psi_2^b) + \bar{\psi}_2^a(\nabla_0\psi_1^a + i\nabla_1\psi_1^b) + \bar{\psi}_1^b(\nabla_0\psi_2^b - i\nabla_1\psi_2^a)$$
$$+ \bar{\psi}_2^b(\nabla_0\psi_1^b + i\nabla_1\psi_1^a) + \text{other blocks} \qquad (5.79)$$

Let's examine the discrete differences here in more detail. Consider three sites 1, 2, and 3 in one lattice direction and the definitions of first and second derivatives,

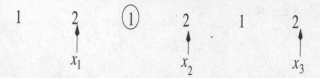

Figure 5.3. Coupled sites in one direction on a two-dimensional staggered-fermion lattice.

$$\nabla f(2) \equiv \tfrac{1}{2}(f(3) - f(1)), \qquad \nabla^2 f(2) \equiv f(3) + f(1) - 2f(2) \quad (5.80)$$

Then various finite differences involving the three sites can be written in these terms as

$$f(3) - f(2) = \tfrac{1}{2}(f(3) - f(1)) + \tfrac{1}{2}(f(3) + f(1) - 2f(2)) = \nabla f + \tfrac{1}{2}\nabla^2 f$$

$$(5.81)$$

and

$$f(2) - f(1) = \tfrac{1}{2}(f(3) - f(1)) + \tfrac{1}{2}(f(3) + f(1) - 2f(2)) = \nabla f - \tfrac{1}{2}\nabla^2 f$$

$$(5.82)$$

Now introduce the "block" derivative which couples corresponding fermion fields between adjacent blocks and does not contain any mixing. The action as it is written at present contains terms that couple degrees of freedom entirely inside a block as well as terms connecting adjacent blocks. For example, the action written above contains differences like $\psi_2^a(x_2) - \psi_2^a(x_1)$, in the notation of Fig. 5.3. However, using the algebra above,

$$\psi_2^a(x_2) - \psi_2^a(x_1) = \tfrac{1}{2}\big(\psi_2^a(x_3) - \psi_2^a(x_1)\big) - \tfrac{1}{2}\big(\psi_2^a(x_3) + \psi_2^a(x_1) - 2\psi_2^a(x_2)\big)$$

$$= \hat{\nabla}\psi_2^a - \tfrac{1}{2}\hat{\nabla}^2\psi_2^a \quad (5.83)$$

Using block derivatives, the variables ψ_i^a and ψ_i^b, $i = 1, 2$ can be thought to live at the centers of each block. The variables are now organized to account for the multiple flavors which appear in the continuum limit.

Introducing the block derivatives into the expression for the action written in terms of ψ's gives

$$S = \bar{\psi}_1^a\big[\big(\hat{\nabla}_0\psi_2^a - \tfrac{1}{2}\hat{\nabla}_0^2\psi_2^a\big) - i\big(\hat{\nabla}_1\psi_2^b - \tfrac{1}{2}\hat{\nabla}_1^2\psi_2^b\big)\big]$$

$$+ \bar{\psi}_2^a\big[\big(\hat{\nabla}_0\psi_1^a + \tfrac{1}{2}\hat{\nabla}_1^2\psi_1^a\big) + i\big(\hat{\nabla}_1\psi_1^b - \tfrac{1}{2}\hat{\nabla}_1^2\psi_1^b\big)\big]$$

$$- \bar{\psi}_1^b\big[\big(\hat{\nabla}_0\psi_2^b + \tfrac{1}{2}\hat{\nabla}_0^2\psi_2^b\big) - i\big(\hat{\nabla}_1\psi_2^a + \tfrac{1}{2}\hat{\nabla}_1^2\psi_2^a\big)\big]$$

$$+ \bar{\psi}_2^b\big[\big(\hat{\nabla}_0\psi_1^b - \tfrac{1}{2}\hat{\nabla}_1^2\psi_1^b\big) + i\big(\hat{\nabla}_1\psi_1^a + \tfrac{1}{2}\hat{\nabla}_1^2\psi_1^a\big)\big] \quad (5.84)$$

We can collect the terms linear in ∇ and identify the u and d fermions to write

$$S = \bar{u}\hat{D}u + \bar{\bar{d}}\hat{\bar{D}}\tilde{d} + \bar{\psi}_1^a\left(-\frac{1}{2}\hat{\nabla}_0^2\psi_2^a + \frac{i}{2}\hat{\nabla}_1^2\psi_2^b\right) + \bar{\psi}_2^a\left(\frac{1}{2}\hat{\nabla}_0^2\psi_1^a - \frac{i}{2}\hat{\nabla}_1^2\psi_1^b\right)$$

$$+ \bar{\psi}_1^b\left(\frac{1}{2}\hat{\nabla}_0^2\psi_2^b - \frac{i}{2}\hat{\nabla}_1^2\psi_2^a\right) + \bar{\psi}_2^b\left(-\frac{1}{2}\hat{\nabla}_0^2\psi_1^b + \frac{i}{2}\hat{\nabla}_1^2\psi_1^a\right) \qquad (5.85)$$

The awkward term $\bar{\bar{d}}\hat{\bar{D}}\tilde{d}$ can be written $\bar{d}\hat{D}d$ using the identities discussed earlier. The last four terms can also be simplified by rewriting them in terms of u and d and by introducing flavor-isospin notation,

$$f = \begin{pmatrix} u \\ d \end{pmatrix} \qquad (5.86)$$

and the flavor matrices,

$$T_0 = \begin{pmatrix} 0 & -1 \\ 1 & 0 \end{pmatrix}, \qquad T_1 = \begin{pmatrix} 0 & -i \\ -i & 0 \end{pmatrix} \qquad (5.87)$$

After some algebra, we have the final result

$$S = \bar{f}\hat{D}f + \bar{f}\gamma_5 T_\mu \frac{1}{2}\hat{\nabla}_\mu^2 f \qquad (5.88)$$

This form of the action has the good features discussed above. The first term is free of flavor mixing and displays the two species of the continuum limit. The second term is an irrelevant boson-kinetic-energy term with the fermion remnants, the matrices $\gamma_5 T_\mu$. We shall see below that the matrix character of this term insures the existence of a continuous remnant of chiral symmetry at all lattice spacings while it breaks explicitly the axial-flavor-neutral symmetries.

5.5 Staggered fermions and remnants of chiral symmetry

One of the best features of Euclidean staggered fermions is that they possess a continuous remnant of chiral symmetry even on the strongly cutoff lattice. This allows one to study some aspects of spontaneous symmetry breaking and light-pion physics even on a coarse lattice. Just as in our discussion of the Hamiltonian, we will be discussing even and odd, alternating, lattice sites.

To begin, reconsider the action written in terms of the χ variables, and imagine a global $U(1) \times U(1)$ invariance group such that the first $U(1)$ acts on even sites and the second $U(1)$ acts on odd sites,

$$\chi(n) \to U_e \chi(n), \qquad n \text{ even}; \qquad \chi(n) \to U_o \chi(n), \qquad n \text{ odd} \quad (5.89)$$

and

$$\bar{\chi}(n) \to \bar{\chi}(n)U_o^+, \qquad n \text{ even}; \qquad \bar{\chi}(n) \to \bar{\chi}(n)U_e^+, \qquad n \text{ odd} \quad (5.90)$$

Looking back at our expression for the action written in terms of χ, we see that this is clearly a symmetry because only even (odd) sites are coupled to odd (even) sites.

It is instructive to write this symmetry in terms of the conventional Fermi fields,

$$\psi_1 \to U_o \psi_1, \qquad \psi_2 \to U_e \psi_2 \tag{5.91}$$

$$\begin{pmatrix} u_1 \\ u_2 \end{pmatrix} = \begin{pmatrix} \psi_1^a + \psi_1^b \\ \psi_2^a + \psi_2^b \end{pmatrix}$$

$$\to \begin{pmatrix} U_o u_1 \\ U_e u_2 \end{pmatrix} = \frac{1}{2}(U_o + U_e) \begin{pmatrix} u_1 \\ u_2 \end{pmatrix} + \frac{1}{2}(U_o - U_e) \begin{pmatrix} u_1 \\ -u_2 \end{pmatrix} \tag{5.92}$$

Similarly,

$$\begin{pmatrix} \tilde{d}_1 \\ \tilde{d}_2 \end{pmatrix} \to \frac{1}{2}(U_o + U_e) \begin{pmatrix} \tilde{d}_1 \\ \tilde{d}_2 \end{pmatrix} + \frac{1}{2}(U_o - U_e)\gamma_5 \begin{pmatrix} \tilde{d}_1 \\ \tilde{d}_2 \end{pmatrix} \tag{5.93}$$

However, this expression can be written in terms of d by recalling that $\tilde{d} = \sigma_1 d$ and $\gamma_5 = \sigma_3$, so

$$\begin{pmatrix} \tilde{d}_1 \\ \tilde{d}_2 \end{pmatrix} = \sigma_1 \begin{pmatrix} d_1' \\ d_2' \end{pmatrix} = U_+ \sigma_1 \begin{pmatrix} d_1 \\ d_2 \end{pmatrix} + U_- \gamma_5 \sigma_1 \begin{pmatrix} d_1 \\ d_2 \end{pmatrix} \tag{5.94}$$

where we have introduced $U_\pm = (U_o \pm U_e)/2$. Now,

$$d' = \sigma_1 U_+ \sigma_1 d + \sigma_1 U_- \gamma_5 \sigma_1 d \tag{5.95}$$

Finally, since $\sigma_1 \gamma_5 \sigma_1 = -\gamma_5$, we can write

$$d' = U_+ d - U_- \gamma_5 d \tag{5.96}$$

which is identical to the transformation equation for u except for the last sign. Using the isospin notation,

$$f = \begin{pmatrix} u \\ d \end{pmatrix}$$

we have

$$f' = U_+ f + U_- \gamma_5 T_3 f \tag{5.97}$$

where T_3 is the third component of isospin,

$$T_3 = \begin{pmatrix} 1 & 0 \\ 0 & -1 \end{pmatrix}$$

This transformation law is a symmetry of the action. We learn that the original $U_e \times U_o$ symmetry operation contains a vector piece, $1_D \times 1_F$, fermion-number conservation, and an axial piece, $\gamma_5 \times T_3$, one component of axial isospin. The second piece forbids a mass term in S. If interactions preserve this symmetry, as they will in our applications, then the mass counterterms cannot be generated by those interactions. This feature means that fine tuning of the fermion action will not be necessary in order to keep the fermion excitations light even in the presence of such interactions. In addition, if the symmetry breaks dynamically (spontaneously), as it will in applications, then, in more than two dimensions, a massless pion will appear in the theory's spectrum in accordance with Goldstone's theorem.

Note that the irrelevant operator in the block-derivative expression for the action breaks the full axial-isospin symmetries to a smaller set of continuous symmetries. It breaks the axial-flavor-neutral symmetries explicitly.

In addition to the continuous $U_o \times U_e$ symmetries, there are discrete symmetries of interest in S. For example, if we translate the system by an odd number of sites in either direction and transform phases appropriately, the action will be unchanged. Therefore, these symmetries are the "square roots of translations" and can be identified with γ_5, T_i, and their products. We refer the interested reader to the literature for more details.

5.6 Exact chiral symmetry on the lattice

Both the original Wilson method and the staggered-fermion method suffer from having too little symmetry on a lattice with a fixed spacing a. However, their continuum limits should have the correct number of species with the correct chiral and flavor symmetries. The approach to the continuum limit can be sped up considerably by improving both actions with next-to-nearest-neighbor couplings whose strengths are engineered to suppress $O(a^2)$ symmetry-breaking effects. With these improvements physical, accurate results from simulations can be expected on modestly sized lattices.

In addition, there are more inventive lattice-fermion approaches that possess more symmetries even on the cutoff lattice, have the correct axial anomaly structure, and may yield nonperturbative formulations of chiral gauge theories such as the standard model of electroweak interactions. Our emphasis here remains on vector theories like QCD, so we will not discuss left–right-asymmetric models, but we shall see that the desire to formulate such models on the lattice in a consistent, nonperturbative fashion is having very significant influence on QCD research.

5.6.1 Domain-wall fermions

Domain-wall methods to decribe light fermions have been known to workers in statistical physics for over 60 years [26], but were rediscovered in lattice theory in the 1990s [27]. Inspiration in high-energy physics came from efforts by physicists interested in grand unification to understand global anomalies within chiral gauge theories for which the renormalizability of the field theory requires that the gauge currents remain anomaly-free. For example, the standard model of electroweak interactions has an anomalous global baryon current, but has anomaly-free local chiral gauge currents. Engineering such a feat on the lattice would be highly nontrivial in the light of the species-doubling phenomena we have already discussed for Wilson and staggered fermions. Here we will be less ambitious and use the domain-wall method to formulate QCD, a vector gauge theory with no left–right asymmetries, with exact chiral and flavor symmetries even at strong coupling on a lattice with a large cutoff length a. Such a formulation has enormous advantages over traditional Wilson and staggered approaches. Having the Goldstone bosons of broken chiral symmetries on strongly coupled lattices for analytic as well as numerical purposes is a great help. One might worry that this is too much of a good thing in the sense that having massless modes on a finite lattice will lead to huge artificial finite-size effects. In fact, when $m_\pi L \approx 1$ (L is the linear extent of the lattice), finite-size effects are considerable, but chiral perturbation theory has precise predictions for them, which is another great help.

To begin, let's outline the strategy used here. In condensed-matter physics one finds models in which there are interfaces between different phases of matter and on the interface conduction electrons can propagate as massless modes. We begin with four-dimensional Euclidean lattice-gauge theory and we append to it a fifth dimension in which a fermion propagates with a spatially dependent mass. Let the mass term in the fifth dimension have the spatial dependence of a kink. We shall see that, in this case, the free Dirac equation will have a massless chiral zero mode bound to the four-dimensional domain wall. There will also be the doublers with opposite chirality on the wall, as we have seen in our discussion of naive fermions above. However, the doublers can be removed by the Wilson term. At this point we have the degrees of freedom of a four-dimensional chiral gauge theory as long as the massive fermion modes which are not bound to the defect decouple and provided that the gauge fields just propagate along the defect in the four dimensions of the original lattice-gauge-theory formulation. On the lattice we must specify the extent of the fifth-dimensional world or give it periodic boundary conditions. For open boundary conditions, there will be an interface at one end of the fifth-dimensional world and an anti-interface at the other end with L_s lattice spacings between them. This means that we can arrange for one handedness of fermion on the interface and the opposite handedness on the anti-interface L_s lattice spaces away in the fifth direction. These components then comprise a single Dirac fermion in which the left- and right-handed components do not mix in the limit $L_s \to \infty$. Therefore, the fermion system has exact chiral symmetry on the lattice! This is perfect for QCD applications.

Extensions of the domain-wall idea to chiral rather than vector gauge theories are controversial and will not be discussed here.

To illustrate how the domain-wall method works, consider the free five-dimensional Dirac equation coupled to a domain wall [28]. Denote the five coordinates $z_\mu = (\vec{x}, s)$, where $\vec{x}$ is the coordinate of our familiar four-space. Taking the fifth dimension to have infinite, open extent, we make a domain wall by considering a spatially dependent mass term with a kink,

$$m(s) \underset{s \to \pm\infty}{\to} \pm m_\pm \tag{5.98}$$

where $m_\pm > 0$. The Dirac operator is then

$$D_5 = \not{\partial} + m(s) = \vec{\gamma} \cdot \vec{\nabla} + \gamma_5 \partial_s + m(s) \tag{5.99}$$

If we choose $m(s)$ to vanish at $s = 0$, we expect massless modes bound to the defect. These zero modes satisfy $D_5 \Psi_0 = \vec{\gamma} \cdot \vec{\nabla} \Psi_0$ and would describe massless propagation along the defect. In detail, we choose the zero modes to have the form,

$$\Psi_0^\pm = e^{i\vec{p} \cdot \vec{x}} \phi_\pm(s) u_\pm \tag{5.100}$$

where the $u_\pm$ are constant four-component chiral spinors, $\gamma_5 u_\pm = \pm u_\pm$, and the functions $\phi_\pm(s)$ satisfy the one-dimensional differential equation

$$[\pm \partial_s + m(s)] \phi_\pm(s) = 0. \tag{5.101}$$

The solutions to this equation are

$$\phi_\pm(s) = \exp\left(\mp \int_0^s m(s') \, ds'\right) \tag{5.102}$$

Note that only one of these solutions, $\phi_+(s)$, is normalizable and admissable. We learn that there is a single, positive-chirality fermion bound to the defect.

This continuum analysis is easily transplanted to the lattice. If we discretize the Dirac operator in the manner of Wilson and use a five-dimensional Wilson term, we can arrange the system to have the same low-energy degrees of freedom as our simple continuum analysis presented above. Again, only a single positive-chirality mode will persist. Of course, on a finite lattice, we would choose periodic or open boundary conditions in the fifth dimension. For example, take a lattice of length L_s is the s direction and choose periodic boundary conditions. Then, given a domain wall at $s = 0$, there must be an anti-domain wall at $s = \pm L_s/2$. For every zero mode at $s = 0$, there is a zero mode of opposite chirality at $s = \pm L_s$. The vector character of the five-dimensional theory is explicit. The zero modes at $s = 0$ and $s = \pm L_s$ have exponentially small overlap, so for finite L_s there is an exponentially small breaking of chirality. This breaking is completely decoupled from the restoration of the spacetime symmetries of the four-dimensional lattice system because one could choose different lattice spacings in the s direction and in the four physical directions. The breaking of chiral symmetry arises only through the finiteness of the fifth dimension.

When gauge fields and the full structure of lattice-gauge theory are incorporated into this framework, we continue to use the same plaquette action in the four-dimensional world, but we do not let the gauge fields extend into the fifth dimension. Now the physical content of the four-dimensional theory brings up some issues challenging our understanding of the Nielsen–Ninomiya theorems. In particular, at low energies it appears that the domain-wall formulation has produced an effective four-dimensional theory of fermion zero modes coupled to gauge fields, which in general must be expected to be anomalous and cannot be simultaneously locally gauge-invariant and chirally invariant. The resolution to this puzzle was, in fact, provided long before the application of these ideas to lattice-gauge theory. In the context of grand unified theories, it was understood that heavy fermions that live away from the domain walls do not fully decouple from the four-dimensional dynamics. They provide the anomalous currents which flow into the fifth dimension and account for the "apparent" nonconservation of the flavor-singlet chiral current in our four-dimensional world [28]. So, the extra dimension is the "loophole" in the Nielsen–Ninomiya theorem that the domain-wall fermions have exploited.

Let's record the domain-wall fermion action as it is used in practical calculations. Some explicit chiral-symmetry breaking is inserted into the model in the form of a conventional bare-quark mass m. The chiral limit of the system is found by first taking $L_s \to \infty$ and then taking $m \to 0$. The fermion action reads

$$S_{\rm f} = - \sum_{x,y,s,s'} \bar{\psi}(\not{D}_{x,y}\delta_{s,s'} + \not{D}_{s,s'}\delta_{x,y})\psi \tag{5.103}$$

where

$$\not{D}_{x,y} = \tfrac{1}{2}[(1 + \gamma_\mu)U_{x,\mu}\delta_{x+\mu,y} + (1 - \gamma_\mu)U_{y,\mu}^\dagger\delta_{x-\mu,y} - (M - 4)\delta_{x,y}] \tag{5.104}$$

and

$$\not{D}_{s,s'} = \begin{cases} P_{\rm R}\delta_{1,s'} - m P_{\rm L}\delta_{L_s-1,s'} - \delta_{0,s'} & s = 0 \\ P_{\rm R}\delta_{s+1,s'} + P_{\rm L}\delta_{s-1,s'} - \delta_{s,s'} & 0 < s < L_s - 1 \\ -m P_{\rm R}\delta_{0,s'} + P_{\rm L}\delta_{L_s-2,s'} - \delta_{L_s-1,s'} & s = L_s - 1 \end{cases} \tag{5.105}$$

where $P_{\rm R,L} = (1 \pm \gamma_5)/2$ are the chiral projection operators. The effectiveness of this action in carrying out the original plan of the domain-wall philosophy is discussed in the literature [29].

Domain-wall fermions have been used for simulations of QCD at nonzero temperature. They could also be used at nonzero chemical potential, following the rules we will introduce in the context of naive and staggered fermions in the next chapter. Since domain-wall fermions are much more computer intensive than staggered fermions, simply because the algorithm runs in five dimensions rather than four and the fifth dimension cannot be taken to be small, their use in applications is still in its infancy. However, for applications in which chiral and flavor symmetries are paramount, we expect that the method will have a bright future.

5.6.2 The Ginsparg–Wilson relation

Long ago, in the early days of lattice-gauge theory, Ginsparg and Wilson, who were engaged in a renormalization-group analysis, suggested that lattice Dirac operators D should be constructed to satisfy the relation [30],

$$\gamma_5 D + D\gamma_5 = aD\gamma_5 D \qquad (5.106)$$

because it guarantees that the fermion propagator anti-commutes with γ_5 at non-zero distances on the lattice with spacing a, thereby respecting an aspect of chiral symmetry. In fact this relation has a more fundamental and useful interpretation [31]. Instead of considering the usual chiral rotation

$$\psi \to e^{i\epsilon\gamma_5}\psi, \qquad \bar{\psi} \to \bar{\psi}e^{i\epsilon\gamma_5} \qquad (5.107)$$

consider a version adapted to a lattice with a given spacing a,

$$\psi \to e^{i\epsilon\gamma_5(1-\frac{1}{2}aD)}\psi, \qquad \bar{\psi} \to \bar{\psi}e^{i\epsilon(1-\frac{1}{2}aD)\gamma_5} \qquad (5.108)$$

If we take ϵ to be infinitesimal, then one can easily check that Eq. (5.106) follows. Equation (5.108) can also be generalized to flavor-non singlet chiral transformations by including a flavor matrix into the transformation law in the usual way.

This "success" looks too good to be true. It appears that we now have a gauge-invariant regularization scheme that has no chiral anomalies. We know from the Nielsen–Ninomiya lattice arguments that this is not possible, assuming that the lattice Dirac operator has reasonable properties that we discussed above. So, can we find a reasonable expression for D that satisfies Eq. (5.106)? A particularly simple candidate is [32]

$$D_N = \frac{1}{a}[1 - A(A^\dagger A)^{-1/2}], \qquad A = 1 - aD_W \qquad (5.109)$$

where D_W is the Wilson form of the lattice Dirac operator, having an irrelevant boson-like piece to suppress fermion doublers at the edges of the Brillouin zones, as discussed and recorded above. We can check that Eq. (5.109) satisfies Eq. (5.106) by direct sustitution. However, D_N looks pathological because it involves the reciprocal of the square root of an operator. However, D_N is not nonlocal. If we transform to momentum space, we can see this explicitly. Use the abbreviations $\hat{p} = (2/a)\sin(ap_\mu/2)$ and $\tilde{p} = (1/a)\sin(ap_\mu)$ and then calculate the Fourier transform,

$$a\tilde{D}(p) = 1 - \left(1 - \frac{1}{2}a^2\hat{p}^2 - ia\gamma_\mu\tilde{p}_\mu\right)\left(1 + \frac{1}{2}a^4\sum_{\mu<\nu}\hat{p}_\mu^2\hat{p}_\nu^2\right)^{-1/2} \tag{5.110}$$

This expression has the right properties. It is analytic and periodic in p_μ within the Brillouin zone, it reduces to $i\gamma_\mu p_\mu$ for low momenta, it is invertable for all nonzero momenta, and in x space it falls off exponentially at a rate proportional to $1/a$. Since it does not anti-commute with γ_5, it does not have the usual chiral symmetry appropriate to continuum field theory and the Nielsen–Ninomiya no-go theorem does not apply to it. Nonetheless, since the symmetry it does have is a continuous chiral transformation relevant to the lattice, we need to understand the fate of the anomaly [31].

Consider calculating an expectation value of an operator O involving fermion fields,

$$\langle O \rangle_F = \int \prod_x \mathrm{d}\psi(x)\,\mathrm{d}\bar{\psi}(x)\, O\, e^{-S_F} \tag{5.111}$$

Now apply the lattice chiral symmetry Eq. (5.108) for infinitesimal ϵ, and find that the variation in the expectation value is not zero because the measure in the fermion integral is not invariant,

$$\langle \delta O \rangle_F = -a\, \mathrm{tr}\{\gamma_5 D\}\, \langle O \rangle_F \tag{5.112}$$

where the trace goes over all fermion fields. In the case of free fermions, the trace vanishes and the symmetry is exact, as expected. In addition, for flavor-nonsinglet chiral rotations, we obtain zero because of the presence of the group generator in the trace. However, in other interacting theories $-a\,\mathrm{tr}\{\gamma_5 D\}$ is the chiral anomaly and it will not vanish in general. A formal, but brief, derivation of the anomaly equation can be given in this context [31]. More explicit derivations also exist.

Begin by writing the left-hand side of Eq. (5.106) for the operator $z - D$ in the case that D satisfies Eq. (5.106),

$$a(z - D)\gamma_5(z - D) = z(2 - az)\gamma_5 - (1 - az)[(z - D)\gamma_5 + \gamma_5(z - D)] \tag{5.113}$$

Since we want an expression for the anomaly, multiply this equation by $(z - D)^{-1}$, taking z outside the spectrum of D, and trace it,

$$-a\, \mathrm{tr}\{\gamma_5 D\} = z(2 - az)\, \mathrm{tr}\{\gamma_5(z - D)^{-1}\} \tag{5.114}$$

Finally, divide throughout by $z(2 - az)$ and integrate over an infinitesimal circle that encloses only the spectral value 0 of D,

$$-a \, \text{tr}\{\gamma_5 D\} = 2 \, \text{tr}\{\gamma_5 P_0\} = 2N_f \times \text{index}(D) \qquad (5.115)$$

where

$$P_0 = \oint \frac{dz}{2\pi i}(z - D)^{-1} \qquad (5.116)$$

We identify Eq. (5.115) as a formal statement of the axial anomaly, expressed in terms of zero modes of the Dirac operator, which is particularly appropriate for topological analyses, and was touched on above in our discussion of instantons.

This very curious fermion method is being studied theoretically and numerically and it looks promising on all fronts. The action associated with D_N involves direct couplings between arbitrarily separated sites and is not local. The importance of these direct couplings will depend on the gauge fields involved, so little can be said of a general, generic nature. However, results of some studies suggest that it will be possible to effect the inversions of the lattice Dirac operator to implement simulation studies based on this fermion method. Several more orders of magnitude of computer time will be needed compared with the use of staggered fermions. The method could be applied to finite-temperature studies of QCD and QCD with a chemical potential, if all these preliminaries work out. The usefulness of having another fermion method with good chiral properties cannot be underestimated in this field.

Looking beyond vectorlike theories such as QCD, there is the hope that the Ginsparg–Wilson-fermion method will allow precise and practical formulations of the standard model of electroweak interactions as well as other chiral gauge theories. Then their nonperturbative physical content could be analyzed. In the case of the standard model, such an advance would allow simulations of the early universe in which electroweak symmetry breaking was restored. Perhaps early-universe studies of the baryon asymmetry in the universe would yield to simulations. Perhaps.

5.7 Chiral-symmetry breaking on the lattice

Lattice-gauge theory provides a good framework for studying the dynamical breaking of chiral symmetry and soft-pion phenomenology. Condensation phenomena and the calculation of order parameters such as $\langle \bar{\psi}\psi \rangle$ are practical jobs for computer simulations of QCD and much has been learned about this subject from numerical simulations over the years.

Perhaps the simplest argument for the spontaneous breaking of chiral symmetry was abstracted by A. Casher from models in $1 + 1$ dimensions [33]. The argument motivates the belief that vector theories that confine quarks necessarily spontaneously break chiral symmetry. Begin with a theory whose Lagrangian is chirally symmetric and whose quarks have vanishing bare masses. Suppose also that the confining force is spin-independent and that bound $q\bar{q}$ pairs exist in an *s*-wave state. View these states semi-classically as in the quark model. Consider a turning point in the worldline of the bound quark. The quark must turn around, but its spin does not because the confining force does not cause spin flips. Therefore, its chirality (chirality equals helicity for massless quarks) must change sign. So chirality is *not* a good quantum number and we must have dynamically induced chiral-symmetry breaking!

What is the physical mechanism behind this miracle? When the quark turns around, it must pick up a unit of chirality from the vacuum. Therefore, the vacuum must have indefinite chirality, as a condensate of $\bar{q}q$ pairs would. Chirality must be spontaneously broken in order to support the bound states assumed.

In fact, the formation of bound states, rather than confinement, is central to Casher's argument. In QCD it appears that chiral-symmetry breaking occurs at intermediate length and coupling scales on which the instanton liquid describes the dynamics. Confinement and its accompanying linear potential work on larger length scales.

The Casher physical picture of chiral-symmetry breaking is rather different than that presented by the model which began the field of soft pions and chiral-symmetry breaking, the Nambu–Jona Lasinio model [34]. These models were inspired by BCS superconductivity in which short-range forces between electrons cause the formation of an ee condensate giving rise to superconductivity in various low-temperature metals. The original BCS models of superconductivity were essentially $(1 + 1)$-dimensional Gross–Neveu (four-Fermi) models which were asympotically free and led to superconductivity even in the weak-coupling limit. The point was that arbitrarily weak short-range couplings between electrons at the top of a filled, sharp Fermi surface are always sufficient to destabilize the free-electron state and lead to a lower-energy state in which the electrons are paired up (Cooper pairs) into charge-two loosely bound states that can condense. The four-dimensional four-Fermi models considered by Nambu and Jona Lasinio to model chiral-symmetry breaking required a critical coupling: only above the critical coupling, $g^2 > g_c^2$, would chiral symmetry be broken. Unlike the BCS model, there is no filled Fermi sea in the quantum vacuum, just a filled Dirac sea, so there is no inherent instability for a weak interaction to play upon and cause condensation. Instead, the short-range four-Fermi attractive potential must be increased in strength before quark–anti-quark pairs will condense into the vacuum and spontaneously break chiral symmetry. Unfortunately, the four-Fermi model in four dimensions is unrenormalizable, so the Nambu–Jona Lasinio model suffers from cutoff dependence and various other pathologies. However, there is a great deal of truth in their enormously important work.

There is an important rigorous relationship between the spectrum of the quark propagator and the chiral condensate that was inspired by Casher's qualitative argument. Consider $\rho(\lambda)\,d\lambda$, the mean number of eigenvalues of the fermion propagator contained in the interval $d\lambda$, per unit volume. The theorem of Banks and Casher relates this density of zero modes to the chiral condensate [35],

$$\langle 0|\bar{q}q|0\rangle = -\pi\rho(0) \qquad (5.117)$$

To derive this result, consider the fermion Green function S_F in a background gauge field F,

$$S_F(x, y) = \langle q(x)\bar{q}(y)\rangle_F = \sum_n \frac{u_n(x)u_n^\dagger(y)}{m - i\lambda_n} \qquad (5.118)$$

where $u_n(x)$ and λ_n are eigenfunctions and eigenvalues of the Euclidean Dirac operator,

$$\slashed{D}u_n(x) = \lambda_n u_n(x) \qquad (5.119)$$

Note that, except for zero modes, the eigenfunctions occur in pairs, u_n and $\gamma_5 u_n$, with opposite eigenvalues. So, setting x equal to y and integrating over x, we have,

$$\frac{1}{V}\int d^4x\,\langle 0|\bar{q}(x)q(x)|0\rangle_F = -\frac{2m}{V}\sum_{\lambda_n>0}\frac{1}{m^2+\lambda_n^2} \qquad (5.120)$$

Averaging over all gauge-field configurations and taking the thermodynamic limit $V \to \infty$,

$$\langle 0|\bar{q}q|0\rangle = -2m\int_0^\infty d\lambda\,\frac{\rho(\lambda)}{m^2+\lambda^2} \qquad (5.121)$$

This is a formal expression since the integral is not well behaved at large λ. However, we are interested just in the possibility of spontaneous symmetry breaking, which is controlled by the low-λ behavior of $\rho(\lambda)$. In fact, if $\rho(\lambda)$ behaves as λ^α, then the right-hand side of this equation is propotional to m^α. Therefore, a nonzero result arises if and only if $\rho(\lambda)$ tends to a nonzero limit for $\lambda \to 0$. On completing the integration over λ in this case, we obtain the Banks–Casher result.

Since the fermion propagator is central to lattice-gauge-theory algorithms and the density of eigenvalues is sensitive to topological excitations in QCD, this result has played an important role in diverse studies of chiral-symmetry breaking.

There are many related ideas about the mechanism of chiral-symmetry breaking in the modern field-theory literature. For example, in gauge theories of grand unified theories (GUTs), one develops a rough picture of spontaneous symmetry-breaking patterns by modeling the theory's complex dynamics with simple, single-gauge-boson exchange between quarks and anti-quarks. On the basis of some experience with dynamical calculations using gap and Schwinger–Dyson equations, one supposes that a condensate forms and the relevant symmetry breaks dynamically if the gauge-boson coupling exceeds a critical value of the order of unity, $g_c^2/(4\pi) \approx O(1)$. In these models with many quark and boson species, one considers all the possible condensates and symmetry-breaking patterns and one considers single-boson exchange and hypothesizes that there will be condensation and symmetry breaking in that channel where single-boson exchange yields the strongest force [36]. Quarks in high representations of the gauge group are, therefore, predicted to condense preferentially. Models addressing the hierarchy problem of GUTs have been proposed on this basis. Since the dynamics of these theories are so rich, other seemingly totally different mechanisms have also been proposed. For example, instantons can lead to condensation and spontaneous symmetry breaking and these effects must, if possible, be accounted for. In the context of QCD in baryon-rich environments, the role of instantons appears to be especially important at moderate densities.

Lattice-gauge theory can add controlled, accurate calculations to all of these scenarios. Large-scale simulations of QCD have given good evidence of chiral-symmetry breaking in fairly realistic settings and these calculations are becoming more sophisticated with time. Soft-pion dynamics can also be simulated, so the low-energy theorems that accompany spontaneous symmetry breaking can be simulated, checked, and understood as well. These calculations must make contact with the continuum limit of the lattice model in order for them to have real value and, for the present generation of simulations, this requires large lattices and weak gauge coupling, according to asymptotic freedom.

It is interesting and instructive to see chiral-symmetry breaking in a simple and short lattice calculation [37]. This can be done for strong coupling using expansion methods, as we will illustrate now. These calculations offer a realization of Casher's argument: the lattice model of QCD confines quarks at strong coupling and it also breaks chiral symmetry.

Consider the lattice action for a "naive"-fermion formulation of QCD,

$$S = \frac{1}{2}\Sigma_{r,v}\bar{\psi}(r)\gamma_\mu n_\mu U(r, n_\mu)\psi(r + n_\mu) + m\Sigma_r\bar{\psi}(r)\psi(r)$$

$$+ \frac{1}{2g^2}\Sigma_{\text{plaq}}\,\text{tr}\,UUUU + \text{h.c.} \tag{5.122}$$

where the sum over the unit vector n_μ includes all the sites $r + n_\mu$ neighboring r. We will use this naive-fermion action (species "doubling" would produce 16 continuum quarks) because the notation is very simple and clear, and it will be adequate to illustrate several points.

Now suppose that the lattice coupling is large, $g^2 \gg 1$. Then the tr $UUUU$ term can be dropped. The U matrices lose their real dynamics and become random matrices whose functional integrals are easy. We will discuss this aspect of the calculation below after we have seen that the model breaks chiral symmetry dynamically.

We have added a small fermion-mass term as a symmetry-breaking field to the action so that we can search for spontaneous symmetry breaking in the usual fashion – by calculating the condensate and other quantities at nonzero m and then letting $m \to 0$ and seeing whether the condensate survives.

We organize the action into two terms,

$$S = S_0 + S_{\text{int}} \tag{5.123}$$

where

$$S_0 = m \Sigma_r \bar{\psi}(r)\psi(r), \qquad S_{\text{int}} = \tfrac{1}{2} \Sigma_{r,v} \bar{\psi}(r) \gamma_\mu n_\mu U(r, n_\mu) \psi(r + n_\mu) \tag{5.124}$$

The hopping term, S_{int}, will have to be treated to all orders if we are to obtain useful results. The calculation will resemble a self-consistent mean-field theory expressed in graphical terms.

In order to evaluate the graphs, we will need two basic contractions. The first deals with the fermion piece of the action,

$$\langle \psi_\beta^b(r) \bar{\psi}_\alpha^a(r) \rangle = \frac{1}{m} \delta_{\alpha\beta} \delta^{ab} \tag{5.125}$$

where α and β are color indices ranging from 1 to N for SU(N) color, and a and b are the spinors' Dirac indices ranging from 1 to $2^{d/2}$ (d is the dimensionality of spacetime). The expectation value in the equation is taken with respect to S_0. When S_{int} is treated as a perturbation, i.e. $e^{S_0 + S_{\text{int}}}$ is expanded in powers of S_{int}, local gauge invariance implies that only closed fermion loops contribute. On reading off S_{int}, we see that each link contributes a factor $-\tfrac{1}{2} \gamma_\mu n_\mu U(r, n_\mu)$ to the amplitude.

We will meet the contraction

$$\int [dU] U_{ij} U_{kl}^\dagger = \frac{1}{N} \delta_{il} \delta_{jk} \tag{5.126}$$

in these strong-coupling expansions. The final graphical rule we need is the prescription, from Fermi statistics, that each closed fermion loop is accompanied by a factor of -1.

Now we can begin constructing graphs. The zeroth-order contribution for $\langle \bar{\psi}(0)\psi(0) \rangle$ is NC/m, where we have summed over the indices on the Fermi fields and have defined $C = 2^{d/2}$ for clarity.

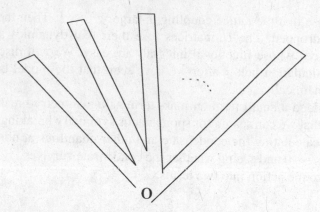

Figure 5.4. Dominant graphs contributing to chiral-symmetry breaking on the four-dimensional lattice for strong coupling, large N, and large spacetime dimensionality.

To second order in S_{int}, the fermion can hop to a nearest neighbor and back, so it contributes $(NC/m)[2d/(4m^2)]$, where the factor $2d$ counts the number of nearest neigbors. Note that the fermion path encloses zero area and the integral $\int[dU]U_{ij}U_{kl}^\dagger = (1/N)\delta_{il}\delta_{jk}$ has been used.

To proceed to higher orders we must take another limit to control the topologies of the graphs which dominate the calculation. In the limit of a large number of colors, $N \to \infty$, we will argue that only graphs with many zero-area "petals," as shown in Fig. 5.4, dominate. A graph with n petals contributes

$$A_n = \frac{NC}{m}\left(\frac{d}{2m^2}\right)^n = \frac{NC}{m}\left(\frac{d}{2m}\frac{\langle\bar\psi\psi\rangle_0}{NC}\right)^n \quad (5.127)$$

where we have identified the zeroth-order estimate for the chiral condensate, $\langle\bar\psi\psi\rangle_0$, on the right-hand side of this expression.

At this point we could just select this set of graphs, let $n \to \infty$, and compute $\langle\bar\psi(0)\psi(0)\rangle$. However, a more interesting, dynamical, and self-consistent calculation can be done. Recognize that the end of each "petal" of each graph is the source of an additional "flower." When this is done, the perturbative expansion at $N \to \infty$ becomes a fermion mean-field theory.

With this improvement the graphs A_n become, simply on replacing the zeroth-order $\langle\bar\psi\psi\rangle_0$ by its "exact" value determined by self-consistency,

$$A_n = \frac{NC}{m}\left(\frac{d}{2m}\frac{\langle\bar\psi\psi\rangle}{NC}\right)^n \quad (5.128)$$

Summing all these graphs for $\langle \bar{\psi}\psi \rangle$ gives

$$\langle \bar{\psi}\psi \rangle = \frac{CN}{m} \sum_{n=0}^{\infty} \left(\frac{d}{2m^2} \frac{\langle \bar{\psi}\psi \rangle}{NC} \right)^n$$

$$= \frac{CN}{m - \dfrac{d}{2NC}\langle \bar{\psi}\psi \rangle} \qquad (5.129)$$

which expresses the self-consistency of this mean-field approach. In the limit $m \to 0$, we find a nonsingular result, $\langle \bar{\psi}\psi \rangle = -iCN\sqrt{2/d}$. Finally, mapping back to Minkowski space, $\psi \to \psi_{\mathrm{M}}$, we have

$$\langle \bar{\psi}_{\mathrm{M}}\psi_{\mathrm{M}} \rangle = -i\langle \bar{\psi}\psi \rangle = -CN\sqrt{\frac{2}{d}} \qquad (5.130)$$

which gives spontaneous chiral-symmetry breaking.

Let's end with a few qualitative comments about this calculation. First, note the mean-field character of the calculation: the order parameter $\langle \bar{\psi}\psi \rangle$ is determined at one site in terms of $\langle \bar{\psi}\psi \rangle$ at its nearest-neighbor sites self-consistently. Note that the "petals" in A_n do not overlap, so our counting factors are good only for large dimensions d, another approximation underlying most mean-field calculations. One should consider this calculation as the leading nontrivial term in a $1/d$ expansion. Another technicality: vacuum fluctuations – disconnected graphs – did not appear in the calculation. One can check, however, that such graphs cancel out between the numerator and denominator in the formal expression for $\langle \bar{\psi}\psi \rangle$ at large d.

The calculation respects the spirit of Casher's intuitive argument. In particular, the random character of the U matrices at strong coupling was essential here and it was also the ingredient in showing confinement, the area law for Wilson loops. Also, at strong coupling quarks will be bound into s-wave bound states.

Another illuminating calculation is the verification of the Goldstone theorem here. We want to check that the spontaneous breakdown of chiral symmetry is accompanied by a massless pion and that the PCAC relation holds. It is interesting that we can obtain these results easily on the lattice at strong coupling. Although the lattice badly violates spacetime symmetries, the staggered-fermion method has enough elements of chiral symmetry to get some qualitative physical effects correct. Of course, these results must be obtained in the theory in the weak-coupling, continuum limit in order to make contact with real physics. That is being done through extensive computer simulations at present, and quantitative agreement between the lattice results and hadronic data is emerging.

We need to calculate the pion propagator,

$$D(r - r') = \langle (\bar{\psi}(r)\gamma_5\psi(r))(\bar{\psi}(r')\gamma_5\psi(r')) \rangle \qquad (5.131)$$

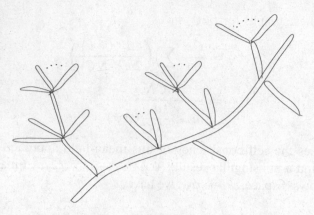

Figure 5.5. Graphs contributing to the pion propagator at strong coupling.

in the same fermion mean-field scheme as that in which we calculated the order parameter $\langle \bar{\psi}\psi \rangle$. The graphs that dominate are shown in Fig. 5.5 and they can be summed by the same methods as those used to obtain the order parameter $\langle \bar{\psi}\psi \rangle$, yielding

$$D(k) \underset{k \to 0}{\sim} \frac{4NC}{k^2 + \frac{4d}{NC}m\langle \bar{\psi}\psi \rangle} \tag{5.132}$$

We can let the explicit chiral-symmetry-breaking effect m vanish, and see that the pion becomes massless. For m small but nonzero, we can also read off a strong-coupling version of the current-algebra PCAC relation,

$$f_\pi^2 m_\pi^2 = m\langle \bar{\psi}_M \psi_M \rangle \tag{5.133}$$

in the form

$$\frac{NC}{4d}m_\pi^2 = m\langle \bar{\psi}_M \psi_M \rangle \tag{5.134}$$

Because of these results we can expect a smooth approach to continuum-limit soft-pion physics starting from lattice calculations. In addition, we can easily experiment in lattice calculations with $\langle \bar{\psi}\psi \rangle$ and the pion propagator in high-temperature and high-chemical-potential environments to discover new phenomena.

We have left one detail undone. We need to check that the graphs shown in Fig. 5.5 actually dominate the calculation when $N \to \infty$. The argument

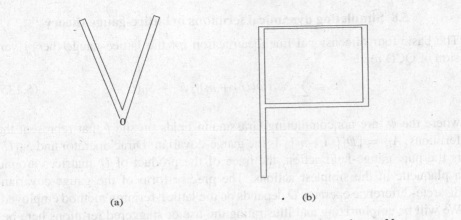

(a) . (b)

Figure 5.6. Dominant and subdominant graphs at large N.

here is essentially the same as the more familiar argument in the continuum $N \to \infty$ limit of Yang–Mills theory. For example, consider Fig. 5.6(b) with the subdominant graph having a hole in the quark worldline and compare it with Fig. 5.6(a) without the hole. The dominant graph without the hole behaves as N, which simply counts the number of quark colors circulating in the closed infinitely narrow fermion loop. The graph with the hole behaves as $O(1)$ in N. This can be checked explicitly using the U-matrix integral discussed above. Alternatively, one can note that both branches of the internal loop in Fig. 5.6(b) with the hole must be local color singlets separately so there is an additional constraint in this graph, which eliminates one power of N.

One more observation here concerns the quenched limit of the theory. This is the limit in which the number of fermion species N_f is taken to zero so that the dynamics becomes that of just the gauge fields. It is clear in the "flower" graphs of the mean-field calculation that the fermion effects consist of a single continuous worldline. It is also clear that the graphs with loops, such as that shown in Fig. 5.6(b), are subdominant in the quenched $N_f \to 0$ limit. So this strong-coupling calculation suggests that the quenched limit of the theory, which is relatively easy to study by simulation methods since there are no quark loops in the dynamics, should be at least a good qualitative guide to the physics of spontaneous chiral-symmetry breaking. This has certainly proven to be the case, even though the quenched limit has some pathologies that ultimately limit its usefulness. In particular, the quenched limit ignores the anomaly and suffers from too much symmetry, a light η', and attendant specious infrared singularities.

5.8 Simulating dynamical fermions in lattice-gauge theory

The basic four-dimensional Euclidean action for the lattice-gauge-theory version of QCD reads

$$S = \sum_{ij} \bar{\psi}_i [\not{D}(U) + m]_{ij} \psi_j + S_0(U) \tag{5.135}$$

where the ψ_i are noncommuting Grassmann fields on site i that represent the fermions, $A_{ij} = [\not{D}(U) + m]_{ij}$ is the gauge-covariant Dirac operator and $S_0(U)$ is the pure gauge-field action, the trace of the product of U matrices around a plaquette in the simplest actions. The precise form of the gauge-covariant discrete-difference operator $\not{D}$ depends on the lattice-fermion method employed. We will be emphasizing and illustrating the use of staggered fermions here because of their good chiral-symmetry properties, their practical advantages in simulations, and their wide use in thermal and chemical-potential studies of lattice QCD. So in this case ψ_i will be a one-component Grassmann field and the first term of the action reads

$$\sum_n \bar{\psi}(n) \left(\frac{1}{2} \sum_\mu^4 \eta_\mu [U_\mu(n)\psi(n+\mu) - U_{n-\mu}^\dagger(n)\psi(n-\mu)] + m \right) \psi(n) \tag{5.136}$$

where the η_μ are the phase factors guaranteeing that ψ satisfies the four-dimensional Dirac equation and carries spin $\frac{1}{2}$. The reader will recall the appearance of these phases η_μ in our explicit two-dimensional discussion of staggered fermions. The four-dimensional generalization can be found in the literature. The $U_\mu(n)$ in the formula are the gauge-field SU(3) rotation matrices that reside on the links of the lattice between sites n and $n + \mu$.

A lattice-gauge-theory simulation of QCD would calculate expectation values of gauge- and fermion-field observables in the partition function,

$$Z = \int \prod_i d\psi_i \prod_j d\bar{\psi}_j \prod_{n,\mu} dU_\mu(n) \exp(-S) \tag{5.137}$$

Since the ψ are anti-commuting numbers, a direct simulation of this equation through a Metropolis or heat-bath algorithm familiar from classical statistical mechanics is not practical on a classical computer that can manipulate only ordinary binary numbers. Luckily, the Grassmann variables can be integrated out since they appear only in a quadratic form,

$$Z = \int \prod_{n,\mu} dU_\mu(n) \det[\not{D}(U) + m] \exp(-S_0(U))$$

$$= \int \prod_{n,\mu} dU_\mu(n) \exp[-S_0(U) + \text{tr}\ln(\not{D} + m)] \tag{5.138}$$

where we see the infamous fermion determinant. It is not clear that integrating out the fermions has done us much good, since the final expression is a nonlocal effective interaction among all the $U_\mu(n)$ matrices. At least one can check that the fermion matrix $\not{D} + m$ has a positive-semi-definite determinant, so the measure does constitute a conventional probabilistic measure. The fermion determinant contains the physics of all closed fermion loops. In particular, it contains the Fermi-statistic rule that associates a factor of minus 1 with each closed fermion loop.

Various approaches to evaluating the partition function by importance sampling have been suggested. We will consider the microcanonical, Langevin, and hybrid approaches here because they have been particularly productive in QCD simulations. Many results have been obtained for QCD spectroscopy at zero temperature, as well as the thermodynamics at nonzero temperature. Unfortunately, as we will discuss in greater detail below, the fermion determinant is not real at nonzero chemical potential for the gauge group SU(3). Therefore, most simulation studies have been performed for the gauge group SU(2) or the isospin chemical potential for the gauge group SU(3) and, as we will discuss, the most interesting physics has not been studied by lattice methods. It is to be hoped that improvements will be made, new algorithms will be invented, and this situation will improve. Presumably these developments will grow out of the known successful but limited methods which will be introduced now.

5.9 The microcanonical ensemble and molecular dynamics

We begin with the molecular-dynamics approach to problems in equilibrium statistical mechanics because this framework will produce a means to simulate the fermionic partition function above. To begin, we review the idea of this approach. Suppose that we had a sample of material and wanted to measure its bulk thermodynamic properties to high precision. We would prepare a sample of the material and make many sequential measurements of the properties of interest to us. We need many measurements in order to obtain averages with small statistical-error bars because the system is experiencing fluctuations due to its internal dynamics and, perhaps, its external environment. Our averaging procedure consists of taking a time average over the system's intrinsic dynamics. Under many circumstances we will discuss below, this time average will be identical to a configurational average of the same properties. The configurational average is an idea due to the inventors of statistical mechanics, Maxwell and others, whereby one imagines preparing a very large ensemble of identical copies of the physical system and taking averages by concatenating measurements of each particular member of the ensemble. For the same given external conditions, the time average of the properties of one system can be shown to be identical to the ensemble-average. In most theoretical discussions, one focuses on the

ensemble-average approach because of its mathematical clarity, but in most experimental situations, one measures the time averages of the observables and
relies on the molecular dynamics of the system to sweep the one experimental
system through the available phase space sufficiently that the idealized ensemble
average and the actual time average are identical.

Let's illustrate the molecular-dynamics approach with a simple problem. Consider a boson field ϕ defined on a lattice with sites labeled i. The theory has
an action $S(\phi)$ that determines its partition function and equilibrium statistical-
mechanical properties. The system, however, does not have a natural dynamics
to describe its approach to equilibrium. We are free to invent dynamics as long
as the resulting scheme produces the correct equilibrium statistical mechanics.
Two schemes that have been studied and analyzed extensively are the molecular-
dynamics approach to the microcanonical ensemble and the Langevin equation
for the canonical ensemble.

In the molecular-dynamics approach, we associate the action $S(\phi)$ with a
"potential" $V(\phi) \equiv \beta^{-1} S(\phi)$ and construct a fictitious "Hamiltonian,"

$$H = T + V = \sum_i \frac{1}{2} p_i^2 + V(\phi) \tag{5.139}$$

where i labels the lattice sites and p_i will be interpreted to be the momenta
conjugate to the field ϕ_i. However, using this Hamiltonian we could consider the
canonical ensemble and the classical statistical mechanics based on the phase
space $\prod_i dp_i \, d\phi_i$ and the Boltzmann factor $\exp(-\beta H)$. Since the p_i integrals
are trivial, this formulation reduces to the original path-integral formulation of
the boson-field theory, $Z = \int \prod_i d\phi_i \exp S(\phi)$.

To construct a molecular-dynamics approach to the same problem, identify p_i
with the momenta conjugate to ϕ_i by introducing a "time," τ, into the problem,
$p_i = d\phi_i / d\tau$. In the context of computer simulations, τ is essentially computer
time. Now the ensemble given by the phase-space measure $\prod_i dp_i \, d\phi_i$ and the
Boltzmann factor $\exp(-\beta H)$ defines the usual canonical ensemble of classical statistical mechanics. There is still nothing new or of any advantage here
until one passes to the microcanonical ensemble. In this approach the energy
of the physical system is fixed, $H = E$, and the measure in phase space is
$\prod_i dp_i \, d\phi_i \, \delta(H - E)$. Observables in the system are functions of p_i and ϕ_i and
have expectation values

$$\langle F \rangle = \frac{1}{Z} \int \prod_i dp_i \, d\phi_i \, \delta(H - E) F(p_i, \phi_i) \tag{5.140}$$

If F is just a function of the field ϕ_i, then standard arguments show that $\langle F \rangle$
calculated in the microcanonical ensemble is identical to $\langle F \rangle$ calculated in the
canonical ensemble in the thermodynamical limit of large volumes, $V \to \infty$.

However $\langle F \rangle$ can also be calculated from the evolution with "time" of the classical system. This is the molecular-dynamics approach to the problem. Let $(\phi_i(\tau), p_i(\tau))$ label the phase-space point of the physical system. Then a "time" average of F would be

$$\langle F \rangle = \lim_{T \to \infty} \frac{1}{T} \int_0^T F(p_i(\tau), \phi_i(\tau)) \, d\tau \qquad (5.141)$$

This "time" average reproduces the expectation value in the microcanonical ensemble stated above if the ergodic hypothesis works for this physical system. Roughly speaking, one must assume that the molecular dynamics of the system carries the phase-space point $(\phi_i(\tau), p_i(\tau))$ uniformly over the entire energy shell $H = E$. One says that the degrees of freedom "mix" under their dynamical evolution. Exceptions to the ergodic hypothesis are afforded by free fields, for which there is, of course, no mixing at all, and weakly coupled systems, for which the mixing is inadequate or incomplete. In computer simulations of molecular-dynamical systems, one must carefully check that adequate mixing occurs.

Since molecular-dynamics simulations are done at fixed "energy" rather than fixed "temperature," we must read off the system's "temperature" from the results. (In most of our applications, "temperature" is actually "coupling.") The precise correspondence follows from the equipartition theorem for the kinetic energy, E,

$$\langle KE \rangle = \tfrac{1}{2} \beta^{-1} N^* \qquad (5.142)$$

where N^* is the number of independent, physically excited degrees of freedom in the system.

The molecular-dynamics approach to the boson system represents a clear alternative to standard Metropolis and heat-bath methods for simulation of the canonical ensemble. Those approaches move the phase-space point of the system over the entirety (it is hoped) of the phase space at given temperature in small, local jumps. In the molecular-dynamics approach, "time" averages of observables are accumulated over long computer runs. The molecular-dynamics approach simulates the microcanonical ensemble at given "energy" and it is (1) fully deterministic, (2) involves ordinary, coupled differential equations, and (3) generalizes to a practical, local algorithm for noncommuting fermions.

The molecular-dynamics approach to simulating QCD, whose action was written above, consists of inventing a scheme of "time"-dependent SU(3) gauge-field matrices, and pseudo-fermion fields, so that the "time" averages of these variables reproduce the ensemble averages of observables made of gauge fields and fermions in the original QCD lattice action.

The molecular-dynamics scheme is as follows. Let the gluon link variables $U_\mu(n)$ have a "time" dependence τ and let there be boson fields ("pseudo-fermions") whose evolution with "time" will generate the fermion determinant. One Lagrangian that will accomplish this reads

$$L = -S_0(U) + \frac{1}{2} \sum_{n,\mu} \dot{U}_\mu^\dagger(n) \hat{P} \dot{U}_\mu(n) + \sum_{ij} \dot{\phi}_i^\dagger [A^\dagger A]_{ij} \dot{\phi}_j - \omega^2 \sum_i \phi_i^\dagger \phi_i$$

$$(5.143)$$

where $A_{ij} = [\not{D}(U) + m]_{ij}$ is the lattice Dirac operator introduced earlier and $\hat{P}$ is a projection operator diag$(1, 1, 0)$ which picks out independent variables in the SU(3) matrix $\dot{U}_\mu$.

L consists of kinetic-energy terms both for the gauge field and for the "pseudo-fermion" field as well as potential terms for both. Each kinetic-energy term has special features. In the gauge-field piece we must include the projector $\hat{P}$ to account for the fact that one cannot treat each element in the 3×3 matrix U as independent because of the constraints imposed by its SU(3) character. One has a great deal of freedom in the parametrization of these matrices as well as freedom in how the constraints are treated. In the pseudo-fermion piece of the kinetic energy, we have placed the positive operator $A^\dagger A$ in the role one might have expected a "mass-squared" term. We shall see that the appearance of $A^\dagger A$ here will generate the fermion determinant in the *numerator* of the partition function. Finally, note that L is local because the Dirac operator couples only nearest-neighbor sites.

It is useful to construct this system's Hamiltonian in order to discuss its molecular dynamics and thermodynamics. We have momenta canonically conjugate to the gauge and pseudo-fermion degrees of freedom,

$$p_\mu = \dot{U}_\mu(n), \qquad P_i = [\dot{\phi}^\dagger A^\dagger A]_i \qquad (5.144)$$

The Hamiltonian then reads

$$H = \frac{1}{2} \sum_i p_i^2 + \sum_i P_i^\dagger (A^\dagger A)_{ij}^{-1} P_j + S_0(U) + \omega^2 \sum_i \phi_i^\dagger \phi_i \quad (5.145)$$

and the Hamiltonian equations of motion are

$$\dot{P}^\dagger = \frac{d}{d\tau}(A^\dagger(U)A(U)\dot{\phi}) = -\omega^2 \phi \qquad (5.146)$$

$$\dot{p} = \ddot{U} = -\frac{\partial}{\partial U^\dagger} S_0(U) + \dot{\phi}^\dagger \frac{\partial}{\partial U^\dagger}(A^\dagger(U)A(U))\dot{\phi} \qquad (5.147)$$

These equations are meant to give the reader an idea of the molecular-dynamics equations for this system. In order to implement them numerically, a convenient parametrization for the U matrices must be chosen and one must be careful that the special unitary character of the matrices is preserved by the evolution with "time". These matters of detail are nicely covered in the literature, but the point to be stressed here is that the Hamilton equations constitute a tractable (well, tractable for a computer) set of ordinary differential equations for which there are well-studied algorithms (such as second-order Runge–Kutta, or conjugate gradient, algorithms) that can solve them with controlled errors. The fermions do introduce an added complication that makes this system of coupled equations more difficult to simulate than a standard molecular-dynamics system that consists of classical, commuting variables. In particular, we see that, to advance the system of equations forward in "time" by one step, one must solve a linear set of equations for $\dot\phi$ of the form $A^\dagger(U)A(U)\dot\phi =$ source. Since A is a sparse matrix, this can be done efficiently by standard algorithms, such as the conjugate-gradient method. The number of operations in such an algorithm grows proportionally to the number of degrees of freedom on the finite lattice, so the computer time needed for the full simulation scales with the size of the system in a conventional fashion and it is practical, although demanding, to simulate large lattices of QCD. Computer timing and issues of the best algorithm etc. all change with hardware and software issues and the interested reader should consult the most up-to-date research articles to appreciate the state of the art.

Our last task is to check that the Lagrangian L invented above actually gives the original path integral. The canonical ensemble based on the Hamiltonian above reads

$$Z = \int DU\, Dp\, D\phi\, D\phi^\dagger\, DP\, DP^\dagger \exp(-H/T) \qquad (5.148)$$

All the integrals except U enter H quadratically, so the integrals can be done,

$$Z \propto \int DU\, \det^2 A(U) \exp(-S_0(U)/T) \qquad (5.149)$$

which is the required answer, except for the second, rather than the first, power of the fermion determinant. This can be cured by realizing that $A^\dagger(U)A(U)$ does not couple nearest-neighbor ϕ fields in the staggered-fermion method. One can think of the four-dimensional lattice of sites as two interleaving lattices of "even" and "odd" sites, and the field ϕ_i can be set to zero on one of the sublattices and the dynamics will leave it zero forever. When this is done, $\det^2 A$ is replaced by $\det A$, and we have our final, acceptable scheme.

In retrospect we can appreciate the character of the tricks that allowed a purely bosonic system of equations to simulate a fermionic field theory, complete with the exclusion principle and all of the subtle correlations that are implied. The pseudo-fermion kinetic-energy term in L is $\frac{1}{2}mv^2$ with $m \propto A^\dagger(U)A(U)$. When the Hamiltonian is constructed, we have $p^2/(2m)$ and the resulting $(A^\dagger(U)A(U))^{-1}$ here is responsible for the *positive* power of $\det(A^\dagger(U)A(U)) = \det^2 A(U)$ in the partition function. A crucial feature of this scheme is how it avoids the full, nonlocal character of the fermion determinant. The full fermion determinant is built up in the molecular-dynamics algorithm wherein, for each "time" step, the linear system $A^\dagger(U)A(U)\dot{\phi} = $ source is solved for $\dot{\phi}$. This step is a local operation since $A^\dagger(U)A(U)$ couples only nearby degrees of freedom. Only by iterating this procedure over many "time" steps does one solve the molecular-dynamics equations and build up the full, nonlocal fermion determinant.

Ths last ingredient in the algorithm is the calculation of the coupling constant β. This is done through the equipartition theorem, which states that each term in the system's kinetic energy has an average value of $\frac{1}{2}kT$,

$$\frac{1}{2}\beta^{-1}N^* = \langle T \rangle = \left\langle \dot{\phi}A^\dagger A\dot{\phi} + \frac{1}{2}\sum_{n,\mu} \dot{U}_\mu^\dagger(n)\hat{P}\dot{U}_\mu(n) \right\rangle \qquad (5.150)$$

The numerical value of N^* depends on the particular parametrization of the U matrices.

5.10 Langevin and hybrid algorithms

Molecular-dynamics methods have several drawbacks. The method simulates the microcanonical ensemble of given "energy" and it works only if the ergodic hypothesis applies. It is very inconvenient to have to read off the coupling constant of a simulation after it is complete rather than before. However, this is just an annoyance in comparison with the possibility that the dynamics might not sample all of available phase space and produce equilibrium expectation values. In computer simulations with a finite set of weakly interacting degrees of freedom, violations of the ergodic hypothesis are virtually guaranteed.

Luckily the molecular-dynamics algorithm can be improved by learning some lessons from the Langevin equation, a stochastic differential equation that simulates the canonical ensemble. The molecular-dynamics method and the Langevin equation can be merged into a "hybrid" algorithm that will simulate lattice QCD with any number of quarks.

First, let's introduce the Langevin equation, which is famous from Brownian motion, in a simple context of a single Bose variable q. We want to calculate an expectation value,

$$\langle F(q) \rangle = \frac{1}{Z} \int dq \, F(q) \exp(-S(q)) \qquad (5.151)$$

In the microcanonical/molecular-dynamics approach, we have seen that this system would be given by the dynamics,

$$\ddot{q} = -\partial S(q)/\partial q \qquad (5.152)$$

and expectation values would be replaced by "time" averages,

$$\langle F(q) \rangle = \lim_{T \to \infty} \frac{1}{T} \int_0^T F(q(\tau)) \, d\tau \qquad (5.153)$$

By contrast, in the Langevin approach the system is subject to external, explicit white noise,

$$\dot{q} = -\partial S(q)/\partial q + \eta(\tau)$$
$$\langle \eta(\tau)\eta(\tau') \rangle = 2\delta(\tau - \tau') \qquad (5.154)$$

and average quantities are calculated using "time" averaging as above. The stochastic differential equation of Langevin causes $q(\tau)$ to execute a forced random walk such that it covers phase space with the Boltzmann weight $\exp(-S(q)) \, dq$.

The molecular-dynamics equations and the Langevin equations look rather contrary, but the fact that there is an intimate relation between them is best shown by replacing each differential equation by a discrete difference equation, as a computer programmer would do. The Langevin equation becomes, denoting χ_n as noise,

$$q_{n+1} = q_n + \Delta \chi_n - \tfrac{1}{2}\Delta^2 S'(q_n)$$
$$\chi_n \chi_{n'} = \delta_{nn'}$$
$$\tau_{n+1} - \tau_n = \tfrac{1}{2}\Delta^2 \qquad (5.155)$$

whereas the molecular-dynamics equations read

$$q_{n+1} = 2q_n - q_{n-1} - \Delta^2 S'(q_n)$$
$$\tau_{n+1} - \tau_n = \Delta \qquad (5.156)$$

which can be written more suggestively as

$$q_{n+1} = q_n + \tfrac{1}{2}(q_{n+1} - q_n) - \tfrac{1}{2}\Delta^2 S'(q_n) \qquad (5.157)$$

So we see that "noise" in the Langevin scheme corresponds to "velocity" in the molecular-dynamics scheme. The "time" differences correspond as $\tau_{n+1} - \tau_n = \tfrac{1}{2}\Delta^2$ in the Langevin scheme and $\tau_{n+1} - \tau_n = \Delta$ in the molecular-dynamics scheme.

These two schemes have the following features: 1. The Langevin equation has explicit noise, so it is ergodic by construction. Unfortunately, it samples phase space very slowly in many cases because $q(\tau)$ executes a forced random walk, which, if the noise dominates, fills space at a rate $\propto \sqrt{N}$, where N is the number of "time" steps. 2. Molecular dynamics uses the classical equations of motion, so it is as efficient as possible at probing the important regions of phase space locally. Also, its "time" step is large, so it moves along its trajectories rapidly. However, it need not be ergodic. For example, for long "times" the algorithm may become trapped in small regions of phase space and be unable to explore the entire energy shell. By comparison, the Langevin method could experience less of a problem in a similar situation because its explicit noise term "kicks" it through phase space.

These considerations, qualitative as they are, suggested to early workers in the field of lattice simulations of fermionic problems that a "hybrid" method combining the best features of each algorithm might be worthwhile [38]. In the hybrid scheme, one states that a time step will be either Langevin (with probability $p\Delta$) or molecular dynamic (with probability $1 - p\Delta$):

$$q_{n+1} = q_n + \Delta v_n - \tfrac{1}{2}\Delta^2 S'(q_n) \tag{5.158}$$

with

$$v_n = (q_{n+1} - q_{n-1})/(2\Delta), \qquad \text{with probability } 1 - p\Delta \tag{5.159}$$
$$v_n = \chi_n, \qquad \text{with probability } p\Delta \tag{5.160}$$

$$\tau_{n+1} = \tau_n + \Delta \tag{5.161}$$

One can optimize this algorithm by choosing the best probability per unit "time" to "refresh" it, i.e. replace the velocity with a random number. Experience has shown that p should be set to approximately twice the frequency of the "slowest mode" in the system. The idea here is that the fields in the system can be characterized by their variations on the four-dimensional spacetime lattice. Modes, with the slowest variations limit the rate at which the system can change and move through phase space. If one "drives" the system with noise at just this frequency, then one is encouraging the slowest modes, which hold back the system, to change efficiently. This is the best one can do within this framework.

When these ideas are implemented in lattice-gauge theory, one begins with the Lagrangian,

$$L = \frac{1}{2} \sum_{n,\mu} \dot{U}_\mu^\dagger(n) \hat{P} \dot{U}_\mu(n) + \sum_{ij} \dot{\phi}_i^\dagger [A^\dagger A]_{ij} \dot{\phi}_j$$
$$- \omega^2 \sum_i \phi_i^\dagger \phi_i - \beta \sum (\text{tr } UUUU + \text{h.c.}) \tag{5.162}$$

Notice that the "velocitites" $\dot{U}$ and $\dot{\phi}$ appear only in quadratic terms, so, at a chosen "time" step in the evolution of the molecular-dynamics equations, one can replace the $\dot{U}$ fields, say, by a completely new field configuration chosen from the Boltzmann distribution, $\exp(-\tfrac{1}{2}\dot{U}_\mu^\dagger(n) \hat{P} \dot{U}_\mu(n))$. A similar procedure can be followed for $\dot{\phi}$ but a row in the inverse of the matrix $A^\dagger A$ will be required in order to follow standard formulas and one must choose $\dot{\phi}$ in the Boltzmann distribution $\exp(-\dot{\phi}_i^\dagger [A^\dagger A]_{ij} \dot{\phi}_j)$. Finally, since even the pseudo-fermion field ϕ appears only as a quadratic form in L, it can be "refreshed" according to the Boltzmann distribution $\exp(-\omega^2 \phi^\dagger \phi)$.

There are variations on this early hybrid algorithm that are very useful. In one case, the pseudo-fermion field is replace by noise. We can see that this is possible because ϕ enters L just as a quadratic form. The advantage of this algorithm is its apparently greater efficiency as well as the possibility of tuning and treating the number of flavors as a continuous parameter [38]. This is important because in QCD applications we want to simulate two almost massless "up" and "down" quarks, a heavier "strange" quark, etc. This noisy fermion algorithm can effectively treat the power of the fermion determinant in the path integral as a free parameter. This algorithm plays an important role in the chemical-potential simulations, as we will discuss later.

Another very important variation is the so-called hybrid Monte Carlo algorithm [39]. Instead of just arbitrarily "refreshing" the variables in the molecular-dynamics evolution from "time" to "time," this algorithm executes a Monte Carlo step with a "guidance" Hamiltonian that has an accept–reject criterion. A consequence of this procedure is that the algorithm now becomes exact, like a traditional Monte Carlo algorithm. This means that the "time"-step errors found on discretizing the molecular-dynamics differential equations are completely removed. Of course, large "time" steps will not work well here because the probability of rejection of the Monte Carlo step will approach unity in that case and the algorithm will not push the system through phase space efficiently. In addition, this algorithm cannot take fractional powers of the fermion determinant because it cannot use noisy pseudo-fermion fields. However, if this algorithm can simulate a model of interest, its lack of "time"-step errors affords it clear advantages.

6
The Hamiltonian version of lattice-gauge theory

6.1 Continuous time and discrete space

Another form of lattice-gauge theory that is very intuitively appealing is the theory's Hamiltonian form. In many problems it is very instructive to make energy estimates, look at wavefunctions, and estimate the masses, sizes, and shapes of bound states etc. that a quantum-mechanical formulation with a Hamiltonian gives. Sometimes these simple things are hard to extract from a path-integral formulation. Mechanisms of confinement, flux-tube dynamics, and chiral-symmetry breaking are also amenable to Hamiltonian analysis. Deconfinement at finite T and transitions as functions of chemical potential will also be discussed in this formulation below. Since there is much more pioneering physics to be done in QCD in extreme environments, it is important to have many approaches available.

In the Hamiltonian formulation one uses a spatial lattice and leaves the time variable continuous, as we have illustrated in several $1 + 1$ examples above. The Hamiltonian is the generator of time evolution and acts on quantized states. The Hamiltonian could be obtained methodically from the path-integral formulation by calculating the system's transfer matrix and then taking the time continuum limit [9]. It is more instructive, however, to construct the Hamiltonian from scratch. In fact, we can think about just two spatial points, write the Hamiltonian for that system, and then consider a full three-dimensional spatial lattice.

As in the Euclidean formulation, the key to the construction is the requirement that local color-gauge invariance be an exact symmetry for any lattice spacing a. Let there be two sites with the link between them. The fermion field will live on the sites $i = 1, 2$ and it will be denoted $\psi_\alpha(i)$, where α is a color index. We will leave the color index implicit in many cases where its use is clear from the context. A local color transformation is implemented through an $SU_c(3)$ rotation matrix $V(i)$ in the fundamental representation

$$\psi(i) \to V(i)\psi(i) \tag{6.1}$$

Therefore, we can think of independent color frames of reference at each site i. Each term in the Hamiltonian H must be invariant with respect to independent rotations of each frame. Clearly a single-site mass term satisfies this criterion trivially,

$$m \sum_i \psi^\dagger(i)\psi(i) \tag{6.2}$$

However, the construction of a discrete form of the kinetic energy is not so simple because the "obvious" hopping Hamiltonian,

$$H_q = i[\psi^\dagger(2)\psi(1) - \psi^\dagger(2)\psi(1)] \quad \text{(tentative)} \tag{6.3}$$

is not locally gauge-invariant. Following Yang and Mills, we introduce an $SU_c(3)$ rotation matrix that relates the color frames at sites 1 and 2. In the fundamental representation it reads

$$U_3 = \exp\left(i\frac{\lambda^\alpha}{2}B_\alpha\right) \tag{6.4}$$

where B_α ($\alpha = 1, 2, 3, \ldots, 8$) is an angular variable. The geometrical meaning of U_3 determines its color-transformation law,

$$U \to V(1)UV(2)^{-1} \tag{6.5}$$

Using U we can write the hopping Hamiltonian with local color symmetry,

$$H_q = i[\psi^\dagger(2)U\psi(1) - \psi^\dagger(2)U^{-1}\psi(1)] \tag{6.6}$$

The U matrices are essential ingredients here, so let's study them in isolation first. These degrees of freedom relate adjacent color frames, so they occupy the links between frames. The color reference frames at each link can be rotated in color space and, to describe the invariance of the theory with respect to local color-gauge symmetry, we need the generators $E_\alpha(i)$, $\alpha = 1, 2, 3, \ldots, 8$, for these rotations at each site i. They are determined by their commutation relations,

$$[E_\alpha(i), E_\beta(j)] = if_{\alpha\beta\gamma}E_\gamma(j)\delta_{ij} \tag{6.7}$$

where $f_{\alpha\beta\gamma}$ are the structure constants of SU(3). Later we will relate the generators $E_\alpha(i)$ to non-Abelian electric flux, which will play an important role in the dynamics of pure gauge fields. However, our emphasis here is on geometrical features of gauge fields. In particular, the transformation law of U under rotations V of the color reference frames gives us the commutation relations of $E_\alpha(i)$ with U. Consider an infinitesimal local color rotation written to first order in the generators,

$$V(j) \approx 1 + \frac{i\epsilon^\alpha(j)}{2} E_\alpha(j) \tag{6.8}$$

where $\epsilon^\alpha(j)$ is an infinitesimal color vector of generalized angles. Substituting this expansion into the transformation relation for U gives

$$[E_\alpha(1), U_3] = \frac{\lambda^\alpha}{2} U_3, \qquad [E_\alpha(2), U_3] = -U_3 \frac{\lambda^\alpha}{2} \tag{6.9}$$

Since $E_\alpha(i)$ is a color vector and since the U operator relates adjacent color frames of reference, we can show that $E_\alpha(1)$ and $E_\alpha(2)$ are simply related,

$$E(2) = -U_8 E(1) \tag{6.10}$$

where U_8 means that U has been expressed in the eight-dimensional adjoint representation of SU(3). This result follows from the relations for commutation of $E_\alpha(i)$ with U_3 via a standard exercise in Clebsch–Gordon coeffients, making an **8** from the product of a **3̄** and a **3**. The important point about this result is that it means that color electric flux maintains its magnitude across links,

$$E(2)^2 = E(1)^2 \tag{6.11}$$

Now consider the space of states of the gauge field. There is an SU(3) group at each site, so we must label the states by their irreducible representations. In SU(3) each multiplet is labeled by two invariants, a quadratic and a cubic Casimir operator, C_2 and C_3. In addition, within each multiplet we need two additional quantum numbers to label a state uniquely and we will do that with a two-dimensional vector $\vec{m}$. (All this should be familiar from ordinary particle physics, in which we assign particles to a representation of $SU_{flavor}(3)$, perhaps an octet or decuplet, and then label the particular particle by its third component of isospin I_3 and its hypercharge Y.) A state in our world of two sites would be written

$$\prod_{i=1}^{2} |\vec{m}(i); C_2(i), C_3(i)\rangle \tag{6.12}$$

However, the quantum numbers on one site are not independent of those on the other. In particular, since the quadratic Casimir operator is just E^2 and we noted that $E(1)^2 = E(2)^2$, it must be that

$$C_2(1) = C_2(2) \tag{6.13}$$

In addition, C_3 is a second invariant constructed from the group generators, so

$$C_3(1) = C_3(2) \tag{6.14}$$

Now we have enough preliminaries to construct the space of states explicitly. We begin with the ground state $|0\rangle$ which should be annihilated by the generators,

$$E(1)|0\rangle = E(2)|0\rangle = 0 \tag{6.15}$$

To make states in higher multiplets, we apply matrix representations of the U matrix. In particular, to fill out the $(\bar{3}, 3)$ representation of SU(3) $\times$ SU(3), consider

$$(U_3)_l^k |0\rangle \tag{6.16}$$

To check this claim, measure the square of the color electric flux in this state, using the transformation laws given above,

$$E^2(1)U_3|0\rangle = \left(\tfrac{1}{2}\lambda^\alpha\right)\left(\tfrac{1}{2}\lambda_\alpha\right)U_3|0\rangle = C_2 U_3|0\rangle \tag{6.17}$$

where C_2 is the value of the quadratic Casimir operator in the fundamental $\mathbf{3}$ representation, $C_2 = \tfrac{4}{3}$.

Before developing more aspects of the lattice theory, let's predict some of its continuum correspondences. We did this exercise in detail for the Euclidean four-dimensional version of the theory, so the results should not be surprising. First, the generators of local color rotations are related to the Yang–Mills field-strength tensor,

$$\vec{E}_\alpha(\vec{r}) \cdot \hat{n} = \frac{a^2}{g} F_\alpha^{0k}(\vec{r}) n_k \tag{6.18}$$

where $\hat{n}$ is a unit vector pointing between lattice sites, F_α^{0k} is the Yang–Mills field strength, and g is the coupling constant.

The rotation matrix is related to the vector potential $\vec{A}_\alpha(\vec{r})$,

$$U_3(\vec{r}, \hat{n}) = \exp\left(iag\vec{A}_\alpha(\vec{r}) \cdot \hat{n} \frac{\lambda^\alpha}{2}\right) \tag{6.19}$$

if we choose the class of temporal gauges, $A_\alpha^0(\vec{r}) = 0$.

It is interesting that constructions of this sort pre-date lattice-gauge theory and even Yang–Mills theory by many years. As discussed in the context of path-integrals above, when J. Schwinger was first formulating QED he considered point-split fermion currents like

$$\lim_{\vec{\epsilon} \to 0} \bar{\psi}(\vec{r} + \vec{\epsilon})\gamma_\mu\gamma_5 e^{(ig\vec{A}(\vec{r}) \cdot \vec{\epsilon})} \psi(\vec{r}) \tag{6.20}$$

By considering local gauge transformations, Schwinger showed that this point-split fermion bilinear is gauge-invariant even for nonzero $\vec{\epsilon}$. This proved essential in Schwinger's discussions of the axial anomaly, the lack of conservation of the axial-vector current of QED due to quantum fluctuations. In terms closer to our present discussion, we can check that Schwinger's construction satisfies Gauss' law and respects the fact that a positively charged fermion is the source of one unit of electric flux and that a negatively charged fermion is the sink of that flux. Consider the state

$$|s\rangle = e^{ig\vec{A}\cdot\vec{\epsilon}}|0\rangle \tag{6.21}$$

Since $\vec{E} = \dot{\vec{A}}$ in the temporal gauge ($A_0 = 0$), the exponential $\exp(ig\vec{A}\cdot\vec{\epsilon})$ is a shift operator for $\vec{E}$. We compute

$$\langle s|\vec{E}|s\rangle = g\vec{\epsilon} \tag{6.22}$$

The close analogy of these results to our equations above involving the color-group generator $E(j)$ and the U matrix will prove important.

Now let's put quark degrees of freedom onto the two-site world. The quarks reside in the fundamental representation of the color group, so they transform under color rotations as

$$\psi(j) \to V(j)\psi(j) \tag{6.23}$$

For an infinitesimal rotation, we again write

$$V(j) \approx 1 + \frac{i}{2}\epsilon^\alpha G_\alpha^{\rm f} + \cdots \tag{6.24}$$

where $G_\alpha^{\rm f}$ is the fermionic piece of the generator. The infinitesimal form of the rotation equation reads

$$\delta\psi = [G_\alpha^{\rm f}, \psi] = \frac{i}{2}\epsilon^\alpha \lambda_\alpha \psi \tag{6.25}$$

which indicates, after some standard textbook algebra in second quantization, that the fermion piece of the generator at site i is

$$G^{\rm f}(i) = \psi^\dagger(i)\frac{\lambda_\alpha}{2}\psi(i) \tag{6.26}$$

This means that the total generator at the site i is

$$G_\alpha = \psi^\dagger(i)\frac{\lambda_\alpha}{2}\psi(i) + E_\alpha(i) \tag{6.27}$$

Local color-gauge invariance can now be stated in a precise, convenient form. A physical state should be invariant with respect to local color rotations and therefore must be annihilated by G_α,

$$G_\alpha|0\rangle = 0 \qquad (6.28)$$

It is instructive to make some gauge-invariant states and interpret them physically. The vacuum is locally invariant with respect to color rotations and is annihilated by $\psi(i)^-$, $\psi^\dagger(i)^-$, and $E_\alpha(i)$. The state of a quark and an anti-quark at site i is gauge-invariant if the color indices are contracted into a singlet,

$$\sum_j \psi_j^\dagger(i)\psi_j(i)|0\rangle \equiv \psi^\dagger(i)\psi(i)|0\rangle \qquad (6.29)$$

However, if we place a quark on one site and an anti-quark on a neighboring site, then gauge invariance requires the presence of a U_3 operator between them,

$$\sum_j \psi_j^\dagger(1)U_{ij}^{(3)}\psi_j(2)|0\rangle \equiv \psi^\dagger(1)U_3\psi(2)|0\rangle \qquad (6.30)$$

where the U operator enforces Gauss' law and places one unit of electric flux between the fermion and the anti-fermion. Note that the state also has all of its color indices locally contracted into color singlets, so it is clearly invariant under local color rotations.

Our last task in the context of the two-site model is the construction of a gauge-invariant Hamiltonian that will generalize to a full three-dimensional, infinite spatial lattice. We have discussed fermion-mass and hopping terms. We will need to elevate the gauge field to the status of a dynamical variable. Since $E_\alpha(i)$ is the gerator of color-frame rotations and since its square, the quadratic Casimir operator for the gauge variables, is locally gauge-invariant, a sensible candidate term should involve $\sum_{i\alpha} E_\alpha(i)E_\alpha(i)$. We must pick the appropriate prefactor here so that the Hamiltonian reduces to the known result when the lattice spacing is taken to zero for smooth fields. The result is

$$H = \frac{g^2}{2a}\sum_i E^2(i) + \frac{i}{a}[\psi^\dagger(1)U\psi(2) - \psi^\dagger(2)U^\dagger\psi(1)] + m\sum_i \psi^\dagger(i)\psi(i)$$

$$(6.31)$$

This is as far as we can go with two sites. H is clearly missing magnetic effects in the gauge field. Recall from our discussion of the Euclidean version of the theory that these effects come from plaquettes, the product of U matrices around the smallest loops on the lattice. To get the full H we turn to an infinite, three-dimensional lattice. Label sites with a triplet of integers, $\vec{r} = (n_1, n_2, n_3)$ and label directed links with $\vec{r}$ and a unit vector pointing toward one of the six nearest-neighbor sites. The fermion fields will reside on the sites, $\psi(\vec{r})$. The electric-flux variable, $\vec{E}_\alpha(\vec{r}) \cdot \hat{n}$, will also reside on sites but will point along the various links. The gauge fields $U(\vec{r}, \hat{n})$ will live on links. Most of the algebra we did on our two-site world generalizes to the three-dimensional spatial lattice in the obvious fashion. The generator of color rotations now has a single-site fermion term and six contributions from the electric-flux variables on all the links attached to that site. The condition that physical states be locally color-gauge-invariant reads as before,

$$G_\alpha |0\rangle = 0 \qquad (6.32)$$

with

$$G_\alpha(\vec{r}) = \psi^\dagger(\vec{r}) \frac{\lambda_\alpha}{2} \psi(\vec{r}) + \sum_{\hat{n}} E_\alpha(\vec{r}) \cdot \hat{n} \qquad (6.33)$$

It is instructive to consider some physical states made by applying gauge-invariant operators to the vacuum. First consider pure gauge-field examples. We can multiply strings of U matrices together, but, in order to make only gauge-invariant operators, color indices must be contracted locally. In other words, let Γ be a closed path of links and label the links with increasing integers. Then an operator of the form

$$U(\Gamma) = \text{tr}\, U(1)U(2)U(3) \ldots U(n) \qquad (6.34)$$

is locally gauge-invariant and makes a closed string of electric flux when it operates on the vacuum. The shortest closed path on the lattice consists of the square of four links and, as in the Euclidean version of lattice-gauge theory, these quantities will be used to make the "magnetic" terms in the Hamiltonian.

The energy of a closed, nonintersecting loop of n links is proportional to the product of the electric flux on each of the links with n, the number of links. Since the electric flux is quantized according to the possible values of the quadratic Casimir operator of SU(3),

$$[E(\vec{r}) \cdot \hat{n})]^2 = 0, \ \frac{4}{3}, \ 3, \ \frac{10}{3}, \ \frac{16}{3}, \ldots \qquad (6.35)$$

the energy per length of such electric flux tubes is quantized.

Next consider states that contain fermions in addition to gauge-field excitations. A meson state might be

$$\sum_k \bar{\psi}_k(i) \gamma \tau \psi_k(i) |0\rangle \equiv \bar{\psi}(i) \gamma \tau \psi(i) |0\rangle \qquad (6.36)$$

where the color indices k are summed locally into color singlets. The γ and τ matrices are short for an appropriate Dirac matrix and a flavor matrix that acts on the flavor indices of the fermion field operator that we have left implicit. Baryon states can also be made using fermion operators on a single site,

$$\sum_{klm} \epsilon_{klm} \psi_k^\dagger(i) \psi_l^\dagger(i) \psi_m^\dagger(i) |0\rangle \tag{6.37}$$

where the symbol ϵ constructs a color singlet from three indices.

Excited meson and baryon states can be made using operators in which fermion fields are at different sites and strings of U are inserted to satisfy Gauss' law.

Now we should write down the Hamiltonian for the three-dimensional lattice system. It reads

$$H = \frac{g^2}{2a} \sum_{\vec{r},\hat{n}} [\vec{E}_\alpha(\vec{r}) \cdot \hat{n}]^2 - \frac{1}{g^2 a} \sum_{\text{plaquettes}} (\text{tr} \, UUUU + \text{h.c.})$$

$$+ \frac{1}{2a} \sum_{\vec{r}} [(-1)^y \chi^\dagger(\vec{r}) U(\vec{r}, n_z) \chi(\vec{r} + n_z) + (-1)^z \chi^\dagger(\vec{r}) U(\vec{r}, n_x) \chi(\vec{r} + n_x)$$

$$+ (-1)^x \chi^\dagger(\vec{r}) U(\vec{r}, n_y) \chi(\vec{r} + n_y) + \text{h.c.}] \tag{6.38}$$

where we have not included a bare-fermion-mass term. The elaborate fermion-hopping term contains the phases of Kogut–Susskind (staggered) fermions in three dimensions [40]. We recorded the Euclidean version of these phases above. With these phases the Dirac equation for a massless iso-doublet is retrieved in the continuum limit. The relative strength of the electric and magnetic gauge terms in the first line of the equation also guarantees that the appropriate, conventional continuum Hamiltonian of QCD will be obtained. The verification of these points follows our earlier demonstration for the Euclidean formulation and that algebra will not be repeated here. The important point, however, is that, for strong coupling on a coarse lattice, the terms are clearly ranked in order of importance in a strong-coupling expansion in terms of the variable $x = 1/g^2$. We write H in dimensionless form,

$$W = \frac{2a}{g^2} H = W_e + x W_q + 2x^2 W_m \tag{6.39}$$

where

$$W_e = \sum_{\vec{r},\hat{n}} [\vec{E}_\alpha(\vec{r}) \cdot \hat{n}]^2 \tag{6.40}$$

$$W_q = \sum_{\vec{r}} [(-1)^y \chi^\dagger(\vec{r}) U(\vec{r}, n_z) \chi(\vec{r} + n_z) + \cdots \tag{6.41}$$

$$W_m = -\sum_{\text{plaquettes}} (\text{tr} \, UUUU + \text{h.c.}) \tag{6.42}$$

This Hamiltonian has been used for QCD spectroscopy calculations as well as formal developments in the field, such as large-N expansions and mean-field approximations.

Let's gain some familiarity with this Hamiltonian. Suppose that a, the spatial lattice spacing, is fixed and the coupling constant g is large. Suppose also that the dynamical Fermi fields χ are absent. In other words, we are working in the "quenched" approximation in which all the dynamics resides in the gauge fields which provide an environment in which "external" fermions can propagate. Then the static term, W_e, dominates the energetics. If we place a static source of one unit of electric flux at a site $\vec{r}$, then one unit of electric flux must emanate from it, which means that a $U(\vec{r}, \hat{n})$ must occupy a link radiating from site $\vec{r}$. However, this operator has an open color index associated with the nearest-neighbor site $\vec{r} + \hat{n}$, so it must be connected to another $U(\vec{r} + \hat{n}, \hat{m})$, etc. Each U operator costs an energy $[g^2/(2a)]C_2(3)$ and there is an infinite number of links in the string attached to the color source. More simply, the fixed source gives rise to a flux tube of infinite extent and infinite energy (linearly divergent). A simple generalization from this exercise is that the only finite-energy states in the theory will be color singlets.

To illustrate the importance of the magnetic term in the Hamiltonian, consider a state with a source of color in the 3 representation at one site and a sink of color in the $\bar{3}$ representation at another site. A gauge-invariant operator making such a state would have a string of U operators extending from the source to the sink. In other words, a straight thin flux tube holds the source and sink together by a linearly confining potential. Now treat the effects of the magnetic piece of the Hamiltonian in ordinary perturbation theory in the expansion parameter $x = 1/g^2$. According to Schrödinger perturbation theory, the perturbation will act on the unperturbed state to produce a correction. In particular, the magnetic term of the gauge-field action can be applied to the unperturbed string state with one of the links in W_m hitting the same link on which there is a U from the unperturbed state. It is clear that the result is a "fluctuating string." In addition, the quark-hopping term can act as a perturbation and, if it acts on a link on which there is a U from the original state, the hopping term can annihilate that bit of string and place a quark and an anti-quark down so that the original string is broken. This effect can also be interpreted as "screening" because the perturbation has now produced extended color-singlet "meson" states that do not experience a long-range confining potential. This calculation illustrates a general effect: when dynamical quarks are included in calculations, they screen the long-range linear potential, leaving just color-singlet states and shorter-range interactions.

Systematic calculations of light- and heavy-quark spectroscopy, the heavy-quark potential, electric-flux distributions, etc. have been done using these methods carried to much higher order in a strong-coupling expansion. Some illustrations will be included below.

6.2 Quark confinement in Hamiltonian lattice-gauge theory and thin strings

One of the goals of lattice-gauge theory is the development of a better understanding of quark confinement. We have considered it briefly above and will return to it several times again in the context of extreme environments. Although computer simulations of SU(3) and simpler lattice-gauge theories provide good evidence for flux-tube formation and confinement, a simple continuum understanding of the phenomenon is lacking.

As we have discussed, quark confinement is easy to understand on a coarse, strongly coupled lattice. Of course, such a structure breaks our cherished spacetime symmetries, so the significance of such a result is suspect. Nonetheless, we will discuss spacetime symmetries and their restoration in the weak-coupling limit of lattice-gauge theory, and gain more insight into flux-tube dynamics.

Begin with the on-axis heavy-quark potential. We place a Q and a $\bar{Q}$, created by operators $\psi^{\dagger}(\vec{n})$ and $\psi(\vec{n}+\vec{R})$, at the sites $\vec{n}$ and $\vec{R}$ with a string of U matrices between them,

$$\psi^{\dagger}(\vec{n}) \left\{ \prod_{\text{path}} U \right\} \psi(\vec{n}+\vec{R}) \tag{6.43}$$

where the "path" extends from $\vec{n}$ to $\vec{n} + \vec{R}$, but is otherwise arbitrary.

It is easy to calculate the energy of such states for strong coupling, $g^2 \gg 1$. When $g^2 \gg 1$, the leading term in H is

$$H_{\text{o}} = \frac{g^2}{2a} \sum_l E_l^2 \tag{6.44}$$

The vacuum, into which we will place the $Q\bar{Q}$ pair, must minimize H_{o}, so each of its links must be a color-singlet state,

$$E_l^{\alpha}|0\rangle = 0 \tag{6.45}$$

So the $Q\bar{Q}$ state of minimal energy should have the fewest excited links in its $\prod_{\text{path}} U$ term. Each excited link will have the energy, in SU(2) gauge theory,

$$E^2 U|0\rangle = \tfrac{1}{2}\tau^{\alpha} \cdot \tfrac{1}{2}\tau^{\alpha} U|0\rangle = \tfrac{3}{4} U|0\rangle \tag{6.46}$$

which one proves from the SU(2) commutation rule,

$$[E^{\alpha}, U] = \tfrac{1}{2}\tau^{\alpha} U \tag{6.47}$$

So, if the Q and $\bar{Q}$ lie along an axis with R/a links between them, the energy of their state is

$$V(R) = \frac{g^2}{2a} \cdot \frac{3}{4} \cdot \frac{R}{a} = \sqrt{\sigma} R \tag{6.48}$$

which gives us the strong-coupling limit of the string tension in this Hamiltonian formulation of lattice-gauge theory,

$$\sqrt{\sigma} = \frac{3}{8}\frac{g^2}{a^2} \tag{6.49}$$

This formula illustrates the renormalization program of lattice-gauge theory in a very primitive fashion. The string tension is a physical quantity that is independent of the cutoff a. However, this formula for the string tension will be independent of a only if $g \sim a$: if the string tension were computed on two different lattices with different spacings but the same H, then the coupling on the coarser lattice would have to be larger and proportional to a, in order that σ would be independent of the cutoff. This characterizes confinement without screening: the coupling grows without bound at large distances in order to guarantee that only color-singlet states have finite, bounded energies.

This strong coupling behavior, $g \sim a$, would have to match with asymptotic freedom, $g \sim 1/\ln a$, on a fine lattice so that the physical, continuum theory resides in a single phase. The continuum limit of the Hamiltonian lattice theory would then experience a single phase: confinement at large distances and asymptotic freedom and a parton description at short distances.

6.3 Relativistic thin strings, delocalization, and Casimir forces

How does confinement emerge from asymptotic freedom at short distances? How does the flux law change over length scales, so that at small length scales the flux spreads out into the familiar three-dimensional pattern, while at large distances it is squeezed into a thin flux tube? Is a thin tube really a possible configuration of flux?

To gain some insight into these longstanding questions, it is instructive to consider a very simple model: a thin structureless string in continuum, relativistic spacetime. Pin down the ends of the string at positions $\vec{x} = -\frac{1}{2}\vec{R}$ and $\vec{x} = \frac{1}{2}\vec{R}$. What are the qualitative characteristics of the spatial distribution the string can assume? Will it resemble a "straight flux tube" that could be responsible for confinement through a linearly rising heavy-quark potential?

We will see that, as R grows large, translation invariance implies that the string wanders off the axis between its pinned ends and that its wandering transverse to the axis grows without bound at the rate $\ln R$.

Let the string be described by a two-component vector field, which gives its location off the axis between its ends, $\vec{\xi}(t, z)$ for z in the range, $-\frac{1}{2}R \leq z \leq \frac{1}{2}R$. The string is pinned at its ends, as shown in Fig. 6.1, $\vec{\xi}(t, -\frac{1}{2}R) = \vec{\xi}(t, \frac{1}{2}R) = 0$.

We want an effective action to describe the low-frequency modes of this string. Make the following assumptions about it.

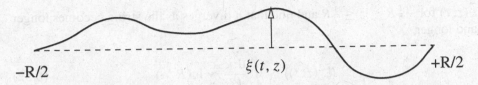

Figure 6.1. The fluctuating thin string with pinned ends.

1. S_{eff} must be local, so it can be written as an integral over a density L, which can depend on $\vec{\xi}(t, z)$ and a finite number of its derivatives at (t, z), $S_{\text{eff}} = \int dz\, dt\, L$.

2. L itself should be invariant with respect to the following spacetime symmetries:

   (a) Poincaré transformations in the (t, z) plane, and

   (b) O(2) rotations and translations of the string $\vec{\xi}$.

Now we can write down the terms that can enter the effective Lagranian L. We are interested only in terms that can contribute to its low-energy behavior. Property 2(b) precludes a "mass" term that would be proportional to $\vec{\xi}^2$. So L must be made up of derivatives, $\partial_\mu \vec{\xi}$ and $\partial_\mu \partial_\nu \vec{\xi}$ etc., where the derivatives ∂_μ refer to t and z,

$$S_{\text{eff}} = \int dz\, dt\, [\partial_\mu \vec{\xi} \cdot \partial^\mu \vec{\xi} + b\, \partial_\mu \partial^\mu \vec{\xi} \cdot \partial_\nu \partial^\nu \vec{\xi} + c(\partial_\mu \vec{\xi} \cdot \partial^\mu \vec{\xi})^2 + \cdots] \quad (6.50)$$

where the integral extends over the range $-\frac{1}{2}R \leq z \leq \frac{1}{2}R$ and all t. If we are interested only in the low-momentum properties of the string, on the scale of an ultraviolet cutoff a, then only the first term in the expression for S_{eff} is significant. The other terms in the expression are irrelevant, i.e. the parameters b and c have dimensions of length to various positive powers, which will become suppression factors by powers of the ultraviolet cutoff in applications. In condensed-matter physics this is the character of a "spin-wave" expansion. So, for studying phenomena with characteristic wavelengths $\lambda \gg a$, S_{eff} is well approximated by the action of two massless fields in two dimensions,

$$S_{\text{eff}} = \int dz\, dt\, \partial_\mu \vec{\xi} \cdot \partial^\mu \vec{\xi} \quad (6.51)$$

We know that massless modes in two dimensions have severe infrared problems that will make some features of the string sensitive to its length R. For example, the string cannot be straight, because this would mean that $\langle \vec{\xi}(z, t) \rangle = 0$ with small fluctuations. In fact, using S_{eff} it is easy to calculate the variance in

$\vec{\xi}(z, t)$ for $-\frac{1}{2}R \leq z \leq \frac{1}{2}R$ and find that it diverges as the string becomes longer and longer,

$$\langle \vec{\xi}^2(z, t) \rangle \sim \int \frac{d^2 k}{k^2} \sim \ln(R/a) \tag{6.52}$$

where the integral over wavelengths is cut off by R in the infrared and a in the ultraviolet. For a physical fluctuating string, its thickness would replace the ultraviolet cutoff a. Since the transverse fluctuations in the position of the string diverge as R grows large, the string is "delocalized." This calculation is an example of the Mermin–Wagner theorem that we discussed above in the context of two-dimensional spin systems. If $\langle \vec{\xi}(z, t) \rangle = 0$ with small fluctuations, then translation symmetry would be broken. However, the theorem states that continuous global symmetries cannot break down spontaneously in two dimensions in a theory with only local couplings.

We learn from this that, although the flux tube can maintain a thin intrinsic thickness, it must "snake" out into the transverse plane in an unbounded fashion, such that it eventually loses memory of its boundary conditions at $\pm\frac{1}{2}R$.

A second physical effect of these massless modes, capillary waves in the language of interfaces in condensed-matter physics, is the existence of a universal $1/R$ potential in the string channel of the theory [41],

$$V(R) = \sqrt{\sigma}R + c - \frac{(d - 2)\pi/24}{R} + \cdots \tag{6.53}$$

The $1/R$ potential here is a universal one-dimensional Casimir effect and will be derived below. It is a Casimir effect because it records the energy in the transverse fluctuations in the string of length R that depend on the boundary conditions, namely the fact that the string is pinned down at its ends. It is universal because its existence just depends on continuous translation symmetry in spacetime. In fact, its strength depends just on the dimensionality of spacetime; $d - 2$ counts the number of spatial dimensions transverse to the string.

Let's do the calculation following Casimir's original method for obtaining his famous effect for a black box immersed in the electromagnetic vacuum. In this case each transverse mode contributes $\frac{1}{2}\hbar\omega$ to the ground-state energy,

$$\Delta V(R) = \frac{1}{2} \sum_n \epsilon_n, \qquad \epsilon = \pi n/R, \qquad n = 0, 1, 2, 3, \ldots \tag{6.54}$$

where we have counted massless normal modes for a field with fixed boundary conditions, $\vec{\xi}(-R/2, t) = \vec{\xi}(+R/2, t) = 0$. The sum over normal modes diverges – the fluctuations shift the ground-state energy. There must be a term that is extensive (proportional to R) and will renormalize the string tension additively, and we know from experience with the Casimir effect in three-dimensional cavities that there will be other terms with weaker dependences on R that produce attraction and/or repulsion between elements of the boundary. To find and separate these effects, we must regulate the sum with a convergence factor $e^{-\alpha n}$ and let $\alpha \to 0$ at the end of the calculation. In order to think physically about this procedure and give α a physical interpretation, suppose that the string has a small intrinsic thickness of a, a natural ultraviolet cutoff. Then the largest physically sensible value for n is $\sim R/a$, the largest energy ϵ is $\leq \pi/a$, and the minimal value for α is $\sim a/R$.

With this regularization procedure, the change in the heavy-quark potential becomes,

$$\Delta V(\alpha, R) = \frac{\pi}{2R} \sum_n n e^{-\alpha n} = -\frac{\pi}{2R} \frac{d}{d\alpha} \sum_n e^{-\alpha n} \tag{6.55}$$

$$= -\frac{\pi}{2R} \frac{d}{d\alpha} \sum_n \left(\frac{1}{e^\alpha - 1} \right) = -\frac{\pi}{2R} \sum_n B_n \frac{(n-1)\alpha^{n-2}}{n!} \tag{6.56}$$

where B_n are the Bernoulli numbers.

Only two terms survive in this result. There is the $n = 0$ term which is extensive and contributes to the string tension through an additive renormalization. (Since the minimum physical value of α is $\sim a/R$, the $n = 0$ term is proportional to R/a^2, as expected of the divergent vacuum energy of a scalar field in a $1 + 1$ dimensional box of extent R.) The second term comes from $n = 2$, and is insensitive to α, the regulator,

$$\Delta V(R) = -\frac{\pi}{2R} \frac{1}{2} B_2 = -\frac{\pi/24}{R} \tag{6.57}$$

We obtain a term of this value for each transverse dimension, so, if there are $d - 2$ of them, the final result is

$$\Delta V(R) = -\frac{(d-2)\pi/24}{R} \tag{6.58}$$

It is very interesting that results from the most accurate lattice-simulation studies of the heavy-quark potential confirm the presence of the universal $1/R$ term of the simplest bosonic string model. Simulations in three and four Euclidean dimensions require the $d - 2$ counting factor as well. The $1/R$ effect is evident in the lattice data even for relatively "short" distances, of the order of half a fermi. It appears that the thickness of the confining flux tube is quite small, $\sim \frac{1}{3}$ fm, and the confining geometry is physically significant at separations $\geq \frac{1}{2}$ fm. These scales in the physics of confinement have not been assimilated into a first principles understanding of the dynamics of QCD at this time.

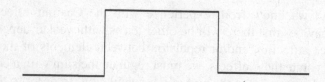

Figure 6.2. A kink and an anti-kink on a thin string.

6.4 Roughening and the restoration of spatial symmetries

How does the delocalization of the string show up in lattice-gauge theory? For strong coupling, the cubic symmetry of the spatial lattice stabilizes the straight string and inhibits the transverse fluctuations that could delocalize it. There is a nonzero gap in the spectrum of the transverse fluctuations of the string.

Consider an on-axis string consisting of N links. The string has an energy of $[g^2/(2a)]N$ in the standard Hamiltonian version of the theory. However, if we put a transverse wave into the string, we have an extra energy of $2 \cdot [g^2/(2a)]\frac{3}{4}$, in the strong-coupling limit. The excited string has a "kink" and an "anti-kink", as shown in Fig. 6.2. The strong-coupling energy of the single-kink state is $m_k = [g^2/(2a)]\frac{3}{4}$. The kink should be thought of as a particle – its wavefunction, energy, and momentum are localized and it can travel along the string. It spreads the transverse distribution of the electric flux of the string, but, for strong coupling, the transverse distribution is bounded because the wandering of the string is inhibited by the finite energy barrier, m_k. However, as we consider weaker coupling g^2 and approach the continuum limit, m_k will decrease. In fact, this mass gap will vanish at a nonzero value of the coupling, g_R, before the continuum limit is reached. This critical coupling g_R is called the "roughening" point, and has been studied very well in condensed-matter investigations into the physics of interfaces.

In the context of Hamiltonian lattice-gauge theory, one can calculate the kink's mass in strong-coupling perturbation theory and estimate the critical coupling g_R at which m_k vanishes. In such calculations, the magnetic terms in H smear out the kink state and reduce its energy. At the roughening point g_R, there is no barrier to transverse wanderings of the on-axis flux tube. In fact, at this point, the transverse profile of the string widens like $\sim \ln(R/a)$, as predicted by the continuum effective-Lagrangian model.

All of this has been studied very well in lattice-gauge theory and in condensed-matter physics of interfaces. It is interesting to consider the roughening transition in the context of two-dimensional interfaces. We might be considering the three-dimensional Ising model at low T in the magnetized phase and choose boundary conditions to form an interface for which all the spins above the x–y plane are "up" and all those below the plane are "down." We are interested in how the interface at $z = 0$ fluctuates as the temperature is increased. The fluctuations are frequently studied in the "solid-on-solid" approximation.

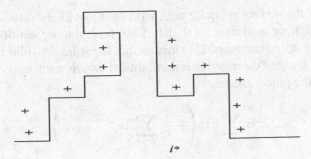

Figure 6.3. An interface configuration with an "overhang."

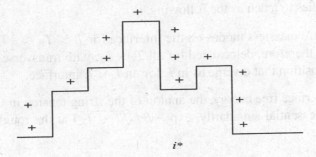

Figure 6.4. An interface configuration of the "solid-on-solid" variety.

One considers the flat, planar interface at low temperature T and neglects all bulk fluctuations that are not connected to the interface. One makes the simplifying assumption that one can ignore those rare fluctuations with overhangs, as shown in Fig. 6.3. More likely fluctuations are shown in Fig. 6.4, where we see that the height of each column of "down" spins above or below the x–y plane is a single-valued function. Call it n_{i^*}, where i^* labels a dual site of the two-dimensional x–y lattice. Since only broken bonds contribute to the action, the partition function of the interface is

$$Z = \sum_{n_{i^*}} \exp\left(-\frac{4}{T} \sum_{\langle i^* j^* \rangle} |n_{i^*} - n_{j^*}|\right) \qquad (6.59)$$

Considering sequential slices across the interface, this is clearly a natural lattice version of the thin string discussed above.

It is interesting to ask how well this "solid-on-solid" model describes the real interface in the three-dimensional Ising model. This is an "experimental" question, so we turn to computer simulations. The simulations indicate that the roughening temperature T_R is about half the bulk transition temperature T_c, the Curie point between bulk magnetization and bulk disorder. Since $T_R \ll T_c$, it is reasonable to ignore bulk fluctuations – the bulk Ising model is quite "cold" and ordered at T_R and vacuum fluctuations are rare. In addition, for $T \approx T_R$, fluctuations of the surface are quite rare, with the most likely values of $n_{i*} - n_{j*}$, for nearest-neighbor dual sites, of $0, \pm 1$. Therefore, the neglect of "overhangs" is a compelling approximation. This means that the solid-on-solid model is well approximated by the "discrete Gaussian" model which we found earlier in our discussion of the planar model,

$$Z = \sum_{n_{i*}} \exp\left(-\frac{4}{T} \sum_{\langle i*j* \rangle} |n_{i*} - n_{j*}|^2\right) \tag{6.60}$$

Since this model is dual to the planar model, we know all about interfaces with no additional work. What luck! In particular, the classic work of Kosterlitz and Thouless applies to teach us the following.

1. There are massless modes on the interface for $T > T_R \approx \frac{1}{2} T_c$. The interface is, therefore, delocalized for all $T > T_R$, with transverse fluctuations in its position that diverge as $\ln L$ for an $L \times L$ interface.

2. The interface free energy, the analog of the string tension in QCD, develops an essential singularity, $\exp(-B/\sqrt{T - T_R})$ at the roughening temperature.

We learn some general results from these exercises. The string sector of QCD has some very curious features. It does not have a mass gap simply because of the geometry of the string and the fact that spacetime is homogeneous. Although the underlying dynamics of QCD which generate linear confinement are rather special, we believe that there are several consequences, such as delocalization and the universal $1/R$ potential, that are simple, geometrical, and universal.

The roughening transition has additional practical significance for lattice-gauge theory. For strong couplings above the roughening coupling g_R, on-axis flux tubes are smooth, with unlikely fluctuations. At g_R the string becomes rough and obtains some of the features of a thin string embedded in a homogeneous, continuum spacetime. Therefore, this sector of the theory and the dynamics of the string, in particular, decouple from the lattice with its cubic symmetry. The string begins to behave as if the underlying spacetime had full rotational and translational symmetry at the roughening point.

It is interesting to see how this works out in Hamiltonian lattice-gauge theory. For the purposes of illustration consider $(2 + 1)$-dimensional SU(3) gauge theory. Let the Q and $\bar{Q}$ be placed off axis as shown in Fig. 6.5. This general configuration will help us discuss the restoration of spatial symmetries. At strong coupling the flux must travel from the quark to the anti-quark by a route of minimal distance. The leading order, $g \gg 1$, heavy-quark potential is then

$$V(x, y) = \frac{g^2}{2a} \frac{3}{4} (|x| + |y|) \tag{6.61}$$

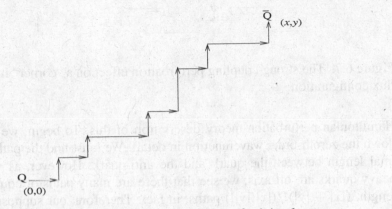

Figure 6.5. An off-axis flux configuration of minimal energy.

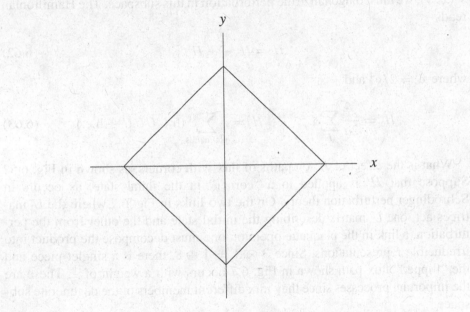

Figure 6.6. An equipotential surface of the heavy-quark potential at strong coupling.

which has the equipotentials shown in Fig. 6.6. $V(x, y)$ has the expected discrete rotational and translational symmetries inherited from the cubic lattice. The rotational symmetry of a continuum calculation has been distorted by the discrete lattice.

It is interesting to see how this situation changes as fluctuations develop at weaker coupling and g approaches the roughening transition g_R. Consider a

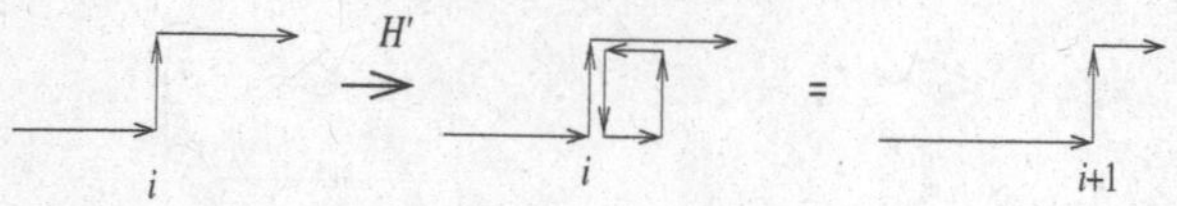

Figure 6.7. The strong-coupling perturbation effect on a "corner" in an off-axis flux configuration.

Hamiltonian perturbation-theory description of this. To begin, we must write down the zeroth-order wavefunction in detail. We must find the path(s) of minimal length between the quark and the anti-quark. However, as soon as the heavy quarks are off axis, we see that there are many paths of equal, minimal length, $(|x| + |y|)!/(|x|!|y|!)$ paths, in fact. Therefore, our supposedly simple perturbation-theory calculation becomes a tricky exercise in degenerate perturbation theory. To calculate the first-order correction to the heavy-quark potential $V(x, y)$, we must diagonalize the perturbation in this subspace. The Hamiltonian reads

$$H = H_{\mathrm{o}} - x_{\mathrm{s}} H' \tag{6.62}$$

where $x_{\mathrm{s}} = 2/g^4$ and

$$H_{\mathrm{o}} = \frac{g^2}{2a} \sum_l E_l^2, \qquad H' = \sum_{\text{plaquettes}} (\mathrm{tr}\, UUUU + \text{h.c.}) \tag{6.63}$$

What is the effect of H' on paths of flux with corners, as shown in Fig. 6.5? Suppose that H' is applied to a "corner" in the inital state, as occurs in Schrödinger perturbation theory. On the two links in Fig. 6.7 where the U matrices act, one U matrix describing the initial state and the other from the perturbation, a link in the plaquette operator, one must decompose the product into irreducible representations. Since $3 \otimes \bar{3} = 1 \oplus 8$, there is a singlet piece and the "flipped" flux path shown in Fig. 6.7 occurs with a weight of $\frac{1}{3}$. These are the important processes since they mix different members of the degenerate subspace.

Recognize this as a problem in electronic conductivity – fermions hopping along a chain. To make this point quantitative and useful, make the definitions

$$y \text{ link with flux} \rightarrow \text{fermion at site } i$$
$$x \text{ link with flux} \rightarrow \text{absence of a fermion at site } i$$

So, imagine a box of length $L = |x| + |y|$ containing $|y|$ fermions. It is clear from the perturbation-theory effects discussed above that the perturbation allows a "fermion" to hop one site to a nearest neighbor *if* that site were initially unoccupied. Therefore, the restriction of H' to the degenerate subspace is equivalent

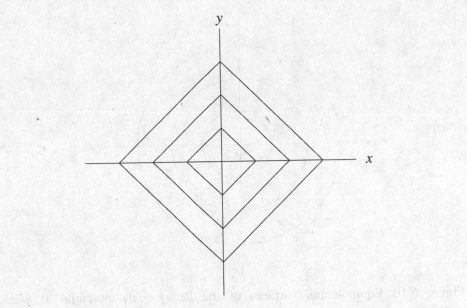

Figure 6.8. Equipotential surfaces of the heavy-quark potential at strong coupling.

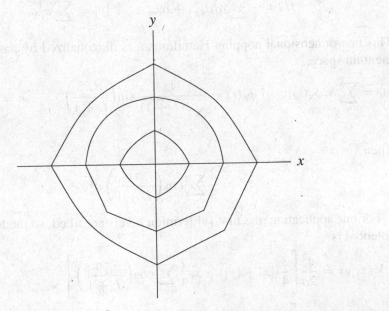

Figure 6.9. Equipotential surfaces of the heavy-quark potential at intermediate coupling.

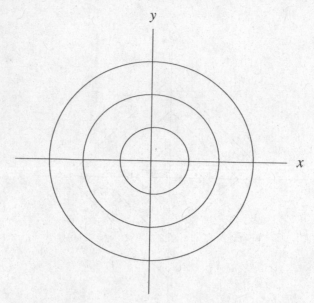

Figure 6.10. Equipotential surfaces of the heavy-quark potential at weak coupling.

to the fermion hopping operator,

$$H' = \frac{1}{3} \sum_i a_i^\dagger a_{i+1} + \text{h.c.}, \qquad |y| = \sum_i a_i^\dagger a_i \qquad (6.64)$$

This one-dimensional hopping Hamiltonian is diagonalized by passing to momentum space,

$$a_i = \sum_n a_n \phi_n(i), \qquad \phi_n(i) = \frac{1}{\sqrt{2(L+1)}} \sin\left(\frac{\pi n i}{L+1}\right), \qquad n = 1, 2, \ldots, L$$

$$(6.65)$$

Then,

$$H' = \frac{2}{3} \sum_n \cos\left(\frac{\pi n}{L+1}\right) a_n^\dagger a_n \qquad (6.66)$$

For our application, the first $|y|$ fermion levels are filled, so the heavy-quark potential is

$$V(x, y) = \frac{g^2}{2a} \left[\frac{3}{4}(|x| + |y|) - x_s \frac{2}{3} \sum_{n=1}^{|y|} \cos\left(\frac{\pi n}{L+1}\right) \right] \qquad (6.67)$$

$$= \frac{g^2}{2a} \left[\frac{3}{4}(|x| + |y|) \right.$$

$$-\frac{4}{3g^4}\left(\frac{\sin\left(\dfrac{\pi(|y|+1)}{2(|x|+|y|+1)}\right)\sin\left(\dfrac{\pi(|x|+1)}{2(|x|+|y|+1)}\right)}{\sin\left(\dfrac{\pi}{2(|x|+|y|+1)}\right)}\right)-1\right]$$

$$(6.68)$$

It is interesting to plot the equipotentials of this $V(x, y)$, especially since the formula is rather opaque(!). The equipotentials are shown in Figs. 6.8–6.10. The restoration of rotational symmetry is surprisingly compelling for such a short calculation. This calculation is sensible only for couplings stronger than g_R because it assumes analyticity. g_R turns out to be rather small in these gauge models [42]. Higher-order corrections can also be calculated and they do not destroy the good agreement of this low-order calculation. Computer simulations also show that restoration of rotational symmetry occurs. It is also amusing that V can be expanded for large L and power-law corrections to the linear potential can be found. These are nonuniversal because this calculation starts in the strong-coupling region where continuous spacetime symmetries are lost. Nonetheless, the fluctuating string of degenerate perturbation theory illustrates how power-law corrections to the linear potential occur because of the geometry of the confining string while local excitations in the bulk, such as glueballs, baryons, and mesons, all have considerable gaps in their spectra of states.

7

Phase transitions in lattice-gauge theory at high temperatures

7.1 Finite-temperature transitions at strong coupling

When QCD is heated sufficiently it will undergo a phase transition from a state of colorless strongly interacting particles to a quark–gluon plasma. The properties of this phase transition and properties of the high-temperature plasma are the subjects of a great deal of modern research in QCD.

We have seen that pure gauge-field lattice dynamics confines quarks at strong coupling. If we use the quenched approximation in which there are no internal quark loops, then external static quarks experience a linear confining potential and the only finite-energy states must be color singlets consisting of confined quarks and/or anti-quarks. It is not difficult to show that, if this theory is heated, it will experience a transition to a quark–gluon plasma [43]. The point is that, as the temperature is increased, thermal fluctuations will become large enough to melt the linear flux tube which had produced confinement. At high temperature the vacuum itself becomes a tangle of closed flux loops, bubbling and fluctuating. When a static quark is placed into this vacuum, the extra flux line emanating from it blends into a vacuum already rich in flux and the resulting state does not cost an energy that linearly diverges. In the hot gauge-field state, there is no barrier to generating more or less flux and complete screening of the external, static quark color can occur, no matter what representation of SU(3) it lies in.

This physical picture has been confirmed in computer simulations of lattice-gauge theory of pure gauge fields and of full QCD. It has been confirmed in the scaling window of the continuum limit as well as for strong coupling, for which simple statistical-mechanics models indicate its correctness.

To illustrate the argument with a minimum of analysis, consider the pure gauge theory based on the group U(1). This model is called "PQED," periodic QED, and is especially easy to analyze. It confines at strong coupling, as we have shown above, because the periodicity of the U(1) gauge variable quantizes the model's electric flux on each link. This was the crucial element in the demonstration of confinement at strong coupling for lattice-gauge theories, not the non-Abelian nature of the gauge group. The non-Abelian character of the gauge group is important in taking the continuum limit using asymptotic freedom and finding that the theory's continuum limit is a unique confining one. The continuum limit of PQED is more problematical, but there are thought to be two distinct alternatives: (1) for weak lattice coupling the theory reduces to ordinary continuum QED in the continuum limit and has massless photons and no confining forces, and (2) for strong lattice coupling the periodicity of the U(1) variables is relevant and the theory contains massless photons and magnetic monopoles that lead to confinement. It may be that the monopoles survive the continuum limit and produce a strong-coupling theory that confines strongly interacting "electrons."

We are content here to consider the U(1) theory at strong coupling and show that it loses its confining nature at high temperature. In this case the gauge variables are just phases,

$$U(\vec{r}, \hat{n}) = \exp(i\phi(\vec{r}, \hat{n})) \tag{7.1}$$

The generator of $U(1)$ gauge rotations at sites, $E(\vec{r}, \hat{n})$, satisfies

$$[\phi(\vec{r}, \hat{n}), E(\vec{r}, \hat{n})] = i \tag{7.2}$$

The periodicty of the gauge variables U implies that the spectrum of possible eigenvalues of the generator $E(\vec{r}, \hat{n})$ consists of the integers.

At strong coupling the Hamiltonian is dominated by its electric-flux term,

$$H = \frac{g^2}{2a} \sum_{\vec{r}, \hat{n}} \vec{E}^2(\vec{r}, \hat{n}) \tag{7.3}$$

and the theory confines fermions that carry charge.

The partition function describing the thermodynamics at temperature T, $\beta = 1/(kT)$, is

$$Z(\beta) = \sum{}' \exp(-\beta H) \tag{7.4}$$

and the only subtlety here is the sum over physical states $\sum'$. These states must satisfy Gauss' law, as stated above, which means that, in the absence of matter-field sources (static charges) of flux, the sum of the generators on the links emanating from each site on the lattice must vanish,

$$\sum_{\hat{n}} E(\vec{r}, \hat{n})|0\rangle = 0 \tag{7.5}$$

where the sum goes over the six lattice directions from site $\vec{r}$. In order to incorporate this constraint into the expression for the partition function, we exponentiate it,

$$\delta\left(\sum_{\hat{n}} E(\vec{r}, \hat{n})\right) = \frac{1}{2\pi} \int_{-\pi}^{\pi} \exp\left(i\alpha \sum_{\hat{n}} E(\vec{r}, \hat{n})\right) \tag{7.6}$$

Now the partition function reads, up to unimportant constants,

$$Z(\beta) = \sum_{E} \prod_{\text{links}} \exp(-\beta H) \prod_{\text{sites}} \int_{-\pi}^{\pi} d\alpha(\vec{r}) \exp\left(i\alpha(\vec{r}) \sum_{\hat{n}} E(\vec{r}, \hat{n})\right) \tag{7.7}$$

We can write this in a more useful form as

$$Z(\beta) = \int_{-\pi}^{\pi} \prod_{\vec{r}} d\alpha(\vec{r}) \prod_{\text{links}} \sum_{E} \exp\left(-\frac{\beta g^2}{2a} E(\vec{r}, \hat{n})^2\right.$$

$$\left. + i[\alpha(\vec{r}) - \alpha(\vec{r} + \hat{n})] E(\vec{r}, \hat{n})\right) \tag{7.8}$$

The sum over the integers E defines the periodic Gaussian function, an old friend. To evaluate the partition function we need one form of the Poisson summation formula, which reads

$$\sum_{n} \exp(cn^2 + i\alpha n) = \left(\frac{\pi}{c}\right)^{1/2} \text{pexp}\left(-\frac{1}{4c}\alpha^2\right) \tag{7.9}$$

where pexp denotes the periodic Gaussian, defined by

$$\text{pexp}(-\gamma\alpha^2) \equiv \sum_{m} \exp[-\gamma(\alpha + 2\pi m)^2] \tag{7.10}$$

It is a superposition of Gaussians with period 2π. It is a very convenient function because Gaussians are easy to integrate, while its periodicity in $\phi \to \phi + 2\pi$ is manifest. In addition, it has the general features of $\exp(-2\gamma)\exp(2\gamma\cos\phi)$ for large γ.

Using this identity, the partition function becomes

$$Z(\beta) = \int_{-\pi}^{\pi} d\alpha(\vec{r}) \prod_{\text{links}} \text{pexp}\left(-\frac{a}{2\beta g^2}(\alpha(\vec{r}) - \alpha(\vec{r} + \hat{n}))^2\right) \tag{7.11}$$

where again we have dropped unimportant constants of integration.

What are the features of this model? It is a system of nearest-neighbor Abelian spins coupled together through the periodic Gaussian and closely resembles the planar Heisenberg model of statistical-mechanics fame,

$$Z(\beta) = \int_{-\pi}^{\pi} d\alpha(\vec{r}) \prod_{\text{links}} \exp\left(-\frac{a}{\beta g^2}\cos(\alpha(\vec{r}) - \alpha(\vec{r} + \hat{n}))\right) \tag{7.12}$$

In fact, the periodic Gaussian expression is the so-called Villain approximation to the planar Heisenberg model. Note that the temperature of the original problem in quarks and gluons maps onto the inverse temperature of the Villain model. This is yet another duality relationship. In other words, as the temperature of the lattice-gauge theory increases, the temperature of the Villain model decreases. High-temperature features of the gauge model map onto low-temperature features of the spin model and vice versa.

The Villain model is known to have a low-temperature magnetized phase and a high-temperature disordered phase with a second-order phase transition between them. In particular, (1) for small $a/(\beta g^2)$, the system has a short-range, exponential spin–spin correlation function,

$$\left\langle e^{i\alpha(\vec{0})} e^{i\alpha(\vec{r})} \right\rangle \underset{r\to\infty}{\sim} e^{-\mu r} \qquad (7.13)$$

and (2) for large $a/(\beta g^2)$, the system is ordered and has a spin–spin correlation function that approaches a constant at large separations as

$$\left\langle e^{i\alpha(\vec{0})} e^{i\alpha(\vec{r})} \right\rangle \underset{r\to\infty}{\sim} \exp\left(\frac{\beta g^2}{r}\right) \qquad (7.14)$$

The lattice theory will inherit a second-order phase transition from its dual spin system.

To obtain a physical interpretation of the transition, let's place static charges into the gauge system and calculate their force law. This means that we place charges $\pm g$ on sites $\vec{r} = \vec{0}$ and $\vec{r} = \vec{R}$ and recalculate the partition function. The sources are introduced into the formalism by changing Gauss' law appropriately at the two sites,

$$\sum_{\hat{n}} E(\vec{0}, \hat{n})|0\rangle = |0\rangle \qquad (7.15)$$

and

$$\sum_{\hat{n}} E(\vec{R}, \hat{n})|0\rangle = -|0\rangle \qquad (7.16)$$

When we redo the partition-function analysis above, these changes at the two sites alter the δ function at $\vec{r} = \vec{0}$ and $\vec{r} = \vec{R}$, with the result that an extra factor of

$$\exp(i\alpha(\vec{0})) \exp(-i\alpha(\vec{R})) \qquad (7.17)$$

is introduced everywhere. Our partition function is now proportional to a spin–spin correlation function,

$$Z(\beta, \vec{R}) = \int_{-\pi}^{\pi} d\alpha(\vec{r}) \prod_{\text{links}} \text{pexp}\left(-\frac{a}{2\beta g^2}(\alpha(\vec{r}) - \alpha(\vec{r} + \hat{n}))^2\right) e^{i\alpha(\vec{0})} e^{-i\alpha(\vec{R})}$$

$$= Z(\beta)\left\langle e^{i\alpha(\vec{0})} e^{-i\alpha(\vec{R})} \right\rangle \qquad (7.18)$$

As reviewed above, we know the behavior of these correlation functions in the dual-spin model, so we can obtain the extra free energy of the lattice-gauge system due to the sources,

$$F(\beta, \vec{R}) = -[\ln Z(\beta, \vec{R}) - \ln Z(\beta)]/\beta = \frac{1}{\beta} \ln\left(\langle e^{i\alpha(\vec{0})} e^{-i\alpha(\vec{R})}\rangle\right) \quad (7.19)$$

Using Eqs. (7.13) and (7.14), we see that the extra energy due to the static charges in the gauge theory at low gauge temperatures is

$$F(\beta, \vec{R}) \underset{R \to \infty}{\sim} \frac{\mu}{\beta} R \quad (7.20)$$

whereas for high temperatures it is

$$F(\beta, \vec{R}) \underset{R \to \infty}{\sim} \frac{g^2}{R} \quad (7.21)$$

So we learn that the gauge theory experiences a deconfining phase transition at some critical temperature at which the heavy-quark potential switches from linear confinement to ordinary Coulomb behavior.

This is an interesting, nontrivial result. However, it is not the whole story for several reasons. First, the calculation was done for strong coupling, far from the lattice theory's continuum limit. Results from additional analytic and numerical work indicate, however, that the important conclusion that there is deconfining transition in the model survives greater sophistication. Second, these Abelian models should be replaced by non-Abelian gauge fields. This complication can also be handled analytically at strong coupling and the deconfinement transition remains. Third, the gauge model does not contain matter fields that could screen the color of static sources and replace the linear confining potential at low temperatures by a short-range Yukawa potential and the Coulomb potential at high temperatures by a short-range Debye screened potential. Introducing simple matter fields into the strong-coupling considerations above confirms these expectations. In fact, dynamical matter fields act as external symmetry-breaking fields in the dual-spin model and smooth out the phase transition illustrated here and replace it with a crossover from Yukawa to Debye behavior. The sharp transition is eliminated. However, as we will discuss below, there is a phase transition in full QCD at nonzero temperature, but it is associated with chiral-symmetry breaking rather than confinement and the heavy-quark potential.

7.2 Simulations at nonzero temperature

When four-dimensional lattice-gauge theory is formulated at nonzero T, one considers asymmetric lattices. Let the spatial box have N sites in the x, y, and

z directions and denote the lattice spacing a. The physical volume of the box is then $V_s = (aN)^3$. The temporal direction is special and, as we know from conventional continuum field theory, has an extent proportional to the reciprocal of the temperature T. In fact, if we denote the number of lattice sites in the temporal direction N_t and the lattice spacing in this direction a_t, then the inverse temperature is

$$T^{-1} = a_t N_t \tag{7.22}$$

Simulation studies are done using periodic boundary conditions imposed on the bosonic degrees of freedom and anti-periodic boundary conditions for the fermions. Of course, it is crucial to use such boundary conditions in the temporal direction in order to get the spin-statistics relations correct.

In most simulations one chooses the spatial box to be large and varies the temporal extent to change T. There are many ways to do this and the best choice might depend on the goals of a particular simulation. For example, most simulations are done with the physical lengths of the lattice links the same in the spatial and temporal directions, $a = a_t$, but with $N_t \ll N$. Then T is varied by varying the coupling $\beta = 6/g^2$ of the SU(3) lattice action. As β is taken larger (weaker coupling), the lattice system finds itself closer to the continuum limit where the correlation length of the model diverges in units of the lattice spacing a. This means that $a_t N_t$ is effectively smaller when it is measured in physical units of the correlation length, so the temperature, the reciprocal of $a_t N_t$, has effectively been increased. Therefore, in many publications, authors plot observables against β and state that increasing β effectively heats the system.

To make this connection quantitative we must return to the Callan–Symanzik function and calculate the dependence of the lattice coupling on the cutoff a.

Recall that the idea of the Callan–Symanzik equation is the realization that, in a renormalizable field theory, the coupling defined at a certain length scale must depend on that length scale in order that physical quantities will, in fact, turn out to be cutoff-independent. For example, suppose that we calculate the proton's mass M in a lattice version of QCD with a lattice coupling g. Then, by dimensional analysis, the mass would have the form

$$M = \frac{1}{a} f(g^2) \tag{7.23}$$

because a, the lattice spacing, is the only dimensional parameter in the theory. However, since M must be independent of a, this equation implies that g must depend on a in a precise fashion. The Callan–Symanzik equation makes this connection precise in the realm where perturbation theory applies.

Through two loops in perturbation theory, the explicit lattice perturbation-theory calculations yield

$$a \frac{\partial g}{\partial a} = \beta_0 g^3 + \beta_1 g^5 \tag{7.24}$$

where the coefficients are

$$\beta_0 = \frac{1}{16\pi^2} \left(\frac{11N}{3} - \frac{2N_f}{3} \right)$$

$$\beta_1 = \frac{1}{(16\pi^2)^2} \left(\frac{34N^2}{3} - \frac{10NN_f}{3} - \frac{(N^2 - 1)N_f}{N} \right) \tag{7.25}$$

for SU(N) gauge fields and N_f flavors of light quarks. The important point is, of course, that both β_0 and β_1 are positive if the number of quark species is small enough. This guarantees that the theory is weakly coupled at short distance – as a is taken smaller, then g must also be taken smaller, in accord with the Callan–Symanzik equation. Stated in another fashion, in order to keep the physics independent of the cutoff, for example, to keep the proton mass fixed at 938 MeV/c^2, the coupling must be tuned to the lattice spacing a according to the Callan–Symanzik differential equation.

One can formally integrate the Callan–Symanzik equation,

$$a = \exp\left(-\int^g \frac{dg'}{\beta(g')} \right) \tag{7.26}$$

and use the perturbative expression over that region where g is small,

$$a = \Lambda^{-1} \exp\left(-\frac{1}{2\beta_0 g^2} \right) (\beta_0 g^2)^{-\beta_1/(2\beta_0^2)} \tag{7.27}$$

where the integration constant Λ has the dimensions of a mass and sets the scale of lengths. Since the temperature T has the dimensions of mass, it must be proportional to Λ,

$$T = C\Lambda \tag{7.28}$$

where the numerical constant C relates the two scales.

Other physical quantities with the dimensions of mass should satisfy analogous relations, but each will have its own particular dimensionless constant C. This scaling result guarantees that ratios of quantities with the same dimensions are simple numbers, independently of the cutoff procedure. So we could imagine, as is done in practice, a lattice simulation that determines T and m_ρ in terms of the cutoff a at a lattice coupling small enough that the Callan–Symanzik equation applies. Then the ratio of T to m_ρ would be a property of the continuum limit of the lattice theory and, knowing m_ρ from experiment, we could predict the unknown temperature T.

In practice, it is convenient to consider other more-elegant cutoff procedures and relate the lattice coupling g and its associated scale parameter Λ to those of other schemes, such as Pauli–Villars and dimensional regularization. These intricate details can be found in the literature. In the course of these developments, one proves that the coefficients β_0 and β_1 are identical in each regularization scheme.

We learn from this summary several lessons that are particularly important for lattice calculations and simulations. To make contact with continuum physics, lattice calculations using the simple actions discussed here must involve calculations at weak coupling, in the "scaling window" where the two-loop Callan–Symanzik equation is accurate. Only then can we expect the cutoff a to decouple from the physical quantities of interest. At finite temperature T, we anticipate the need for (1) weak coupling, to find the scaling region, (2) large temporal extents, $N_t \gg 1$, so that discreteness effects are under control, and (3) much larger spatial extents than temporal extents, $N \gg N_t$, so that finite-size effects associated with the size of the spatial box do not interfere with the desired temperature effects coming from N_t.

It is a sobering fact that few, if any, simulations to date satisfy these criteria.

There are many schemes for dealing with these potential problems. One can "improve" the lattice action so that the scaling window expands and continuum physics can be obtained from simulations at larger couplings on smaller lattices. The idea here, as we have seen in past chapters, is that there are $O(a^2)$ deviations of the lattice action from its continuum counterpart even for smooth fields and weak (or vanishing) coupling. By using more sophisticated finite-differencing formulas these errors can be eliminated. Although the lattice action becomes more complicated and the differencing schemes will necessarily involve next-to-nearest-neighbor couplings, the computational gains have been found to be very substantial. The lattice-spacing dependences of quantities such as the heavy-quark potential are dramatically reduced. By including perturbative corrections in this improvement scheme errors of order $O(g^2 a^2)$ can also be removed both from the pure-gauge piece and from the fermion-hopping piece of the action. Such improvements should make simulations aimed at finding the universality classes of various finite-T transitions practical and reliable.

7.3 Pure gauge-field simulations at nonzero temperature

Simulations of pure glue are interesting, practical, and relevant. In the early days of lattice-gauge theory, theorists noted that there are eight adjoint colored gauge fields in QCD while there are only two light fundamental quarks. So "most" of the dynamics of QCD is due to gauge-field effects. In fact, asymptotic freedom is , a property of non-Abelian gauge fields whereas Fermi fields tend to screen color and counter the unique effects of the interacting gauge fields. Therefore, it was felt sensible to generate equilibrated gauge-field configurations at various couplings and then study quark-matrix elements in this background. This approach constitutes the "quenched" approximation to the theory. Light-quark bound-state spectroscopy, chiral-symmetry breaking, and hadronic matrix elements are relatively easily studied in this context and much has been learned. Of course, we have also learned that this "approximation" has some serious limitations. These include the fact that it is misleading at nonzero-baryon-number chemical potential, as we will discuss in Chapter 10. It cannot be used to study QCD's phase diagram in the T–μ plane. In addition, it suffers from other pathologies that derive from the fact that its configurations omit the axial anomaly and therefore display more symmetry (a massless η') than they should. Unfortunately, an "approximation" that has more symmetry than the real theory can have qualitative errors and this occurs in the quenched approach in the limit of light quarks.

With these caveats, it is interesting and useful to address the thermodynamics of pure gauge fields.

At low T we expect that the theory confines quarks through the formation of electric-flux tubes that produce a linear confining potential, as described earlier. In addition, the pure gauge-field theory is expected to generate a mass gap dynamically and have a rich spectrum of color-singlet excitations. The idea is that the flux tube has an energy per unit length, the square root of the string tension σ, so the heavy-quark potential reads

$$V(R) = \sqrt{\sigma}\,R - \frac{\pi/12}{R} + \cdots \qquad (7.29)$$

The string tension σ is generated dynamically and it can be related to the theory's scale parameter Λ, $\sqrt{\sigma} = C\Lambda$. The constant of proportionality C has been measured in lattice simulations. Since glueballs are created from the vacuum by applying operators consisting of closed loops of U operators, they can be thought of as rings of electric flux in the strong-coupling limit, g large, discussed earlier. Their masses and properties are, therefore, closely tied to the formation and scale of the flux-tube mechanism of confinement. Estimates of the glueball masses are made in lattice-simulation studies. They are relatively heavy states that are also expected to exist as fairly sharp resonances in QCD, in which they would be able to decay into lighter mesons. Their experimental discovery would be a triumph both for QCD in general and for lattice-gauge theory in particular.

As the pure gauge theory is heated, we expect a phase transition to a gluonic plasma. In this phase there should be complete color screening. For example, if a quark in the fundamental representation of color SU(3) were placed into the hot vacuum, its color field should be short-ranged and screened exponentially. In particular, the string tension should vanish at the critical point between the low- and high-temperature phases. We obtained these results at strong coupling above and they are expected to be true in the continuum limit of that analysis. Computer simulations in the scaling window support this view.

These qualitative considerations suggest an order parameter for the transition: at low T the theory is expected to confine fundamental quarks through flux-tube formation, whereas at high T, for which asymptotic freedom applies and predicts that regions of fixed volume should have free-field properties decorated by small perturbative corrections that vanish logarithmically as the temperature grows large, quarks should propagate in a color plasma. This difference can be made precise by considering the excess free energy for a static quark charge. The worldline for such a quark would be purely timelike and it would couple to the gauge-field background through a U matrix, as in the hopping fermion actions studied earlier. So, an order parameter would be the Wilson line (called the Polyakov loop in this context when the trace has been taken) [44],

$$L(\vec{x}) = \text{tr}\left(\prod_{x_0=1}^{N_t} U_0(\vec{x}, x_0)\right) \tag{7.30}$$

As discussed above in the context of quark confinement, since this quantity represents a heavy, static quark impurity in the system, the expectation value of the Wilson line, suitably averaged over the spatial box, records the excess free energy due to the quark,

$$e^{-F_q(T)/T} \propto \langle L \rangle = \left|\left\langle \frac{1}{N^3} \sum_{\vec{x}} L(\vec{x}) \right\rangle\right| \tag{7.31}$$

This quantity is an order parameter for the transition to the plasma because it vanishes identically in the low-T phase in the thermodynamic limit, while it is nonzero in the plasma. At low T, the excess free energy that occurs when a static quark is placed in the vacuum diverges as the linear dimension of the spatial box, because the quark is the source of a flux tube. However, in the high-temperature phase the plasma of gluons can screen the color of any impurity, so the excess free energy is finite.

In order to classify the phase transition, we need to determine the global symmetry group that is broken. The order parameter is locally gauge-invariant, so we seek a global aspect of color SU(3) rotations that will survive the trace in the definition of the order parameter. The triality of the color matrix is a natural candidate that is familiar from the old-fashioned quark model. In particular, consider the center of the color SU(3) group which consists of the three complex roots of unity, Z_3. If we make a Z_3 transformation on all the timelike links on a given temporal hyperplane,

$$U_0(\vec{x}, x_0) \rightarrow z U_0(\vec{x}, x_0) \ (x_0 \text{ specified}), \qquad z \in Z_3 \tag{7.32}$$

Note that this transformation leaves the pure gauge-field action unchanged because the action contains only closed loops of U matrices that pierce the special
hyperplane twice (once in the $+t$ direction giving a z factor and once in the $-t$
direction giving a z^* factor) or not at all (purely spatial loops). The transformation affects the Wilson line,

$$L \rightarrow zL \tag{7.33}$$

so, when L develops a ground-state expectation value, the Z_3 symmetry is broken.

The significance of the center of the global gauge group has been demonstrated nicely in computer simulations that have plotted the Monte Carlo "time"
evolution of the phase of the Wilson line and shown that it tunnels among the
three different possible values for z. In fact, the first-order character of the deconfining transition has been confirmed in such studies.

More insight into the order of the finite-temperature deconfinement transition
for SU(3) gauge fields follows from effective-Lagrangian descriptions of the
transition. Conventional wisdom suggests that this transition can be described
by a local, bosonic spin model in three dimensions with an order parameter that
experiences the same breakdown of symmetry. Let's motivate this approach. The
modern renormalization-group approach to phase transitions states that the universality class of a phase transition depends only on a few, crucial features of the
system. The dimensionality, symmetries, and range of interactions of the order
parameter are paramount. The relevant dimensionality here is three dimensions
rather than four because we are studying the equilibrium statistical mechanics of
a system with three spatial dimensions. There is no real temporal development.
More formally, when the system is defined in Euclidean space, the "temporal"
direction actually is just a recipe for incorporating the temperature into the thermodynamics. The temporal direction is fixed and "small," $aT = N_t^{-1}$, relative
to the size of the spatial box. Because of the periodic boundary conditions in the
temporal direction, the possible frequencies that can be fit into the system are
quantized. These are the Matsubara frequencies,

$$\omega_n = 2n\pi T \text{ (bosons)}, \qquad \omega_n = (2n + 1)\pi T \text{ (fermions)} \tag{7.34}$$

where $n = 0, 1, 2, 3, \ldots$. For bosonic fields the lowest Matsubara frequency is
zero, producing a field that does not vary in the temporal direction. The other
Matsubara frequencies are nonzero and large at nonzero T. It is reasonable to
suppose that the modes with nonzero Matsubara frequencies are not excited at
a second-order phase transition. This is because the important modes are essentially massless and only their long-wavelength excitations are statistically
significant at a second-order phase transition. In this situation, there is little probability of exciting nonzero Matsubara frequencies. The system is then effectively
three-dimensional and one says that it is "dimensionally reduced."

Back to the deconfinement transition. Given the plausibility of dimensional reduction for the high-temperature behavior of the pure gauge model, we are led to consider three-dimensional bosonic field theories with Z_3 global symmetries as their prototypes. The simplest of these is the three-state Potts model. It is just the obvious generalization of the Ising model (Z_2) to an order parameter with a threefold symmetry (Z_3). It is known that the three-state Potts model has a first-order transition in three dimensions. Explicit simulations have shown that the deconfinement transition in the SU(3) gauge field case is, in fact, of first order. As stated above, threefold metastability has been found in the Monte Carlo "time" evolution of the phase of the Wilson line, so there is considerable confidence that the Z_3 symmetry and the three-state Potts model are relevant.

The bulk thermodynamics of the finite-T transition in the pure gauge model has been studied by simulation methods in considerable detail. Since the simulations involve only bosonic degrees of freedom, they have been done far better than their light-quark counterparts. It has been found that the pressure of the pure SU(3) gauge system stays small for all temperatures below T_c. This is certainly expected, because the only relevant degrees of freedom in the low-T phase are glueballs, which are, as indicated above, quite heavy and lead to an exponential suppression of pressure and energy density at low T. Above T_c simulations show that the pressure rises rapidly and reaches two thirds of its asymptotic ideal-gas value by a temperature of twice the critical value. However, the approach to the limiting free ideal-gas values for the pressure, entropy, and energy density is quite slow from this point onward. The old hope that high temperature perturbation theory could descibe the quark–gluon plasma accurately for temperatures of several times T_c seems too naive.

Even though the energy density ϵ jumps discontinuously at the first-order plasma transition with a very substantial latent heat of about $1.5 T_c^4$, the system is not well approximated by the ideal-gas equation of state, $p = \epsilon/3$, for temperatures expected to be relevant to heavy-ion collisions and controlled lattice-gauge simulations. One finds a delayed rise in the pressure compared with the energy density, so the velocity of sound in the plasma is predicted to be considerably below its ideal-gas value.

According to the Casher argument relating chiral-symmetry breaking to bound-state formation and confinement, it is natural to wonder whether chiral symmetry is restored at the deconfinement transition at nonzero T. This is not necessarily true because the forces of attraction between the quark and the anti-quark may be strong enough even on the hot, plasma side of the deconfinement transition to lead to $\langle \bar{q}q \rangle$ condensation. However, for quarks in the fundamental representation of the gauge group, it does appear that the screened forces are not sufficient to produce a condensate and chiral symmetry is restored when deconfinement occurs. This has been confirmed in computer simulations of the quenched theory, i.e. we study light quarks propagating in the dynamical ensembles of pure gauge fields, as described above. In fact, the deconfinement and chiral-symmetry transitions are coincident and of first order. So the fact that the expectation of the Wilson line jumps from zero to a finite value at the deconfinement temperature T_d, while the chiral condensate jumps from a sizable value to zero at the same $T_c = T_d$, strongly suggests that the gluonic forces which cause confinement, flux-tube formation, and bound-state formation are also the cause of chiral-symmetry breaking.

7.4 Restoration of chiral symmetry and high temperature

When light quarks are put into the dynamics so that realistic QCD with N_f flavors is being simulated, this scenario must change dramatically. In particular the Z_3 symmetry of the pure gauge-field model is not a symmetry at any temperature of the action with fermions. This is obvious because of the factor of U in the quark hopping term. The Z_3 symmetry is explicitly broken in the action, so the Wilson line should be nonzero under all conditions. This means that the excess free energy due to the static quark in the Wilson line is always finite. There is a simple interpretation for this formal result. Once there are light quarks in the dynamics, then they can screen any impurity such as the Wilson line. So the Wilson line is no longer the source of an "infinitely" long flux tube; instead, the light dynamical quarks terminate the tube and render its effects short-ranged. The static quark now gives rise to a Yukawa potential due to the screening provided by the light quarks.

Since the long-range confining forces are eliminated, it is no longer obvious that there is any chiral-symmetry breaking in lattice QCD. In fact, simulations indicate that chiral-symmetry breaking survives the Yukawa screening at vanishing T. Casher's physical picture continues to hold, apparently because the forces are strong enough to make a rich bound-state spectrum. It appears that, for dynamical quarks with small but nonzero masses, screening is weak enough to permit chiral-symmetry breaking to survive. Simulations suggest that the linear potential extends out to sufficient distances from its source, one light quark in a $\bar{q}q$ state, so that binding can occur over that distance scale and make the linear potential relevant to the spectroscopy of the two-body states. The Casher argument applies as in the model with no screening. However, for distances greater than the size of a typical hadronic bound state, screening due to light quarks becomes overwhelming, eliminating the long-range potential and producing color-neutral bound states.

This physical picture, complete with its particular length scales, finds additional support in finite-T simulations. Simulations of lattice QCD with two or three light quarks show that the Wilson line does jump dramatically at the same temperature T_c as that at which chiral symmetry is restored and the $\bar{q}q$ condensate vanishes exactly. This observation can be made more precise by considering the susceptibilities for the Wilson line and the chiral condensate,

$$\chi_L = N^3(\langle L^2 \rangle - \langle L \rangle^2), \qquad \chi_m = \frac{\partial}{\partial m}\langle \bar{\psi}\psi \rangle \qquad (7.35)$$

These quantities peak where the Wilson line and the chiral condensate, respectively, are changing rapidly. Plots of the susceptibilities pick out narrow regions that, within statistical accuracy, correspond to identical critical temperatures.

In the limit of massless quarks, we expect that a good order parameter for the transition from hadronic matter to the quark–gluon plasma would be the chiral condensate $\langle \bar{\psi}\psi \rangle$. The continuum QCD Lagrangian has the algebraic, chiral symmetries of the massless Dirac equation, $SU(N_f)_L \times SU(N_f)_R$, corresponding to independent γ_5 rotations of the left- and right-handed components of N_f Dirac fermions. This symmetry should be broken dynamically by Casher's argument down to the algebraic symmetry $SU(N_f)_{L+R}$ when the quarks develop a mass. This spontaneous symmetry breaking produces $N_f^2 - 1$ Goldstone particles, which, in the case of $N_f = 2$, corresponds to three pions.

In an effective three-dimensional Lagrangian description of this transition, one would want to include the relevant light bosonic degrees of freedom, which, in the case of chiral symmetry and $N_f = 2$, would be the three pions and the σ particle. In the chirally symmetric phase, these four states would be degenerate and would comprise a chiral vector. In the broken phase, the three pions would become massless Goldstone particles and the σ would be a massive excitation.

In the generalization of this picture to arbitrary N_f, the order parameter is conveniently chosen to be an $N_f \times N_f$ matrix, $\Phi \equiv (\Phi_{ij})$, of quark–anti-quark bilinears that parametrize the symmetry breaking, $\Phi_{ij} \sim \langle \bar{\psi}_i(1 + \gamma_5)\psi_j \rangle$. Under an $SU(N_f)_L \times SU(N_f)_R$ transformation of the quarks, $\Phi \to U_L \Phi U_R$. The low-energy dynamics of the bosonic degrees of freedom are controlled by the three-dimensional effective Lagrangian that has the same global symmetries as the fundamental QCD Lagrangian,

$$L_{\text{eff}} = -\frac{1}{2}\text{tr}(\partial_\mu \Phi^\dagger \, \partial^\mu \Phi) - \frac{1}{2}\text{tr}(\Phi^\dagger \Phi) + \frac{\pi^2}{3}g_1(\text{tr}(\Phi^\dagger \Phi))^2$$

$$+ \frac{\pi^2}{3}g_2(\text{tr}(\Phi^\dagger \Phi)^2) + c(\det \Phi + \det \Phi^\dagger) \qquad (7.36)$$

where all the terms should be familiar and expected, except, perhaps, the last determinantal interaction. This term is needed in order to account for the fact that triangle graphs in QCD break the axial $U(1)_A$ anomaly, and with the help of instantons, lift the η' mass to a large value. These subtle points were discussed in Chapter 4 above. A renormalization-group analysis of this Lagrangian predicts that it undergoes a second-order transition for $N_f < 3$, but a first-order transition otherwise [45]. The one caveat concerns the fate of the η meson at $N_f = 2$. If the $U(1)_A$ symmetry breaking, which is due to the triangle anomaly and instantons, becomes rather weak at T_c, then the η' would become another low-lying bosonic excitation whose inclusion in the dynamics could induce a fluctuation-driven first-order transition. Simulations suggest that this accidental "restoration" of symmetry does not occur.

In the case of two light flavors, this effective Lagrangian suggests that the O(4) spin model determines the universality class of the chirality restoring finite-T transition in two-flavor QCD. In this ideal case of massless quarks, measurement of the properties of the QCD phase transition would provide a nontrivial test of the universality hypothesis.[1] Of course, the actual situation in QCD is rather complicated because the theory has two very light flavors, the up and down quarks, whose Lagrangian masses are just a few MeV, and a third, strange quark whose Lagrangian mass is of the order of 100 MeV. Simulations favor the argument that this quark spectrum leads to a chirality-crossover phenomenon rather than a real phase transition. However, since the massless, two-flavor chirality-transition point would lie "close" to the physical situation, its O(4) critical indices might be relevant to the character of the real system's crossover. Lattice-gauge-theory simulations have not determined the critical indices of the two-flavor model. Improved gauge and fermion actions are being pursued with this goal in mind.

7.5 Hadronic screening lengths

A simple and appealing way to study and understand the chiral-symmetry-restoring transition in QCD is to consider its spectrum and screening lengths for temperatures above and below T_c. Since chiral symmetry is restored at T_c, the pion triplet is not expected to be massless in the plasma. The character of the pion excitations above T_c is not known, but it is suspected that they are quasi-particles in the plasma phase, at least for temperatures near T_c. If this is the case then, since chiral symmetry is restored, there should be a "σ" quasi-particle in the plasma phase degenerate with the pion triplet. The fact that the system undergoes a second-order phase transition at T_c indicates that the σ and the pions were massless at T_c and their masses increase continuously as T increases from T_c. This can be seen in lattice simulations by measuring the chiral susceptibility, defined in the equation above, and finding a divergence at T_c.

This perspective suggests that other aspects of restoration of chiral symmetry can be studied on the lattice by measuring susceptibilities for other quark–anti-quark bilinears. One defines hadronic correlation functions in a given

---

[1] Although there is little doubt that it holds, this is by no means a proven fact in a theory like QCD.

quantum-number channel, and their associated susceptibility,

$$\chi_H = \int_0^{1/T} d\tau \int d^3r\, G_H(\tau, \vec{r}) \tag{7.37}$$

where the meson correlation function is

$$G_H(\tau, \vec{r}) = \langle \bar{\chi}(0)\gamma_H\chi(0)\bar{\chi}(\tau, \vec{r})\gamma_H\chi(\tau, \vec{r})\rangle \tag{7.38}$$

where γ_H is the appropriate collection of γ matrices from which to construct the composite operator with the desired quantum numbers. The susceptibilities are essentially the inverses of the squared masses, m_H^2. High-quality calculations of these susceptibilities then expose the symmetries of both phases [46]. For example, the divergence of the π and σ susceptibilities at T_c shows that the σ becomes massless as the critical point is approached from below and chiral symmetry is restored in the quark–gluon-plasma phase. The two susceptibilities are identical in the hot phase, as expected when chiral symmetry is restored. This also suggests that there are π and σ quasi-particles in the hot phase in the vicinity of T_c.

Another approach to understanding the symmetries of both phases is afforded by hadronic screening lengths. Consider two composite hadronic operators, $A(\tau, \vec{r})$ and $B(\tau, \vec{r})$, made out of the appropriate products of fields – a quark-and-anti-quark field with particular spin and flavor for mesons, and three quark fields to make a particular baryon. If we average these operators over the temporal axis τ and the transverse plane,

$$\bar{A}(z) = \frac{T}{4L^2}\lim_{L\to\infty}\int_0^{1/T} d\tau \int_{-L}^L dx \int_{-L}^L dy\, A(\tau, x, y, z) \tag{7.39}$$

and calculate the correlation function,

$$S_{AB}(z) = \langle \bar{A}(z)\bar{B}(0)\rangle - \langle\bar{A}(0)\rangle\langle\bar{B}(0)\rangle \tag{7.40}$$

which we then fit to a screening form,

$$S_{AB}(z) \underset{|z|\to\infty}{\sim} c\exp(-m_{AB}(T)|z|) \tag{7.41}$$

we can extract the screening masses m_{AB}.

This program has been carried out both for the pure SU(3) gauge model and for QCD with several flavors of light dynamical quarks. One finds a spectrum of states characterizing chiral-symmetry breaking in the hadronic phase: the π is essentially massless while its chiral partner the σ is heavy, the ρ and its partner the a_1 are massive with different masses, and the even-parity N_+ and odd-parity N_- baryons are even more massive and split. In the plasma phase, the π and the σ

become massive and degenerate, as do the other chiral partners. Since the computer signals for these correlation functions are large and relatively free of noise, the simulation suggests that the plasma has a rich nonperturbative spectrum of states for T in the vicinity of T_c.

Additional evidence for the nontrivial character of the plasma phase comes from studies of the Debye screening mass, m_D. To leading order in perturbation theory,

$$m_D = \sqrt{1 + \frac{N_f}{6}} g(T) T \tag{7.42}$$

where $g(T)$ is the running coupling of QCD evaluated at the relevant mass scale, the temperature T. In the region of the transition, $g(T)$ is of the order of unity and the accuracy of expansions in $g(T)$ is questionable. In fact, the measured Debye screening mass is roughly three times the perturbative prediction for accessible values of T. We learn that even the short-distance properties of the plasma are highly nonperturbative for accessible temperatures, even though asympototic freedom assures us that, at truly high T, the bulk thermodynamics of the plasma should be given by free-field estimates decorated with small, calculable perturbative corrections.

Sorting out the physics of the QCD plasma is an active field of research for lattice-gauge theorists, high-energy phenomenologists, and experimentalists.

7.6 Thermal dilepton rates and experimental signatures for the quark–gluon plasma

The hadronic correlation lengths discussed in the previous section provide information about the response of QCD to time-independent, static perturbations. In other words, only zero-frequency components of the correlation functions are involved. It is also interesting, and even more relevant from the experimental point of view, to address the question of the response to time-dependent perturbations. The difficulty here, in comparison with static correlation-length measurements, is that one needs information about the correlation function at real time intervals, whereas lattice field theory is formulated in Euclidean time. In the frequency language, we need the correlation function, or the corresponding spectral function, for a continuum of real values of the frequency, while the Euclidean theory provides us only with discrete Matsubara modes. This difficulty can, in some cases, be overcome, and we shall discuss here the example of the spectral function relevant for dilepton rates.

Dilepton rates have been discussed for many years as an informative probe into the dynamics of the quark–gluon plasma. In such a process, a quark and an anti-quark in the plasma annihilate into a virtual photon, which then materializes as a lepton–anti-lepton pair. The probability of such an event is clearly proportional to the thermal expectation value of the product of the electromagnetic current with itself. The expression for the cross section will be given below. The production rate will be very sensitive to the temperature of the plasma (the larger the temperature, the larger the rate), so the dileptons are preferentially produced in the early stages of a heavy-ion collision. Since they are only weakly coupled to nuclear matter through the fine-structure constant, they have long mean free paths, which allow them to escape from the system without significant final-state interactions. These properties make electromagnetic signals particularly good probes into the physics of the early stages of the formation of the quark–gluon plasma.

It is particularly interesting to focus on hadronic-resonance contributions to the dilepton rate. If the Q^2 of the virtual photon (dilepton pair) matches the mass of a vector-meson state of QCD, we expect an enhancement of the dilepton rate. Since this production occurs inside a volume of quark–gluon plasma, the positions of the peaks and their widths depend sensitively on the interaction of the hadronic state with the plasma itself. By varying Q^2 we can concentrate on different length and time scales of the interactions and screening mechanisms within the quark–gluon plasma. The strength of the ρ signal, for example, tells us how the light-quark dynamics is affected by the medium, whereas the charmonium signal is sensitive to the screening of the interactions between heavier charmed quarks. Experimental and lattice-simulation data are converging on decisive results here. In fact, some lattice results can be interpreted as evidence for the experimental generation of the quark–gluon plasma at the RHIC.

The dilepton rate, which is a function of frequency ω and momentum $\vec{p}$, can be written in the form [47]

$$\frac{dW}{d\omega\,d\vec{p}} = \frac{5\alpha^2}{27\pi^2}\frac{1}{\omega^2(e^{\omega/T}-1)}\sigma_V(\omega,\vec{p};T)$$
(7.43)

where the spectral function $\sigma_V(\omega,\vec{p};T)$ is related to the current–current correlation function, $G_V(\tau,\vec{p};T) = \int d\vec{x}\,e^{i\vec{p}\cdot\vec{x}}\langle J_V^\mu(\tau,\vec{x})J_{V,\mu}^\dagger(0,\vec{0})\rangle$, on the periodic lattice as

$$G_V(\tau,\vec{p};T) = \int_0^\infty d\omega\,\sigma_V(\omega,\vec{p};T)\frac{\cosh[\omega(\tau-1)/(2T)]}{\sinh[\omega/(2T)]}$$
(7.44)

On an $N^3 \times N_\tau$ lattice, one can measure the correlation function G_V for each "time" slice, $\tau_k T = k/N_\tau$, $k = 1, 2, \ldots, N_\tau$, and then attempt to invert the relation between the spectral function and the correlation function.

There are two approaches to executing this inversion. First, one might use a model-dependent functional form that depends on several parameters and determine $\sigma_V(\omega,\vec{p};T)$. This approach has had only limited success. Alternatively, one might use the maximum-entropy method (MEM) [48] and find the "most probable" spectral function. Since the lattice simulation gives the correlation function only at a discrete set of points τ_k, each with statistical errors, the problem of finding the continuous function σ_V of the continuous variable ω is ill-posed. The MEM can produce only a "most probable" answer with a measure of its reliability. The method, however, is well–known and quite successful in optics, where it is used for image reconstruction. The MEM is well grounded in traditional probability theory and uses the χ^2 likelihood function to constrain the maximization of the Shannon–Jaynes entropy function in order to produce the "most likely" spectral function.

As one might guess, it requires many data points, each with small errors, to predict a spectral function reliably. At least several dozen values of τ_k are typically needed, so large lattices are essential, which restricts the application of the method to either quenched QCD studies or simple models at present. This is a pity because collision broadening, which is sensitive to dynamical quarks and Fermi statistics, is expected to play an important role in determining the shape of the spectral function in heavy-ion experiments. In addition, the MEM does not predict the spectral function with uniform confidence over its range of ω values. Typically, the low-ω range is poorly determined. This is also a pity because this is frequently the most significant part of the power spectrum in heavy-ion collisions.

Simulations have shown that the ρ and ϕ peaks in the dilepton spectra are suppressed in the presence of the quark–gluon plasma, in qualitative accord with the experimental trends [47]. However, the dilepton signal is still somewhat larger than that found for free but hot quarks and gluons, at least for ω/T between 4 and 8. Simulations have been done above the transition, at $1.5T_c$ and $3T_c$, so measureable attraction in these light-quark states at these high temperatures is somewhat surprising. Results such as these suggest that, although the bulk thermodynamic behavior of the quark–gluon plasma might be close to the Stefan–Boltzmann limit, detailed hadronic structures are not completely washed out.

Additional simulation results at $0.9T_c$ and $1.2T_c$ have been obtained for the charmonium states. The s-wave states had a diminished (perhaps by a factor of two) signal above the transition where potential models predicted that there would be no signal at all because the gluon plasma would screen the heavy-quark potential so thoroughly that the c–$\bar{c}$ state would become unbound [49]. The p-wave signal, however, was reduced by a factor of about seven across the transition, suggesting that screening had removed these charmonium states from the spectrum. Recent data are shown in Figs. 7.1 and 7.2.

7.7 A tour of the three-flavor QCD phase diagram

In Fig. 7.3, a very busy affair, we present a survey of the phase diagram of lattice QCD in temperature, light-quark masses, and the strange-quark mass. For

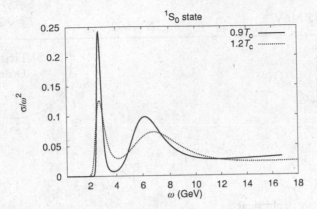

Figure 7.1. S-wave charmonium signals above and below the phase transition in quenched QCD.

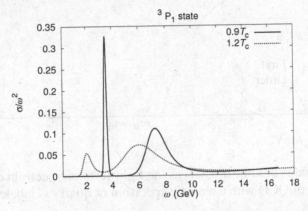

Figure 7.2. P-wave charmonium signals above and below the phase transition in quenched QCD.

experimental purposes and applications in general we are interested in very light "up" and "down" current quark masses and a considerably heavier "strange" quark. To handle a realistic spectrum of quark masses, with "up" and "down" quarks light enough to accommodate a realistic pion, and to make contact with continuum scaling laws will take more computer power and better, improved actions than have been used to date. These improvements are occurring gradually in the field. The presentation here represents the current state of the art and is probably incomplete. Perhaps someone will invent and apply qualitatively better approaches to this problem, which will yield a far more decisive version of our figure.

We will leave to the references discussions of lattice sizes, bare-quark masses, the chosen fermion method, the action used, etc. There are important questions of lattice engineering at play here, but they are best not covered explicitly.

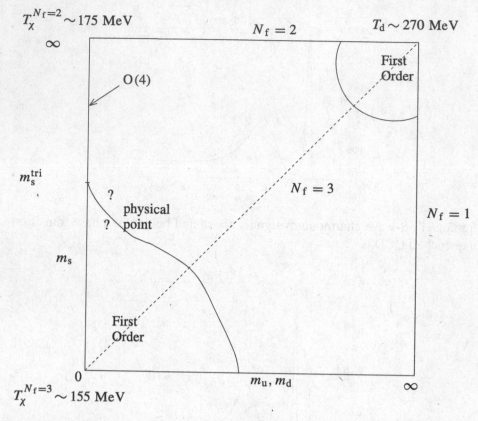

Figure 7.3. Phases of finite-T QCD as the light-fermion spectrum changes. The physical point of QCD with a realistic spectrum of quarks is labeled with question marks.

Anyway, let's take our tour.

Begin at the lower-left-hand corner of Fig. 7.3. Note that the common light-quark mass labels the horizontal axis and the strange-quark mass labels the vertical one. Recall that theory and simulation both indicate that, when the three quarks are massless, there will be a first-order chiral-symmetry-restoration transition between cold hadronic matter and the hot quark–gluon plasma. This result is sensitive to there being three light quarks, as discussed using effective Lagrangians above. In the case of two light quarks, the upper-left-hand side of the phase diagram, the transition is expected to be of second order and in the universality class of the three-dimensional O(4) spin system. The numerical evidence for this universality class is not yet compelling but should improve in the not-too-distant future.

Along the left-hand vertical axis there should be a tricritical point separating the region of first-order transitions from the line of second-order transitions.

Now let's enter the phase diagram along the diagonal from the lower-left-hand corner to the upper-right-hand corner. As the quark masses increase, the first-order chiral transition must weaken. Simulations indicate that the transition becomes a crossover phenomenon and the hadronic state is actually continuously connected to the plasma. The crossover, however, appears quite abrupt in lattice simulations. The boundary between the first-order region and the crossover region is expected to be described by the three-dimensional Ising model, as predicted by effective Lagrangians for analogous liquid–gas transitions. This boundary and its transitions do not have an immediate interpretation in terms of the degrees of freedom of QCD. Instead the transition is governed by the long-wavelength energy and order-parameter fluctuations in the system which are determined phenomenologically. There is some lattice evidence accumulating for the accuracy of this viewpoint and the Ising critical indices [50].

Let's focus more on the following pressing question: where exactly does QCD, with its particular quark masses, fit into this diagram? As stated above, present-day lattice simulations put it into the crossover region, but rather close to the boundary with the first-order chiral region, as shown in greater detail in Fig. 7.4. This figure is labeled by the scale of the light iso-doublet of quark masses, M, and the temperature of the simulation T. The strange-quark mass is taken to be a higher multiple of M, typically $(20–40)M$ in simulation studies. Numerical simulations show that the quantitative features of the change between hadronic and plasma observables are quite clear, and the crossover is quite abrupt. It appears that the high-temperature phase transition relevant to nature will be controlled by the physics of chiral-symmetry breaking rather than by the physics of confinement and deconfinement.

If realistic QCD lies in the crossover region but close to the chiral critical point as shown in Fig. 7.4, then there may be some interesting and curious experimental signatures of the transition. The chiral critical point, labeled "C" in Fig. 7.4, resembles a gas–liquid point in a generic phase diagram and is expected to be described by the three-dimensional Ising model. Therefore, QCD should display its version of "critical opalescence" near that point. An SU(3) flavor-singlet, positive-parity meson should become massless at that point. Perhaps it will be the lightest state in QCD's spectrum in the plasma phase near the critical temperature! Such an odd possibility, a single 0^+ flavor-singlet meson lighter than the familiar pion multiplet, is a matter of speculation and interesting experimental scenarios. We have not heard the last word on this subject and the reader is referred to the literature for additional remarks on this possibility. More simulations with improved actions will be run on the lattice in order to determine where QCD with its particular quark masses actually lies on Fig. 7.4. Smaller light-quark masses and larger lattices, which are needed in order to control finite-size effects, are required.

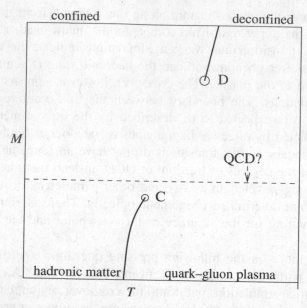

Figure 7.4. A schematic phase diagram of $(2 + 1)$-flavor QCD. The masses of the "up" and "down" quarks are M, the label of the y axis and the strange quark is a much higher multiple of M, such as $(20–40)M$. The temperature T labels the horizontal axis. "C" labels the chiral critical point and "D" labels the deconfinement point. The dashed line, which passes close to point "C," labels the "realistic" QCD situation, we believe.

As we progress up the diagonal in Fig. 7.3 toward the upper-right-hand corner we must meet another boundary to another first-order regime governed by the pure SU(3) gauge-field dynamics. This is the region of heavy-quark masses. As the quark masses become comparable to the reciprocal of the ultraviolet cutoff a, the lattice spacing, they decouple from the low-energy dynamics of the gauge fields and the full theory becomes quenched. The finite-temperature transition is then one of confinement–deconfinement with the Wilson line acting as an order parameter. The global symmetry which is broken in the confined phase and restored as the system is heated and goes through a first-order transition into the unconfined phase is $Z(3)$, the center of the gauge group, as we reviewed in more detail above. Theoretical considerations predicted a first-order phase transition, mapping the transition onto the three-state Potts model in three dimensions. Chiral symmetry is also broken at the first-order transition of the quenched model, with the chiral condensate of massless quarks jumping to zero at that temperature at which heavy quarks would be liberated.

It is interesting to compare the critical temperatures for the three transitions that occur at the corners of the phase diagram Fig. 7.3. In the lower-left-hand corner where three massless quarks experience a chiral-symmetry-restoring transition into the quark–gluon plasma, the critical temperature is the lowest, ~155 MeV. In the upper-left-hand corner where two massless quarks experience a chiral-symmetry-restoring transition into the quark–gluon plasma, the critical temperature is somewhat higher, ~175 MeV. In the upper-right-hand corner where the eight gluons of the SU(3) gauge group experience a deconfining transition from a state of closed, confining string into the gluon plasma, the critical temperature is the highest, ~270 MeV. These systematics, namely that the addition of more and more light quarks leads to more and more screening and a lower critical temperature in each case, seem plausible.

8

Physics of QCD at high temperatures and chemical potentials

8.1 The thermodynamic background

Let's begin our discussion of QCD in extreme environments that are rich in baryons by reviewing some basic concepts of thermodynamics. We want to remind the reader of the partition function, its relation to Euclidean field theory, and the physics and formalism of the chemical potential in simple settings. These discussions will set the stage for more-challenging applications to QCD.

The reader should appreciate that QCD is a remarkable, yet frustrating theory. On the one hand, we know the theory's "first principles" thoroughly, in the sense that we have a well-defined prescription for calculating any quantum-mechanical amplitude governed by strong interactions. The principles are simple and beautiful. However, the phenomena which result from application of these principles are very complicated. For example, asymptotic freedom indicates that a natural language, both for physical ideas and for quantitative estimates, for the short-distance behavior of QCD is that of weakly interacting quarks and gluons. For larger distances and times, this description becomes inadequate in several ways. First, the running coupling grows large and perturbative estimates are no longer reliable. Second, the assumption that physical quantities should be analytic functions of the coupling becomes untenable. Quarks and gluons are not weakly fluctuating degrees of freedom at these length scales. Other dynamical variables, perhaps instantons, color-bearing monopoles, and flux tubes, provide a more relevant description of the physics. These facts indicate that reliable calculations in QCD are beyond our existing mathematical abilities. Most discussions and applications of the theory, starting from the seminal phenomena of color confinement and dynamical mass generation, require physical assumptions inferred from experiments and cannot be derived from mathematical, defensible self-contained arguments. In this book, however, we deliberately limit ourselves to results that can be obtained using purely theoretical means with a minimum of phenomenological input.

As we turn to the properties of QCD in extreme environments, we shall be considering QCD in equilibrium under conditions characterized by finite densities of matter or energy. The theoretical language appropriate for these environments is that of thermodynamics. The primary mathematical object which encompasses all the thermodynamic information in the theory is the thermodynamic partition function. This function can be written as a path integral in QCD. The calculation of this integral from first principles is a very difficult mathematical task, and in the majority of cases it is impossible within our current abilities. There are limiting cases, for example QCD in the limit of large density, for which this calculation can be done, at least to leading order, in an expansion in a controllably small parameter – the effective strong coupling at the relevant energy scale. Another example is the limit of low density and small quark masses (the chiral limit), for which a controllable expansion in momenta and quark masses can be done – chiral perturbation theory.

Lattice simulations and calculations on a computer allow us, in principle, to calculate the partition function (or its derivatives) for any values of parameters. The QCD partition function is an infinite-fold integral that can be rendered finite by limiting the number of quantum-field degrees of freedom through an ultraviolet cutoff (the lattice spacing) and by an infrared cutoff (the lattice volume). Though the results obtained in this way are numerical, in many cases they serve as a basis for our intuition into the behavior of QCD in parameter regimes that are inaccessible analytically. The most basic phenomena, confinement and chiral-symmetry breaking, are prime examples. Lattice QCD simulations often play a role similar to an experiment. Analytic calculations based on controllable expansions as well as lattice calculations based on discretization of the partition function are often referred to as first-principles calculations.

8.1.1 Thermodynamic ensembles

In thermodynamics we do not attempt to describe the state of a system by specifying the values of all its quantum numbers. Rather, we allow the system freedom to occupy probabilistically any one of a set of microscopic states with given macroscopic properties. This set of states is refered to as a thermodynamic ensemble. Such a description is more appropriate for a system with a large number of degrees of freedom, in a situation in which we neither know, nor care to know, the precise behavior of every degree of freedom. Examples of quantities that characterize a system in a thermodynamic description are global, bulk quantities, such as the total volume, total energy, and total value of a conserved charge, such as the baryon number, for example. As long as the system is in a state with given values of such global quantities, we do not care which particular microscopic state it is in. In fact, in an experimental situation, it is even impossible to find out which microscopic state it is in, because of the vast number of degrees of freedom involved. In this situation we describe the state of the system by a probability distribution. At first glance there seems to be considerable ambiguity in the choice of the probability distribution. However, the power of thermodynamics is that this choice is largely insignificant, as long as we remain truly within our macroscopic, or thermodynamic, description.

The most straightforward choice of an ensemble is to pick all the states with given values of the global quantum numbers (energy, charges, etc.) and assign them all the same probability (to be more careful, one should consider a small interval around the given values of global quantities). This is the so-called *microcanonical* ensemble. It corresponds literally to a system that is completely isolated from the environment, so that no conserved quantum numbers in it can change (fluctuate).

Another ensemble, which is more convenient theoretically, is the so-called *canonical* ensemble, which was introduced by Gibbs, in which the energy is allowed to fluctuate in the following way: each state has a probability weight exponentially decreasing with its energy E: $\exp(-E/T)$. The parameter T is the temperature. The higher T, the greater the likelihood that the system is in a higher-energy, excited state. Now, what would the probability distribution for the energy of the system be in this case? This probability is proportional to the number of states with a given energy, $\Gamma(E)$, times the factor $\exp(-E/T)$:

$$P(E) \sim \Gamma(E)e^{-E/T}. \tag{8.1}$$

If the system has a large number of degrees of freedom, the number of states $\Gamma(E)$ should grow exponentially with the energy. This is indicated by simple combinatorics: the total energy is a sum of many small contributions, the number of which grows with the total energy. It should also grow with the total size of the system. As a result, we have two competing factors in Eq. (8.1), one that increases with E and one that decreases. Both factors are exponential, and the numbers in the exponents are large, being proportional to the total number of degrees of freedom. In this situation the probability distribution will be sharply peaked around a value of E at which the rates of change of these two factors balance each other:

$$\frac{d \ln \Gamma(E)}{dE} = \frac{1}{T} \tag{8.2}$$

The width of the peak, ΔE, grows only as the square root of the number of degrees of freedom. Thus the peak becomes relatively narrower and narrower, $\Delta E/E \to 0$, as the volume of the system, V, grows. In the thermodynamic limit, $V \to \infty$, the probability distribution is sharply peaked and is not significantly different from the microcanonical ensemble.

Physically, the canonical ensemble describes a system in a situation in which it is a small part of a huge system (which is called the "thermodynamic reservoir"), which in turn is isolated. The total energy of the huge system is constant, but its energy can be distributed between our subsystem and the rest. The probability of our subsystem having energy E will then be a product of the number of states $\Gamma(E)$ and the number of states of the rest of the huge system $\Gamma_{res}(E_{tot} - E)$. Since the energy of our system is a tiny fraction of the total energy of the reservoir, $E \ll E_{tot}$, we can expand $\ln \Gamma_{res}(E_{tot} - E)$ in powers of E and retain only the linear term. Thus we arrive at the distribution Eq. (8.1) with the temperature now dictated by the reservoir:

$$\frac{1}{T} = \frac{d \ln \Gamma_{res}(E_{tot})}{dE_{tot}} \tag{8.3}$$

The function of energy $\ln \Gamma(E) \equiv S(E)$ is the entropy of the system. We see from Eq. (8.2) that this function determines the relation between the energy of the system and the temperature which is required so that the mean (peak) energy of the canonical ensemble is equal to this energy,

$$\frac{1}{T} = \frac{dS}{dE} \tag{8.4}$$

Since the derivative $d \ln \Gamma_{res}(E_{tot})/dE_{tot} = dS_{res}(E_{tot})/dE_{tot} = 1/T_{res}$ defines the temperature of the reservoir, Eq. (8.3) indicates that the temperatures of the reservoir and of our system are equal – a familiar consequence of thermal equilibrium.

8.1.2 The partition function and Lagrangian

Typically, calculations are much easier using the canonical ensemble rather than the microcanonical ensemble. In particular, the partition function

$$Z \equiv \sum_{\text{all states}} e^{-E/T} \tag{8.5}$$

can be rewritten in terms of a path integral. The path-integral formulation makes it easier to set up perturbative expansions, on the one hand, and lattice discretizations and computer simulations, on the other.

The relation to the path integral arises if we write the partition function in the following way:

$$Z = \sum_{\alpha} \langle \alpha | e^{-H/T} | \alpha \rangle \tag{8.6}$$

where α labels states and, in principle, is a collection of values for all quantum numbers, while H is the Hamiltonian. In each term in the sum we recognize the amplitude of a quantum transition with the following peculiarities: (a) the time Δt during which this transition is allowed to take place is imaginary; $1/T = i\,\Delta t$, and (b) the final and the initial states are the same. Applying standard textbook techniques to rewrite these amplitudes as path integrals, we arrive at

$$Z = \int_{\alpha(0)=\alpha(1/T)} \mathcal{D}\alpha \, \exp\left(-\int_0^{1/T} d\tau \, \mathcal{L}_E\right) \tag{8.7}$$

where the integral in the exponent is over the imaginary time τ and the Euclidean Lagrangian $\mathcal{L}_E$ is obtained from the Lagrangian after substitution of time by $-i\tau$, i.e.

$$\mathcal{L}(t \rightarrow -i\tau) = -\mathcal{L}_E(\tau). \tag{8.8}$$

The minus sign in Eq. (8.8) makes the typical kinetic-energy term $(dq/dt)^2$ transform into a positive term $(dq/d\tau)^2$ in the Euclidean Lagrangian, protecting the path integral from being afflicted by swarms of wiggly trajectories. The boundary conditions on the trajectories in Eq. (8.7) reflect the condition that the initial and final states are the same. A significant advantage of the path-integral representation (8.7) is that the variables of integration may be commuting numbers, so these would be ordinary integrals. Of course there are infinitely many integration variables, so a reduction to a finite number of degrees of freedom, through a lattice's real-space cutoff, for example, is needed before the problem is ready for numerical methods. Another advantage is that it uses the Lagrangian rather than the Hamiltonian, which is often more convenient and powerful, as in the case of gauge theories. In addition, we have seen in earlier chapters that, even in the case of fermion systems in which anti-commuting variables are unavoidable, the path-integral approach is useful and practical for numerical methods.

8.1.3 Conserved charge and chemical potential

So far we have focused on a thermodynamic description that uses the energy as the only parameter that describes the thermodynamic state of the system. There are, however, other global quantities that are of interest to us, which we can and must use to define ensembles. Imagine that we want to describe a volume filled with gas. If we specify only the total energy contained in this volume, the system can achieve such a state by having either many particles with less energy per particle, or fewer but more energetic particles. Experimentally we can distinguish such situations easily (cold versus hot, dense versus dilute), so we would like to have a global parameter that can control this "dimension" in the space of the thermodynamic states the system may occupy. In our example the total number of particles is such a quantity. In this way we can construct an ensemble in which the total number of particles is fixed. Similarly to our discussion of the distribution of energies, we will have the freedom to choose the distribution of the particle number, which again will not affect the thermodynamic properties of a macroscopic system.

In nonrelativistic applications of thermodynamics, it is common to use the total number of particles to label an ensemble. However, in relativistic theories for environments in which particle–antiparticle creation is a common process, the number of particles need not be a quantity that we could fix by isolating the system – this number will still fluctuate due to pair creation and annihilation. For example, in a nonrelativistic plasma the number of electrons is practically constant. This is not true if the plasma is so hot that electron–positron pairs can be created. Even in the relativistic case, however, a conserved charge, such as the number of particles minus the number of antiparticles, can be defined. It does not fluctuate in a closed system and we can always use such a charge to define an ensemble of states with a given value of charge. Of course, in the nonrelativistic limit the charge simply counts the number of particles.

Following our logic in Section 8.1.1 closely, we can either consider the total charge, N, as fixed, or allow states with all values of the charge into the ensemble. This case is called the *grand canonical* ensemble. Here we weight the states according to their charge N with an exponential factor, $e^{\mu N/T}$. The quantity μ, which has the dimension of energy, is the chemical potential of the charge N. This nomenclature can be understood if we consider the chemical-potential factor together with the energy factor:

$$e^{-E/T} \times e^{\mu N/T} = \exp\left(-\frac{E - \mu N}{T}\right) \tag{8.9}$$

One can think that the chemical potential shifts the energy of a given state by an amount proportional to the charge of the state. If the charge is electric, μ plays the same role as electrostatic potential $-\phi$. Positive chemical potential μ favors states with larger positive total charge N.

As in the case of the canonical versus microcanonical ensemble, for which the distribution of the energy was sharply peaked, the distribution of charge is also sharply peaked in a grand canonical ensemble. The combined energy and charge distribution is given by

$$P(E, N) \sim \Gamma(E, N)e^{-E/T}e^{\mu N/T} \tag{8.10}$$

The number (or more properly, the density) of states at a given energy and charge $\Gamma(E, N)$ varies exponentially with changing E or N. CPT invariance tells us that S and Γ can depend only on the absolute value of N. Since we expect that the mean (i.e. peak) value of the charge is zero in a system without a chemical potential, the number of states $\Gamma(E, N)$ must have a peak at $N = 0$ and decrease at large N. The sharp peak in $P(E, N)$ appears when the variation of Γ with E

and N is balanced by the exponential factors in Eq. (8.10):

$$\frac{1}{T} = \frac{\partial S}{\partial E} \quad \text{and} \quad \frac{\mu}{T} = -\frac{\partial S}{\partial N} \tag{8.11}$$

where $S = S(E, N) = \ln \Gamma(E, N)$. As we have just discussed, S decreases with growing $|N|$. Thus the chemical potential has the same sign as the average value of N. We can think of the chemical potential as a parameter that controls the total charge in the system through Eq. (8.11).

The grand canonical ensemble corresponds to a situation in which our system can exchange charge, and energy, with the reservoir. The chemical potential and the temperature of our system are equal to those in the reservoir.

8.1.4 The grand canonical partition function and Lagrangian

The grand canononical partition function reads

$$Z = \sum_{\text{all states}} e^{-E/T} e^{\mu N/T} \tag{8.12}$$

This can also be written as

$$Z = \sum_{\alpha} \langle \alpha | e^{-H/T} e^{\mu N/T} | \alpha \rangle = \operatorname{tr} e^{-(H - \mu N)/T} \tag{8.13}$$

where N in this equation is the charge operator. Note that conservation of charge, i.e. commutativity of N with H, saves us from concerns about operator ordering in the exponent.

We can again use the correspondence between the amplitudes in Eq. (8.13) and Euclidean path integrals to derive the expression for Z in terms of the Lagrangian $\mathcal{L}_E$. How does μ enter? We can answer this question without performing an explicit calculation. As we have already noted, the chemical potential μ plays the same role as an electrostatic potential would have played were N just an electric charge. In addition, we know how the electric charge couples to the electrostatic potential in the Lagrangian. To keep the discussion generic, let us just say that the particles carrying the charge must be described by a *complex* field, which we shall call Ψ (either fermion or boson fields). The conservation of the charge in the Lagrangian description corresponds to a symmetry (by Noether's theorem). This (gauge) symmetry is the phase rotation of the complex field Ψ carrying that charge. The coupling of the chemical (or electrostatic) potential to such a charge in the Lagrangian is obtained by extending the time derivative of the field Ψ,

$$\frac{\partial}{\partial t} \Psi \to \left(\frac{\partial}{\partial t} + i\mu \right) \Psi \tag{8.14}$$

Another intuitively helpful (more mnemonic than rigorous) way to look at it is to note that the replacement

$$\Psi \to \Psi e^{i\mu t} \tag{8.15}$$

produces the same effect as Eq. (8.14), since, due to the invariance of the Lagrangian with respect to global (spacetime-independent) phase rotations of Ψ, the only effect of a time-dependent phase rotation is through the time derivatives. One can then consider the effect of μ as incrementing the frequency of the phase rotation of every state by an amount proportional to the number of Ψ quanta in it. This is equivalent to decreasing the energy of the state, according to the Schrödinger picture in which $|\alpha(t)\rangle = e^{-iEt}|\alpha(0)\rangle$. This is exactly what we want to achieve by introducing μ.[1]

An inquisitive reader might wonder whether the chemical potential could be completely eliminated from the Lagrangian in this way. If so, where does the μ dependence in the partition function in the Lagrangian path-integral formulation come from? Yes, the dependence on μ can be completely eliminated from the Lagrangian. Insofar as processes in the vacuum are concerned, this has to be so, since the origin of the energy scale can be chosen arbitrarily. Because the charge is conserved, the system will never know that the energy changes if the charge is changed. However, in the grand canonical ensemble the relative probability weights of different charge states composing the ensemble depend on μ and the chemical potential is observable. Correspondingly, the dependence on μ, although it can be eliminated from the Euclidean Lagrangian, is transferred to the boundary conditions on the fields in Eq. (8.7). The substitution (8.15), which in Euclidean time is

$$\Psi \to \Psi e^{\mu\tau} \tag{8.16}$$

has to be accompanied by a change in the boundary conditions on the fields in the path integral (8.7), since $\Psi(1/T) \to \Psi(1/T)e^{\mu/T}$. The latter product should be equal to $\Psi(0)$ in the case of bosons, and $-\Psi(0)$ in the case of fermions, as we shall discuss in more detail below.

---

[1] Another intuitive argument. Consider an infinitesimal variation $\delta\mu$ of μ. According to Eqs. (8.14) and (8.15), this is equivalent to an infinitesimal phase change of the field Ψ: $\epsilon = \delta\mu\, t$. The variation of the Lagrangian, to linear order in ϵ and up to total derivatives, should be equal to $j^\mu \partial_\mu \epsilon$, where j^μ is the Noether current, conservation of which is a consequence of the invariance with respect to the phase of Ψ (we are repeating a step from a proof of Noether's theorem). Thus $\delta L = j^0 \delta\mu$. The infinitesimal variation of the Hamiltonian should be equal, up to a sign, to the variation of the Lagrangian and, as we must expect, it is proportional to the density of the charge j^0.

8.1.5 Derivatives of the partition function

The partition function $Z(T, \mu)$ serves as a generating function for all other thermodynamic quantities. Consider some familiar examples. The logarithmic derivatives of Eq. (8.12) with respect to T and μ give the average energy and charge:

$$\langle E - \mu N \rangle = -T^2 \frac{\partial \ln Z}{\partial T}; \qquad \langle N \rangle = T \frac{\partial \ln Z}{\partial \mu} \qquad (8.17)$$

It is convenient to introduce the following *thermodynamic potential*:

$$\Omega(T, \mu) = -T \ln Z \qquad (8.18)$$

For a macroscopic system we can write Ω in the form

$$\Omega = -T \ln\left(\int dE \, dN \, e^S e^{E/T} e^{\mu N/T} \right) = -TS + E - \mu N \qquad (8.19)$$

since the integrand is sharply peaked. Therefore the entropy can be expressed in terms of Ω (and, therefore, in terms of the generating functional Z),

$$S = -\frac{\partial \Omega}{\partial T} \qquad (8.20)$$

A simple way to obtain this relation is to differentiate Eq. (8.19) with respect to T and note that the implicit dependences coming through E and N cancel out because E and N are extremal, i.e. Eqs. (8.11) hold.

At zero temperature, $\Omega = E - \mu N$ is the total energy, corrected by the chemical potential, of the ground state at given μ. The ground state is the state with the minimal value of $E - \mu N$. In general, if Ω depends on an extra parameter (e.g. an order parameter characterizing the structure, or the phase of the system), the value of this parameter in equilibrium will minimize Ω (i.e. the partition function is maximized).

The derivative of $-\Omega$ with respect to the volume produces the pressure

$$p = -\frac{\partial \Omega}{\partial V} \qquad (8.21)$$

For a macroscopic, translation-invariant system, Ω is proportional to the volume V. Therefore, $\partial \Omega / \partial V = \Omega / V$, and Eq. (8.21) implies that

$$\Omega = -pV \qquad (8.22)$$

Equation (8.19) also implies that the energy of the system is

$$E = TS + \mu N - pV \qquad (8.23)$$

which shows that the energy of a system can be increased by supplying heat (with N and V fixed), by adding particles (with S and V fixed), or by squeezing it (with S and N fixed).

8.1.6 An example: fermions

Consider a system of charged, massive noninteracting particles of spin $\frac{1}{2}$, obeying Fermi statistics. The spectrum of such a system is easy to understand by looking at each momentum state separately: each momentum state may be occupied or unoccupied. The difference in the energies between these alternatives is $\omega_{\vec{p}} = \sqrt{\vec{p}^2 + m^2}$. In addition, there is also a twofold degeneracy due to spin: each momentum state $\vec{p}$ can be occupied by a spin-"up" or "down" fermion. Finally, since the particle is charged, there are antiparticle states with the same momentum.

The partition function can be organized in a convenient form. First sum over the states of particles with spin up, and then take the fourth power to account for the same contribution from spin-down particle states and spin-up/down antiparticle states:

$$Z(T) = \prod_{\vec{p}} \left(1 + e^{-\omega_{\vec{p}}/T}\right)^4 \tag{8.24}$$

The chemical potential simply shifts $\omega_{\vec{p}}$ by $-\mu$ for particles and by $+\mu$ for antiparticles:

$$Z(T, \mu) = \prod_{\vec{p}} \left(1 + e^{-(\omega_{\vec{p}}-\mu)/T}\right)^2 \left(1 + e^{-(\omega_{\vec{p}}+\mu)/T}\right)^2 \tag{8.25}$$

The free energy and its derivatives can now be evaluated. For example,

$$N = -\frac{\partial\Omega}{\partial\mu} = 2\sum_{\vec{p}} \left(\frac{1}{e^{(\omega_{\vec{p}}-\mu)/T} + 1} - \frac{1}{e^{(\omega_{\vec{p}}+\mu)/T} + 1}\right) \tag{8.26}$$

has a simple meaning. The first term in the large parentheses is the average number of particles with momentum $\vec{p}$ (the occupation number of this mode), and the second is the average number of antiparticles. The factor of 2 accounts for spin. Note that the occupation numbers decrease exponentially with growing $\omega_{\vec{p}}$, because of the penalty imposed by the Gibbs factor $e^{-E/T}$. When the occupation numbers are small, quantum statistics has no effect, and we shall see below that the Bose particle occupation numbers are the same in this regime. When $\mu = 0$ the average charge N is zero at any temperature.

When the temperature is low compared with μ, the occupation number has the profile of a step function when it is plotted against $\omega_{\vec{p}}$, as shown in Fig. 8.1. This means that modes with $\omega < \mu$ are almost always occupied, whereas the modes with $\omega > \mu$ are almost always empty. At zero temperature, there are no fluctuations, and the ground state of the system at finite μ is very simple: all momentum states within the Fermi surface, $\omega_{\vec{p}} < \mu$, are filled with particles and all momentum states outside the Fermi surface are empty. In our isotropic

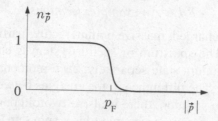

Figure 8.1. The occupation number $n_{\vec{p}}$ of momentum modes (the expression in large parentheses in Eq. (8.26)) of a very cold Fermi gas, $T \ll \mu$. The width of the crossover region from 1 to 0 is of order T/v_F (v_F is the velocity of fermions on the Fermi surface).

example, the Fermi surface is a sphere with radius (Fermi momentum) $p_F = \sqrt{\mu^2 - m^2}$. There are no antiparticles in this ground state.

In the infinite-volume limit the sum over the momentum modes becomes an integral over phase space:

$$\sum_{\vec{p}} \to V \int \frac{d^3\vec{p}}{(2\pi)^3} \tag{8.27}$$

where the factor of $V/(2\pi)^3$ is the density of momentum modes per unit cell in momentum space. At zero or very small T, the density of the charge is therefore proportional to the volume of the Fermi sphere:

$$n \equiv \frac{N}{V} = 2\frac{4\pi p_F^3}{3} = \frac{8\pi}{3}(\mu^2 - m^2)^{3/2} \tag{8.28}$$

Note that, for $\mu < m$, the density of the charge n is zero. There is no Fermi sphere in this case. This is a reflection of a generic property of the partition function at zero T which states that the ground state of the system is the state having the minimal value of the potential $\Omega = E - \mu N$. This state exponentially dominates the sum defining the partition function in the $T \to 0$ limit. At $\mu = 0$ it is the state of lowest energy, the vacuum. At finite μ the ground state cannot change unless another state, with a finite value of charge N can compete for the lowest $E - \mu N$. The ground state will not change unless μ exceeds a value, μ_0, which is the minimum energy per unit charge among all charged states:

$$\mu_0 = \min\left(\frac{E}{N}\right) \tag{8.29}$$

In our example, $\mu_0 = m$, and the state which competes with the vacuum and becomes the ground state is the one with a single fermion at rest: $E/N = m/1$.

In fact, at $\mu = m$ there are two such states due to spin. At higher μ, the degeneracy of states grows rapidly due to the isotropy of momentum space in this example. Adding or removing a fermion on the Fermi sphere $|\vec{p}| = p_F$ does not cost $E - \mu N$. This degeneracy is crucial to the phenomenon of superconductivity. Basic quantum mechanics tells us that even an arbitrarily small perturbation, which has matrix elements between two degenerate states, will have a significant (nonperturbative) effect due to mixing of these two states. In our case, an arbitrarily small attractive interaction will cause a nonperturbative rearrangement of the ground state. The ground state will become a coherent superposition of states with an arbitrary number of fermion pairs of opposite momenta $|\vec{p}| = p_F$, Cooper pairs, added to (or subtracted from) the filled Fermi sphere. This coherent state, the Cooper-pair condensate, is responsible for superconductivity according to the BCS theory. We shall discuss how this phenomenon is realized explicitly in the context of QCD in Section 9.1.

The path-integral (Lagrangian) formulation

We can also calculate the partition function using the path-integral formulation. For the free case it is somewhat more difficult than the combinatoric method, but it is indispensible once interactions become important. It allows us to develop perturbation theory and to set up lattice formulations and simulations. Our free massive fermion should be described by the Dirac Lagrangian, and, including the chemical potential using Eq. (8.14), we have

$$\mathcal{L} = -\bar{\Psi}(i\gamma^\mu \partial_\mu - \gamma^0 \mu - m)\Psi \qquad (8.30)$$

where Ψ is a four-component Dirac spinor, and $\bar{\Psi} = \Psi^\dagger \gamma^0$. Note that the term $\mu\bar{\Psi}\gamma_0\Psi$ is the density of the charge multiplied by the chemical potential, as it should be. The next step is to obtain the Euclidean Lagrangian by substituting $t \to -i\tau$ in Eq. (8.30):

$$\mathcal{L}_E = -\mathcal{L}\big|_{t \to -i\tau} = \bar{\Psi}(\gamma_\mu^E \partial_\mu - \gamma^0 \mu - m)\Psi \qquad (8.31)$$

where we have introduced the notation γ_μ^E for the "Euclidean" Dirac gamma-matrices:

$$\gamma_0^E = \gamma^0, \qquad \gamma_k^E = i\gamma^k \qquad (8.32)$$

All four matrices γ_μ^E are now Hermitian and their anti-commutation relations are also Euclidean-space covariant, $\{\gamma_\mu, \gamma_\nu\} = \delta_{\mu\nu}$. In the following we shall very often omit the index E from the Euclidean matrices, if it is clear from the context that we are working in Euclidean space.

Now we can write the path integral:

$$Z(T, \mu) = \int \mathcal{D}\Psi \exp\left(-\int_0^{1/T} d\tau \int d^3\vec{x}\, \bar{\Psi}(\gamma_\mu^E \partial_\mu - \gamma^0 \mu - m)\Psi\right) \qquad (8.33)$$

The short-hand notation $\int \mathcal{D}\Psi$ hides two important properties of this integral. First, the components of the spinor Ψ are not complex numbers, but are Grassmann numbers (numbers that anti-commute with each other). We shall see that one can always reduce the integral over Grassmann variables to an ordinary Gaussian integral. This will rely on the fact that, even in the interacting case, the Lagrangian is bilinear in fermion fields, or can be reduced to a bilinear by introducing auxiliary bosonic fields. All we need to know about the integrals over Grassmann numbers is that the Gaussian integral is equal to the determinant of the matrix of the bilinear form appearing in the exponent. For our integral in Eq. (8.33) we obtain the determinant of the Dirac operator:

$$Z(T, \mu) = \det\left[\gamma_\mu^E \partial_\mu - \gamma^0 \mu - m\right] \tag{8.34}$$

The second property that our notation in Eq. (8.33) hides is that the boundary conditions on the Grassmann field Ψ are anti-periodic,

$$\Psi(1/T) = -\Psi(0) \tag{8.35}$$

which replaces the periodic boundary conditions for bosonic fields. We can think of this property as a rule that is required in order to obtain the correct partition function for fermions. It correctly reproduces the effects of Fermi statistics.

The integral in Eq. (8.33) is a functional integral and the determinant in Eq. (8.34) is a functional determinant. These quantities are not properly defined until they are regularized and their ultraviolet infinities have been removed. For example, lattice discretization would make the Dirac operator a finite dimensional matrix, whose eigenvalues can be computed by a variety of methods, some numerical.

In the free case we can calculate this determinant analytically. To start, we need to diagonalize the Dirac operator, which can be done in the free case using a Fourier transform. The eigenvalues of a linear combination of Dirac matrices are also easy to find: $a_\mu \gamma_\mu^E = \pm\sqrt{a_\mu a_\mu}$, i.e. two positive and two negative eigenvalues. Thus we find

$$Z = \prod_{\vec{p}} \prod_{p_0} [(p_0 - i\mu)^2 + \vec{p}^2 + m^2]^2 \tag{8.36}$$

The values of $\vec{p}$ are continuous in the $V \to \infty$ limit. However, at finite T, the values of p_0 are discrete: $p_0 = \pm 2\pi T n_{1/2}$, where $n_{1/2} = \frac{1}{2}, \frac{3}{2}, \ldots$, due to the boundary conditions (8.35). These are called Matsubara frequencies. We can take the product over p_0. Let us write

$$Z = \prod_{\vec{p}} \prod_{p_0} (p_0 - i\mu + i\omega_{\vec{p}})^2 (p_0 - i\mu - i\omega_{\vec{p}})^2 \tag{8.37}$$

where $\omega_{\vec{p}}^2 = \vec{p}^2 + m^2$. We can now use the formula[2]

$$\prod_{\pm n_{1/2}} \left(1 + \frac{ia}{n_{1/2}}\right) = \cosh(\pi a) \tag{8.38}$$

to obtain

$$Z = \prod_{\vec{p}} e^{2\omega_{\vec{p}}/T} \left(1 + e^{-(\omega_{\vec{p}}-\mu)/T}\right)^2 \left(1 + e^{-(\omega_{\vec{p}}+\mu)/T}\right)^2 \tag{8.39}$$

up to a constant (infinite) factor. The first exponent on the right-hand side is the contribution of the zero-point energy of each mode, equal to $-\omega_{\vec{p}}/2$, the analog of the zero-point energy for bosons, but with the opposite sign. It is the contribution of the filled Dirac sea, which can be removed by the shift of the origin of the energy. As expected, the final result agrees with Eq. (8.25).

A cold nonideal Fermi gas
The chemical potential has some interesting effects on the ground state of the *interacting* Fermi gas. We briefly discuss two: restoration of chiral symmetry and BCS superconductivity. We shall try to keep our discussion here as generic as possible.

Consider a cold Fermi gas with $\mu > m$. Consider what would happen if the mass of the fermion were an adjustable parameter. How much would be gained in the value of the potential $E - \mu N$ by changing the mass?

$$\Omega_\mu(m) = E - \mu N = 2 \int_{|\vec{p}|<p_F} \frac{d^3\vec{p}}{(2\pi)^3} (\omega_{\vec{p}} - \mu) = \frac{\mu^4}{\pi^2} f\left(\frac{m}{\mu}\right) \tag{8.40}$$

where the dependence on m is in the dimensionless function

$$f(x) = \int_0^{\sqrt{1-x^2}} t^2(\sqrt{t^2 - x^2} - 1)\, dt, \tag{8.41}$$

The integral can be computed easily, but for our discussion we need only some of its qualitative features, which can be seen in Fig. 8.2. Decreasing the mass is advantageous because it decreases the potential Ω_μ. Qualitatively, the radius of the Fermi sphere $p_F = \sqrt{\mu^2 - m^2}$ is larger for smaller m and each extra fermion contributes negatively to $E - \mu N$.

---

[2] A way to derive this formula is to take logarithmic derivatives with respect to a in Eq. (8.38) and then consider the sum rule for the residues of the following function:

$$\sum_{\text{res}} \frac{1}{x^2 + a^2} \tan(\pi x) = 0$$

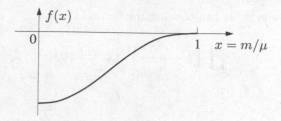

Figure 8.2. The dependence of the potential $\Omega_\mu = E - \mu N$ on the mass at fixed μ: $\Omega_\mu = (\mu^4/\pi^2) f(m/\mu)$.

In fact, to the extent that we understand the origin of fermion masses in our world, the mass is always a dynamical quantity. For example, in the electroweak theory, the mass of the electron is proportional to the expectation value of a dynamical field, the Higgs field. The masses of the baryons in QCD are also of dynamical origin. In other words, chiral-symmetry breaking, which is a necessary companion of fermion-mass generation, occurs via some dynamical mechanism. In the vacuum the value of the mass is a minumum of a certain effective potential. The curvature of this potential depends on the scale of the physics responsible for mass generation. As an example, we can consider a Yukawa theory with small coupling y and an expectation value of the scalar field (the analog of the Higgs) given by σ. The mass of the fermion is then given by

$$m = y\sigma \tag{8.42}$$

The effective potential for the mass of the fermion, $\Omega_m(m)$, is then equal to the effective potential for the Higgs/sigma field $\Omega_\sigma(\sigma)$. Thus the curvature of $\Omega_m(m)$ is proportional to the mass of the Higgs/sigma field:

$$\frac{d^2\Omega_m(m)}{dm^2} = \frac{\partial^2\Omega_\sigma(\sigma)}{y^2 \, \partial\sigma^2} = \frac{m_\sigma^2}{y^2} \tag{8.43}$$

As we have seen above, the chemical potential $\mu > m$ puts "pressure" on the mass to decrease since that decreases Ω_μ. This "pressure" is countered by the dynamical mechanism of mass generation, since it costs energy to shift the mass away from the minimum of Ω_m. The balance is achieved at a smaller value of m, which decreases with growing μ. This new value is given by the minimum of $\Omega_m(m) + \Omega_\mu(m)$ as shown in Fig. 8.3. The size of this effect depends on the curvature of Ω_m. For example, the mass of the electron, given by the Higgs mechanism at a scale of order $100\,\text{GeV}$, will not be moved by any chemical potentials we could hope to achieve. For our practical purposes, we can think of this mass as a constant. However, the mass of a baryon is generated by a mechanism at a scale of the order of $1\,\text{GeV}$, and such chemical potentials are achievable, if not in a laboratory, then in the cores of neutron stars.

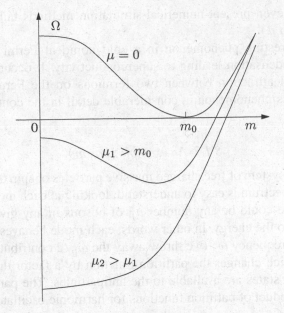

Figure 8.3. The potential $\Omega = \Omega_m + \Omega_\mu$ for the dynamically generated fermion mass for various values of μ. The vacuum value of the mass is m_0. At $\mu = \mu_2$ the mass vanishes.

If μ is sufficiently large, the mass will eventually reach zero, as shown in Fig. 8.3. Chiral symmetry will then be restored. This can happen due to either a second-order (smooth) or first-order (abrupt) transition. The order of the transition depends on the shape of the potential Ω_m. Typically, such a potential has the shape of a double well. If the (negative) curvature of Ω_m at the origin is not large, the potential $\Omega_m + \Omega_\mu$ may develop a second minimum at the origin, which may become lower than the minimum at $m > 0$, causing a first-order transition to occur. We can estimate the critical value of μ in the regime where the Yukawa coupling y is very small. In this regime the transition is of second order since the curvature of Ω_m at the origin is large, proportional to the curvature in the minimum (8.43). The transition occurs when the curvature of Ω_μ at $m = 0$ overcomes the negative curvature of the potential Ω_m at the origin. The latter is proportional to μ^2, while the former is of order m_σ^2 / y^2. Thus

$$\mu_c \sim m_\sigma / y \qquad (8.44)$$

Although this formula cannot be applied for moderate values of y, we can see, on a qualitative level, that, for a theory with $y \sim 1$, we might expect a transition at a value of μ of the order of the mass-generation scale m_σ. In QCD, this scale is of the order of 1 GeV. The transition leads to the restoration of chiral symmetry. It happens, however, in a region of parameters where no controllable calculational methods exist (even present numerical-simulation methods fail in the case of QCD at nonzero μ).

Another interesting phenomenon in a cold nonideal Fermi gas is Cooper pairing and condensation leading to superconductivity. It occurs if there is an arbitrarily small attraction between two fermions on the Fermi surface. We shall discuss this phenomenon in considerable detail in the context of QCD in Section 9.1.

8.1.7 An example: bosons

Consider now a system of free charged massive particles of spin 0, obeying Bose statistics. The spectrum is easy to understand, looking at each momentum mode separately. There could be any number $n_{\vec{p}}$ of bosons in any given mode, contributing $n_{\vec{p}}\omega_{\vec{p}}$ to the energy. In other words, each mode behaves as a harmonic oscillator with frequency $\omega_{\vec{p}}$ (we throw away the $\omega_{\vec{p}}/2$ contribution to the vacuum energy, which changes the partition function by a factor that is irrelevant here). The same states are available to the antiparticles. The partition function factors into a product of partition functions for harmonic oscillators,

$$Z(T) = \prod_{\vec{p}} \left(\frac{1}{1 - e^{-\omega_{\vec{p}}/T}} \right)^2 \tag{8.45}$$

A chemical potential shifts the frequency of each oscillator by $-\mu$ for particles and by μ for antiparticles,

$$Z(T, \mu) = \prod_{\vec{p}} \frac{1}{1 - e^{-(\omega_{\vec{p}} - \mu)/T}} \frac{1}{1 - e^{-(\omega_{\vec{p}} + \mu)/T}} \tag{8.46}$$

Now we can calculate various derivatives. In particular,

$$N = -\frac{\partial \Omega}{\partial \mu} = \sum_{\vec{p}} \left(\frac{1}{e^{(\omega_{\vec{p}} - \mu)/T} - 1} - \frac{1}{e^{(\omega_{\vec{p}} + \mu)/T} - 1} \right] \tag{8.47}$$

The total charge is the difference between the number of particles and the number of antiparticles. The mean occupation numbers of high-frequency modes vanish exponentially when $\omega_{\vec{p}} \gg T$. As we expected, there is no difference between fermion and boson occupation numbers at large $\omega_{\vec{p}}/T$. Of course, $N = 0$ when $\mu = 0$.

Unlike in the free-fermion case, there is a problem when $|\mu| > m$. This can be seen already in the partition function. The sum of probability weights $e^{-n(\omega - \mu)/T}$ over n for modes with $\omega < \mu$ fails to converge. Thus the value of μ for the free gas must be restricted to $|\mu| < m$.[3]

---

[3] In nonrelativistic applications the rest energy of a particle is subtracted from the chemical potential, $\mu_{nr} = \mu - m$. For a free nonrelativistic ideal Bose gas $\mu_{nr} < 0$.

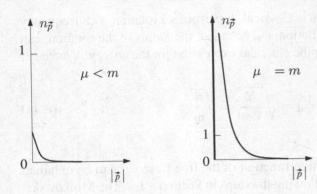

Figure 8.4. Occupation numbers of momentum modes in a cold ideal Bose gas without ($\mu < m$) and with ($\mu = m$) Bose condensation. The occupation of the mode $\vec{p} = 0$ in the latter case is given by $N - N_T$, as discussed in the text.

It is interesting to compare the free Bose gas at zero T with the free-fermion gas. Similarly to the case of fermions, as a consequence of a generic argument, no change in the ground state can occur unless μ (or $-\mu$) exceeds m. However, μ cannot be taken larger than m for the free Bose gas, as we have just noted. So how do we describe a state of a fixed number N of free Bose particles? The character of this state at $T = 0$ is not difficult to understand. Whereas fermions would fill all available momentum states from zero to the Fermi surface in order to minimize the energy at fixed N, all N bosons would simply occupy the single lowest state of $\vec{p} = 0$.

Consider now Eq. (8.47) at finite T. The value of N in Eq. (8.47) increases as μ increases toward its limiting value m. On transforming the sum over $\vec{p}$ into the momentum-space integral (8.27), we can calculate this sum in the limit $V \rightarrow \infty$ and find that it reaches a finite T-dependent value as $\mu \rightarrow m$. For small temperatures $N_T = \text{constant} \times V(Tm)^{3/2}$. Thus $N_T = 0$ at $T = 0$. This means that the integral does not account for the particles in the ground state. What if we consider an ensemble with fixed N (a canonical ensemble)? As long as $N < N_T$ we can find the corresponding μ, which could reproduce this given value of N in the grand canonical ensemble. At $N = N_T$, the value of μ is m. Further increase of N proceeds by putting those extra $N - N_T$ bosons into the $\vec{p} = 0$ state. This process is called Bose condensation. It occurs at fixed T, if the density $n = N/V$ exceeds the critical density $n_T = N_T/V$; or at fixed density, if the temperature is low enough that $n_T < n$. The occupation numbers in a cold Bose gas are illustrated in Fig. 8.4.

When Bose condensation occurs, the occupation number of the $\vec{p} = 0$ mode becomes macroscopic. We can consider this mode as an oscillator with frequency $\omega_{\vec{p}=0} = m$. Since the field ϕ is complex, the oscillator is two-dimensional with O(2) symmetry. Its energy Nm is so much bigger than the level spacing m that its motion is classical. It performs a rotation with frequency m. The amplitude of this oscillation, ϕ_0, or rather, the radius of the rotation, can easily be found by comparing the classical expression for the energy, $Vm^2|\phi_0|^2$, to its quantum value, Nm:

$$|\phi_0| = \sqrt{\frac{N}{Vm}} = \sqrt{\frac{n}{m}} \qquad (8.48)$$

The path integral

The path integral for the partition function of the free Bose gas can be obtained by rather straightforwardly following the steps in Section 8.1.4. The Minkowski-space Lagrangian[4]

$$\mathcal{L} = \phi^*[-(\partial_0 + i\mu)^2 + \vec{\partial}^2 - m^2]\phi \qquad (8.49)$$

leads to the Euclidean Lagrangian,

$$\mathcal{L}_E = \phi^*[-(\partial_\tau - \mu)^2 - \vec{\partial}^2 + m^2]\phi \qquad (8.50)$$

The partition function is given by the path integral

$$Z(T, \mu) = \int \mathcal{D}\phi \, \exp\left(-\int_0^{1/T} d\tau \int d^3x \, \mathcal{L}_E\right) \qquad (8.51)$$

The boundary conditions on the field configurations in the integral are periodic:

$$\phi(1/T) = \phi(0) \qquad (8.52)$$

The field ϕ is a complex number at every point in space and the integrals are taken over its real and imaginary parts (two integrals per point in space). Since the path integral in Eq. (8.51) is Gaussian, it can be evaluated easily,

$$Z(T, \mu) = \det^{-1}\left[-(\partial_\tau - \mu)^2 - \vec{\partial}^2 + m^2\right] \qquad (8.53)$$

The power of the determinant is twice the usual inverse square root obtained from a Gaussian integration because ϕ is complex.

The determinant can be calculated in the same way as the determinant for the fermions. The Matsubara frequencies are equal to $p_0 = 2\pi T n$, where n is now an integer. A helpful trick is to realize that shifting μ by the value $i\pi T$ is equivalent to the needed $2\pi T \times (\frac{1}{2})$ shift of the Matsubara frequences; see Eq. (8.37). Thus we can just pick the final result for the fermions (8.39) and adjust μ, as well as the total power (by a factor of $-\frac{1}{2}$),

$$Z = \prod_{\vec{p}} e^{-\omega_{\vec{p}}/T}\left(1 - e^{-(\omega_{\vec{p}}-\mu)/T}\right)^{-1}\left(1 - e^{-(\omega_{\vec{p}}+\mu)/T}\right)^{-1} \qquad (8.54)$$

---

[4] We added a total derivative to the more common form of the Lagrangian: $|(\partial_0 + i\mu)\phi|^2 - |\vec{\partial}\phi|^2 - m^2$.

which agrees with Eq. (8.46), up to the contribution of the zero-point energies, which were not included in Eq. (8.46).

Note that at $\mu = m$, the constant Fourier mode of the field ϕ has zero curvature in the Lagrangian (8.50). The integral diverges if $\mu = m$ or greater. At $\mu = m$ we can think of the $\phi = $ constant mode as completely unconstrained by the Lagrangian (8.50). This reflects the phenomenon of Bose condensation discussed above.

A cold nonideal Bose gas

Consider the effect of a small interaction on the ground state $(T = 0)$ of the Bose gas. If the interaction is attractive, the system of a macroscopic number of bosons will collapse, i.e. the density will increase, until some repulsive interaction intervenes.

A more interesting case is a weakly repulsive Bose gas. Unlike for the ideal Bose gas, it is now possible to increase μ above m. The number N (or the density) of the particles in the $\vec{p} = 0$ mode will now be controlled by the repulsion. It is easy to derive the leading term in the dependence of N on $\mu - m$. Consider the interaction term $-\lambda|\phi|^4$ in the Lagrangian density, where small and positive λ provides weak (hard core) repulsion. Set $\mu = m$. Without interaction, any occupation number N of the $\vec{p} = 0$ mode will give the same $E - \mu N$. There is a degeneracy of the ground state at $\mu = m$, which, as in the case of fermions, invalidates naive perturbation theory. With the repulsive interaction present, we can calculate the first correction to the energy of the ground state: $E_{\mathrm{I}} = V\lambda|\phi_0|^4$, where ϕ_0 is the amplitude of the $\vec{p} = 0$ mode of the field. As we found above in Eq. (8.48), $|\phi_0| = \sqrt{N/(Vm)}$. Thus we find that the energy now depends on N in the following way:

$$E = mN + E_{\mathrm{I}} = mN + \frac{\lambda}{Vm^2}N^2 \qquad (8.55)$$

Setting the chemical potential to a value $\mu > m$ can balance the energy cost at a given N if we choose μ to satisfy

$$\frac{\partial}{\partial N}(E - \mu N) = \frac{\partial}{\partial N}\left(mN + \frac{\lambda}{Vm^2}N^2 - \mu N\right) = 0 \qquad (8.56)$$

Thus we obtain the relation between N and μ valid for small densities and shown in Fig. 8.5:

$$\frac{N}{V} = \frac{m^2}{2\lambda}(\mu - m), \qquad \mu > m \qquad (8.57)$$

The density grows linearly with the chemical potential. The limit $\lambda = 0$ is singular. The density blows up if μ exceeds m unless $\lambda > 0$.

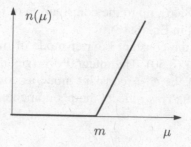

Figure 8.5. The density $n = N/V$ of a cold Bose gas with weak repulsion.

In contrast to the case of fermions, an interaction is required in order to prevent collapse of the Bose gas for $\mu > m$. In the case of fermions this role is played by the exclusion principle, without any interaction, as seen in Eq. (8.28).

8.2 Hadron phenomenology and simple models of the transition to the quark–gluon plasma

It is interesting to make a very simple model of the high-temperature transition to the quark–gluon plasma based on just the composite character of the colorless hadronic states of QCD and counting degrees of freedom [51]. This discussion is probably more useful and reliable at high temperature than at high chemical potential, but we shall apply it equally over the whole phase diagram in the T–μ plane. We will ignore subtleties expected in the physics at low T and high μ just for the sake of orientation. The fact that the μ axis of the phase diagram might be a wealth of physics in phase transition, superconductivity, and exotic states will be addressed below. For the moment just imagine that the world of the T–μ plane is relatively simple (and maybe relatively boring) and that there is a single transition line separating hadronic matter from the quark–gluon plasma, as shown in Fig. 8.6. We shall see that including just knowledge of the number of relevant degrees of freedom in each phase as well as some phenomenology about the formation of hadronic bound states in the low-T and low-μ phase predicts a first-order phase-transition line with very substantial discontinuities.

An underlying assumption in this model is that interactions are weak in an environment of high temperature and/or high chemical potential. Asymptotic freedom suggests this point. For example, at high T, average momentum transfers will be of the order of T, so the relevant running coupling should be characterized by the energy scale T and should be logarithmically small. Similarly, at high chemical potential, the quark Fermi surfaces are high, so μ should characterize their momentum transfers and interactions should again be logarithmically weak. Under closer inspection, we will find limitations in both of these arguments. For example, in an environment of high μ, quarks can engage in

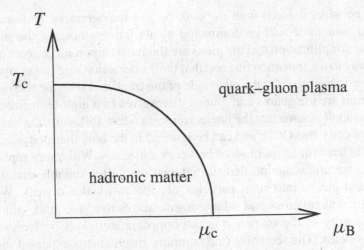

Figure 8.6. The phase diagram of the simple two-component model of QCD.

glancing collisions and these need not involve high momentum transfers. Since sharp Fermi surfaces are unstable against weak, residual interactions, we expect nonperturbative physics to occur even if the coupling is small. Perturbation theory and free-field behavior will give poor estimates and qualitatively wrong physical pictures in these cases.

Some limitations on the applicability of perturbative estimates can be read off expressions for the perturbative quark and gluon masses. For example, the perturbative gluon mass is

$$m_g^2 = N_f \frac{g^2 \mu^2}{6\pi^2} + \left(N_c + \frac{N_f}{2}\right)\frac{g^2 T^2}{9} \tag{8.58}$$

where N_c is the relevant number of colors and N_f is the relevant number of flavors. Certainly perturbation theory is not reliable once $m_g \sim T$ in an environment of vanishing μ. However, this occurs when $\alpha_s \sim 0.18$, a tiny value. We learn that pertubative estimates will be of very limited use at high T. In contrast, at nonvanishing μ and $T = 0$, we find $m_g \sim \mu$ when $\alpha_s \sim 2.40$. This suggests that perturbative estimates are much more valuable in dense quark matter. Of course, this argument overlooks the special kinematics of fermions in the vicinity of a spherical Fermi surface and it ignores nonstatic effects that are unaffected by the Debye screening included in Eq. (8.58). These failures will prove crucial in our discussions of QCD in dense environments.

With these caveats, let's jump into a simple two-component model. Armed with a broarder perspective, we hope that its failures will teach us as much as its successes.

First consider the axis with $\mu_B = 0$. At low temperatures, hadronic matter's thermodynamics should be dominated by its lightest bosons, the pions. Let's make the simplification that the pions are the most important degrees of freedom all the way to the transition line and that their interactions are not as significant as their kinetic energies. On the other side of the transition line, the relevant degrees of freedom are the quarks and gluons themselves in a high-temperature plasma state. We shall assume that the interactions are weak, following the suggestion of asymptotically free QCD, and can be ignored in the bulk thermodynamics of the system, at least for the purpose of ball-park estimates. With more sophistication, perhaps, we are assuming that the interactions, after suitable dressing of the quarks and gluons into quasi-particles, are just residual and weak. We will see from lattice simulations and other considerations that the quark–gluon plasma in the vicinity of the transition is not approximated well by free, essentially massless fields. The screening and damping mechanisms expected in a plasma are certainly at work in this one. Only at truly asymptotic temperatures, if at all, will we be able to approximate the system by free fundamental fields. With all these simplifications, the description of the transition will reduce to an exercise in counting.

The energy density and pressure of the gas of massless pions are

$$\epsilon = 3 \cdot \frac{\pi^2}{30} T^4, \qquad P = 3 \cdot \frac{\pi^2}{90} T^4 \tag{8.59}$$

where the factor of three counts the three types of pions.

The energy density and pressure of the gas of the quark–gluon plasma are estimated similarly,

$$\epsilon = 37 \cdot \frac{\pi^2}{30} T^4 + B, \qquad P = 37 \cdot \frac{\pi^2}{30} T^4 - B \tag{8.60}$$

The number 37 counts the effective number of degrees of freedom including gluons and quarks. For gluons there are eight colors and two spin states, and for quarks there are three colors, two spins, two flavors, and particles and antiparticles. So we find $37 = 2 \times 8 + (\frac{7}{8} \times 2 \times 2 \times 2 \times 3)$. The quantity B, which adds to the energy density and subtracts from the pressure, accounts in an average fashion for interactions that are responsible for the change in the vacuum structure between the low- and high-temperature phases. The constant B was originally introduced in the context of the bag model of hadronic structure which, in one version, envisioned protons as consisting of three essentially massless, confined quarks in a bubble of chirally symmetric vacuum in an otherwise chirally asymmetric environment. For example, the energy of a bag that contains quarks reads

$$E(R) = \frac{4\pi}{3} R^3 B + \frac{C}{R} \tag{8.61}$$

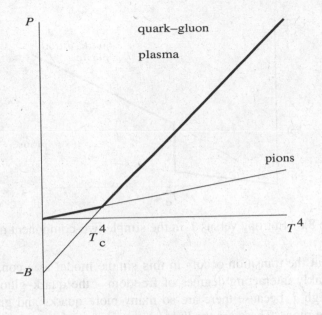

Figure 8.7. Pressure versus T in the simple two-component model of QCD.

where the first term is the bulk energy that accounts for a bubble of radius R of symmetric vacuum and the second term records the energy of the confined, massless quarks in the bubble. The constant C depends on the number of quarks in the particular state of interest. If one considers the proton and minimizes the total energy with respect to R, one finds a proton radius of $R_0 \approx 0.7$ fm and a bag constant $B \approx 175$ MeV fm^{-3}, or $B^{1/4} \approx 192$ MeV. In the equation above for the energy and pressure of the high-temperature vacuum, we now see that B adds to the energy, accounting for the presence of the chirally symmetric phase, and subtracts from the pressure.

We can now compare the pressures in the two phases as a function of the temperature and decide which is the stable state. The relevant plot is shown in Fig. 8.7 and some simple algebra gives the critical temperature in terms of the bag constant,

$$T_c = \left(\frac{45}{17\pi^2} \right)^{1/4} B^{1/4} \approx 0.72 B^{1/4} \tag{8.62}$$

We learn that, for temperatures above T_c the quark–gluon plasma is favored thermodynamically over the pion gas because it has the larger pressure. The numbers work out rather well. For $B^{1/4} \approx 200$ MeV, we have $T_c \approx 150$ MeV, which is near the numerical results coming from large-scale simulations of lattice QCD.

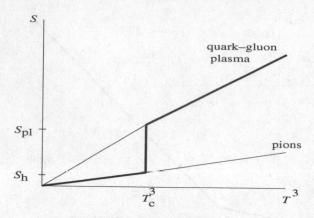

Figure 8.8. Entropy versus T in the simple two-component model.

We see that the transition occurs in this simple model as a consequence of counting weakly interacting degrees of freedom – the quark–gluon plasma is favored at high T because there are so many more quarks and gluon species than there are pions. These are the lightest modes in each phase, and should be the most significant thermodynamically.

This point is made especially well by considering the entropy, $S = \partial P/\partial T$, as a function of T, as shown in Fig. 8.8. The bag constant B does not enter the entropy S, but just affects the estimate of T_c. The jump in the entropy at T_c is directly proportional to the change in the number of light degrees of freedom relevant to the transition at T_c.

Next we want to generalize these considerations into the $T-\mu_B$ plane. This requires additional assumptions, which are almost certainly too naive. Anyway, let's try to obtain a zeroth-order expectation for the phase boundary between the hadronic and quark–gluon-plasma phases throughout the whole $T-\mu_B$ plane. Begin by noticing that the total pressure at T_c is very small in the finite-T considerations above. To a good approximation the transition occurs when the kinetic pressures of the quarks and gluons match the bag constant B. If we take $P = 0$ as the condition for the transition, then our estimate for T_c changes from the value $0.72B^{1/4}$ obtained above accounting for the pions in the hadronic phase to $[90/(37\pi^2)]^{1/4}B^{1/4} \approx 0.70B^{1/4}$. This "success" encourages us to try this simple criterion throughout the $T-\mu_B$ plane. The kinetic pressure of the quarks and gluons at nonzero T and μ_B reads

$$P(\mu_B, T) = \frac{37}{90}\pi^2 T^4 + \frac{\mu_B^2}{9}\left(T^2 + \frac{\mu_B^2}{9\pi^2}\right) \qquad (8.63)$$

Setting this quantity to B, $P(\mu_c, T_c) = B$, predicts a line of transitions with the reasonable shape of Fig. 8.6.

This simple model certainly suggests that counting degrees of freedom plays an important role in the physics of the transition between hadronic matter and the quark–gluon plasma. However, it is not reliable for predicting the order of the transition . . . a first-order transition is almost impossible to avoid in this sort of scenario and we know that the order of the transition in real QCD depends sensitively on the quark content of the theory . . . and it certainly misses a great deal of physics on and near the μ_B axis. Nonetheless, it probably has some truth and utility away from the transition line, especially away from the μ_B axis, as long as one is asking just for estimates of some bulk thermodynamic quantities that are sensitive to counting degrees of freedom and not sensitive to small interaction effects.

In the next section we will try to improve upon this foundation and later in the book we will become much more ambitious as we apply QCD and QCD-like models to the physics in the $T–\mu$ plane.

8.3 A tour of the $T–\mu$ phase diagram

Before we begin our detailed study of QCD and related theories at large chemical potentials, let's review the state of the art of the field's expectations for the $T–\mu$ phase diagram.

The phase diagram we have discussed so far, Fig. 8.6, is simply based on counting degrees of freedom and entropy, and ignores most of the interesting dynamics of QCD. Although counting and comparing degrees of freedom in adjacent phases is very illuminating, we know from other physical systems that these types of arguments miss phases and orders of transitions. It is a sobering thought that "simple" materials such as H_2O have a host of phases, which, in this case, dictate the very character of earth's environment. The symmetry of the H_2O molecule and the tiny residual interactions between polar water molecules lead to the formation of many forms of ice, for example, which influence the earth's weather patterns near its poles. It is a very challenging problem in condensed-matter physics to predict all the phases of ice, their symmetries, and the order of the transitions between them. The problem is simply too hard for our present mastery of computational quantum mechanics. From this perspective, the opinion that the phase diagram of QCD will be "solved" by field theorists any time soon might be too cavalier. A full understanding of the QCD phase diagram will be based on mastering small, residual interactions in QCD in exotic environments. The phase diagram, especially that at low temperatures, will be very rich, and it will be sorted out only through the interplay of theory and experiment. The "experiments" will be those of heavy-ion colliders and physics based on time scales of order 10^{-23} s, as well as observations of distant compact stars over periods of millions of years.

The study of the phases of QCD at moderate μ and small T might lead to a new field: the condensed matter or even chemistry of QCD.

With these "warnings" in mind, let's jump right in!

Our purpose in this section will be to assemble available knowledge, both theoretical and experimental, about QCD and apply it to the construction of the phase diagram in the $T-\mu$ plane. Knowledge of the underlying QCD degrees of freedom allowed us, in the previous section, to make entropy estimates and to draw a rough sketch of the phase diagram based on them. The natural next step is to use the information about the symmetries of QCD, in particular, chiral symmetry, to refine this picture. Many studies of the QCD phase diagram have concentrated on modeling the properties of the chiral phase transition (see, e.g., [52–54]). In this section, we shall discuss the properties of this transition in a more model-independent way. We shall see that chiral symmetry plays a crucial role. We shall also include in our analysis the effects from other phase transitions, such as the nuclear-matter liquid–gas transition, our knowledge of which comes mostly from experiment. The aim of our analysis is to transform this knowledge into the determination of a phase diagram for QCD in the $T-\mu$ plane. Such an analysis is especially important as an extension of Monte Carlo studies, given the technical problems that these encounter with finite baryon charge density.

8.3.1 Symmetry, order parameters, and phase transitions

Perhaps the central theme of our analysis is the use of QCD symmetries to delineate the phases. This does not provide us with much quantitative information about the phase transitions. However, the information this provides is more robust than results obtained from any specific dynamical model of the transition. The transitions which can be studied in this way are transitions between phases, or thermodynamic states of the system, in which the symmetry is realized in one of two different ways: as an exact symmetry or as a spontaneously broken symmetry. Consider as an example the behavior of a ferromagnet. The microscopic interactions respect the global O(3) rotation symmetry. However, the state of the system at zero temperature is not O(3) symmetric – one direction, the direction of spontaneous magnetization, is special. At finite temperature, the thermodynamic state of the system still breaks the O(3) symmetry until the Curie temperature is reached, whereupon the magnetization vanishes, and the microscopic O(3) symmetry becomes an exact symmetry of the thermodynamic state.

The transition between the O(3) symmetric and broken states must correspond to a singularity in the thermodynamic quantities. In order to show that this is the case, another concept becomes useful – the concept of the *order parameter*. For the purpose of this argument, the order parameter is a quantity that must have two properties: (i) it has to vary under the transformations of the symmetry, and

(ii) it must be nonzero in a state with spontaneously broken symmetry. The mean magnetization is such a quantity in the ferromagnet. The symmetry of the thermodynamic state at high T means that all quantities that are not invariant under O(3) rotations must have zero mean (or expectation) values. The magnetization as a function of temperature, therefore, cannot be an analytic function across the Curie temperature T_c. Indeed, a function constant in a finite interval of T must be constant everywhere within the domain of analyticity.

The axial-flavor symmetry of QCD in the limit of two massless quarks, $SU(2)_A$, plays the role of such a symmetry. The order parameter is the expectation value of the chiral condensate $\langle \bar{\psi} \psi \rangle$. This symmetry is restored at finite T, and we must therefore expect a thermodynamic singularity – a phase transition.

What if the symmetry is not exact, as is the case for $SU(2)_A$ of QCD when the quarks are light but not massless? We can go back to the ferromagnet and ask this question. If there is no symmetry, there is no argument based on the order parameter. Any quantity, such as the magnetization, cannot be expected to vanish in any finite region of the temperature – there is no symmetry to protect it. Indeed, the magnetization at $T > T_c$ is simply proportional to the applied magnetic field (for small field), which breaks the O(3) symmetry.

The fate of the transition in this case depends on the order of the transition in the limit of the exact symmetry. In the case of the ferromagnet, the transition at the Curie temperature is of second order and it disappears for arbitrarily small magnetic field. The singularities in the thermodynamic quantities (e.g. in the magnetic susceptibility) round up and become sharp peaks, and the order parameter crosses over from the region where it is large to the region where it is small. Since we expect the transition in QCD with two massless quarks to be of second order, we should expect a similar crossover phenomenon to replace the true phase transition when the small finite quark masses are taken into account.

On the other hand, if the transition was first order when the symmetry was exact, we can expect that the discontinuities which it created in the thermodynamic quantities are smooth functions of the symmetry-breaking parameters. In this case, the first-order transition should persist, at least for small explicit symmetry breaking.

Finally, let us comment on the properties of phase transitions which determine their order. In some sense we shall be giving the definition of second- and first-order phase transitions here. Although for some readers this may constitute a repetition of well-known facts, it may benefit others who might have different notions about the meaning of the order of a transition. We shall define a *first-order* phase transition as a point in the space of thermodynamic parameters (such as T or magnetic field H for the ferromagnet or T and μ in QCD) where two distinct phases, or thermodynamic states, of the system can co-exist. In other words, the system has a choice of thermodynamic state. A typical example is a glass of water with ice. Unless the temperature is exactly $0\,°C$, either all the

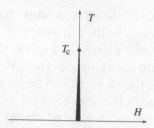

Figure 8.9. The phase diagram of a ferromagnet. The width of the line between $T = 0$ and T_c represents qualitatively the strength of the discontinuity in the magnetization as a function of the magnetic field H across $H = 0$.

water should freeze or all the ice should melt. At zero temperature, water has a choice, to be solid or to be liquid. Let us look at the $T-H$ phase diagram of a ferromagnet in Fig. 8.9. Along the line $H = 0$ below T_c, the ferromagnet has a choice of, in fact, infinitely many states differing by the direction of the magnetization. At any nonzero H the state with M parallel to H is the only truly stable state. Therefore, the line between $T = 0$ and T_c at $H = 0$ is a first-order-transition line. The transition between different phases occurs as a function of the magnetic field.

Note also that each of the co-existing phases is completely free of any singularities at a first-order line. The transition occurs when the system jumps, or rearranges itself from one phase into another. Such a transition can even be prevented by going through the transition point slowly – as in supercooling the water.

In contrast, a *second-order* phase transition is characterized by a singularity in thermodynamic functions. This happens practically always because the second-order transition is a special point on a line of first-order phase transitions – the end-point. It is a point where two or more different phases fuse into one. There is no phase co-existence here, because there is only one phase at a second-order transition. Second-order transitions are also often called *critical points*. Although the same term is often used for the first-order transitions, in this section we try to avoid this confusion and use this term only for second-order, singular, phase transitions. In the $T-H$ plane of the ferromagnet in Fig. 8.9 one can clearly see that phase co-existence ends at T_c. The magnetization vanishes in all phases. There is only one phase at T_c – the symmetric phase. The system has no choice.

8.3.2 Definitions

As we "warned" above, one can expect the phase diagram of QCD to have many subtleties. We can predict some of these by noticing that there are several small parameters, whose interplay may produce subtle, competing effects that will lead to new states of matter: the up- and down-quark masses are small on the natural mass scale of QCD. In addition, the strange-quark mass is comparable to the QCD scale, and it might produce additional exotic states in the phase diagram (see, e.g., Fig. 7.3). In realistic environments, one must also take electromagnetism into account. We shall not even try to attack the full problem. Instead, we shall strip as much from it as we possibly can while still keeping the most basic QCD features in. We shall find that even such a "stripped" version of QCD (F. Wilczek coined the name "QCDLite" for it) presents a theoretical challenge.

Thus we consider pure SU(3) QCD (i) with electroweak interactions turned off and (ii) with two massless quarks. There is then an exact $SU(2)_L \times SU(2)_R \times U(1)_B$ global symmetry of the action, which is spontaneously broken down to $SU(2)_V \times U(1)_B$ at zero and sufficiently low temperatures by the formation of a condensate, $\langle \bar{\psi} \psi \rangle$. Many features of QCD indicate that this is a reasonable approximation, e.g. the lightness of pions and the success of current-algebra relations. (We will comment below on the inclusion of electromagnetic interactions and strange quarks.) This theory is described by a grand canonical partition function, which, written as a path integral, is formally

$$Z \equiv e^{-\Omega(T,\mu)/T} = \int DA \, D\bar{\psi} \, D\psi \, \exp(-S_E). \tag{8.64}$$

The Euclidean action, S_E, is given by

$$S_E = \int_0^{1/T} dx_0 \int d^3x \left[\frac{1}{2g^2} \operatorname{tr}(F_{\mu\nu} F_{\mu\nu}) - \sum_{f=1}^{N_f} \bar{\psi}_f \left(\slashed{\partial} + \slashed{A} + m_f + \frac{\mu}{N_c} \gamma_0 \right) \psi_f \right] \tag{8.65}$$

where $N_f = 2$ is the number of flavors, $N_c = 3$ is the number of colors, and $m_f = m = 0$ is the quark mass. The Euclidean matrices γ_μ are Hermitian. Note that, with our choices of sign, positive m and μ induce positive $\langle \bar{\psi} \psi \rangle$ and $\langle \bar{\psi} \gamma_0 \psi \rangle$. The normalization of μ differs from the normalization customary in lattice calculations by a factor $1/N_c$ (i.e. the baryon charge of a quark). We should be using the symbol μ_B; however, in this section we shall omit the subscript in order to avoid unnecessary clutter. On integrating over the fermion fields we can also write

$$Z = \int DA \exp\left(-\frac{1}{2g^2} \operatorname{tr} (F_{\mu\nu} F_{\mu\nu}) \right) \det\left(\slashed{D} + m_f + \frac{\mu}{N_c} \gamma_0 \right) \tag{8.66}$$

As indicated, this system is characterized by equilibrium values of T and μ. So the system can be viewed as being in thermodynamic equilibrium with a large reservoir of entropy and baryon charge that has these values of T and μ. The total energy and baryon charge of our system fluctuate. Of course, the relative magnitude of these fluctuations is negligible for an open system of macroscopic size. The relation between the chemical potential, μ, and the average number density (per unit volume) of baryons, n, is the same as that between the temperature, T, and the average entropy density (per unit volume), s:

$$nV = \sum_f \langle \bar{\psi}_f \gamma_0 \psi_f \rangle = -\frac{\partial \Omega}{\partial \mu}; \qquad sV = -\frac{\partial \Omega}{\partial T} \qquad (8.67)$$

where Ω is the thermodynamic potential defined in Eq. (8.64). It can also be seen that $\Omega = -pV$, where p is the pressure. In other words, pressure, temperature, and chemical potential are not independent variables for our system. Their variations are related by

$$dp = s\,dT + n\,d\mu \qquad (8.68)$$

Both T and μ (as well as p) are intensive parameters. For a system in thermodynamic equilibrium, these quantities are the same for any of its smaller subsystems. In contrast, the extensive densities s and n can differ for two subsystems even when they are in equilibrium with each other. This happens in the phase-co-existence region, e.g. a glass containing water and ice. It is more convenient to describe the phase diagram in the space of intensive parameters, T and μ. In particular, the first-order phase transition which we shall encounter is characterized by one value of μ but two values of n – the densities of the two co-existing phases. Another reason for working with these coordinates is that first-principles lattice calculations are performed using T and μ as independent variables that can be controlled while the densities are measured. The results of relativistic heavy-ion-collision experiments are also often analyzed using this set of parameters [55].

8.3.3 Zero temperature

We begin by considering the phase diagram as μ is varied along the line $T = 0$. Strictly speaking, we are not dealing with *thermo*dynamics here since the system is in its ground state. This fact leads to a simple property of the function $n(\mu)$ that we have discussed already. Let us write the partition function as the Gibbs sum over all quantum states, α, of the system:

$$Z = \sum_\alpha \exp\left(-\frac{E_\alpha - \mu N_\alpha}{T}\right) \qquad (8.69)$$

where each state is characterized by its energy, E_α, and its baryon charge, N_α. In the limit $T \to 0$, the state with the lowest value of $E_\alpha - \mu N_\alpha$ makes an exponentially dominant contribution to the partition function. When $\mu = 0$, this is the state with $N = 0$ and $E = 0$, i.e. the vacuum or $\alpha = 0$. Let us introduce

$$\mu_0 \equiv \min_\alpha (E_\alpha / N_\alpha) \tag{8.70}$$

As long as $\mu < \mu_0$, the state with the lowest value of $E_\alpha - \mu N_\alpha$ remains the vacuum, $\alpha = 0$. Therefore, we conclude that, at zero temperature,

$$n(\mu) = 0 \qquad \text{for} \qquad \mu < \mu_0 \tag{8.71}$$

What is the value of μ_0? As we have seen, for a theory with a free massive fermion (baryon) this value is equal to the mass of the fermion; $\mu_0 = m$. The function $n(\mu)$ is singular at the point $\mu = m$, Eq. (8.28). The existence of some singularity at the point $\mu = \mu_0$, $T = 0$ is a robust and model-independent prediction. This follows from the fact that a singularity must separate two phases distinguished by an order parameter, e.g. n. The function $n \equiv 0$ cannot be continued to $n \neq 0$ without a singularity, as we pointed out in Section 8.3.1.

What is μ_0 for the case of QCD, and what is the form of the singularity? The answers to these questions are somewhat different in QCD and in the real world (QCD+) which includes other interactions, most notably electromagnetic interactions. Since QCD is our focus here and QCD+ is the ultimate goal of our understanding, we shall consider both cases. It is important to understand their differences if we are to extract physically useful predictions from lattice calculations, which are performed for QCD rather than QCD+.

The energy per baryon, E/N, can also be written as $m_N - (N m_N - E)/N$, where $m_N = m_p \approx m_n$ is the nucleon mass. Therefore, the state which minimizes E/N is that for which the binding energy per nucleon, $\epsilon = (N m_N - E)/N$, is a maximum. Empirically, we know that this state is a single iron nucleus at rest with $N = A = 56$ and $\epsilon \approx 8\,\text{MeV}$. However, in QCD without electromagnetism the binding energy per nucleon increases with A. This is a consequence of the saturation of nuclear forces and can be seen from the Weizsacker formula. Without electromagnetism, only the bulk and surface energy terms are significant for large A:

$$\epsilon(A) \equiv \frac{A m_N - m_A}{A} \approx a_1 - a_2 A^{-1/3} \tag{8.72}$$

with $a_1 \approx 16\,\text{MeV}$ and $a_2 \approx 18\,\text{MeV}$ [56]. As $A \to \infty$, ϵ saturates at the value a_1. This corresponds to the binding energy per nucleon in a macroscopically large sample of nuclear matter as defined by Fetter and Walecka in [56]. We conclude that, in QCD, the density jumps at $\mu = \mu_0 \approx m_N - 16\,\text{MeV}$ to the value of the density of nuclear matter, $n_0 \approx 0.16\,\text{fm}^{-3}$. Therefore, in QCD there is a first-order phase transition, characterized by a discontinuity in the function $n(\mu)$ at $\mu = \mu_0$ (see Fig. 8.10(a)).

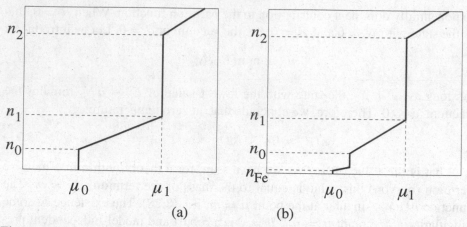

Figure 8.10. Schematic representations of the dependence of the baryon charge density on the chemical potential at $T = 0$ (a) in QCD ($\mu_0 \approx m_N - 16\,\text{MeV}$) and (b) in QCD+ ($\mu_0 \approx m_N - 8\,\text{MeV}$).

In QCD+, the Coulomb forces change the situation near μ_0. The contribution of the Coulomb repulsion to $\epsilon(A)$ is negative: $-0.7\,\text{MeV} \times Z^2/A^{2/3}$, and it is responsible for the experimentally observed maximum in $\epsilon(A)$ at $A \approx 56$. Isospin-singlet nuclear matter ($A = \infty$) is unstable at zero pressure due to Coulomb repulsion. Neutron matter with $Z \ll A$ is also unstable at zero pressure, and we are left to consider a gas of iron nuclei. In order to ensure electric neutrality, we must add electrons. Such a gas is clearly unstable at small densities and forms a solid – iron. Therefore, there is a discontinuity in the value of $n(\mu)$ at $\mu_0 \approx m_N - 8\,\text{MeV}$. This discontinuity is equal to the density of normal matter (i.e. iron) and is about 10^{-14} times smaller than in QCD. For very small $\mu - \mu_0$, $n(\mu)$ has structure, fine on the scale of QCD, which reflects the properties of normal matter under pressure. Then, for $\mu - \mu_0 = \mathcal{O}(10\text{–}200\,\text{MeV})$, we traverse the domain of nuclear physics with the possibility of various phase transitions. In particular, the transition to neutron matter ($Z \ll A$) is probably similar to the transition in QCD at $\mu = \mu_0$. (See Fig. 8.10(b).) In this domain, one may encounter such phenomena as crystallization of nuclear matter [57, 58], superconducting phases of neutron matter and quark matter [21–24], and, due to the strange quark in QCD+, kaon condensation [57, 59] and a transition to strange-quark matter [60, 61]. Moving along the μ axis to the right is equivalent to increasing the pressure: $p = \int n\, d\mu$. Thus, this picture is roughly what one might encounter on moving toward the center of a neutron star from the iron crust at the surface.

Our knowledge of $n(\mu)$ is scanty for densities of order one to ten times n_0 and $\mu - \mu_0 = \mathcal{O}(10\text{–}200\,\text{MeV})$ both in QCD and in QCD+. We can only be sure that $n(\mu)$ is a monotonically increasing function, which follows from the requirement of thermodynamic stability.

The behavior of $n(\mu)$ again becomes calculable in the region of very large $\mu \gg \Lambda_{\text{QCD}}$. In that case, the Pauli exclusion principle forces the quarks to occupy ever-higher momentum states, and, due to asymptotic freedom, the interaction of quarks near the Fermi surface is (logarithmically) weak. The baryon charge density is proportional to the volume of a Fermi sphere of radius $\mu/3$, $n(\mu) \approx N_{\text{f}}(\mu/3)^3/(3\pi^2)$. At low temperatures, only quarks near the Fermi surface contribute to the Debye screening of the gauge fields. The square of the screening mass, m_{D}^2, is proportional to the area of the Fermi surface: $m_{\text{D}}^2 \sim g^2 \mu^2$. This means that color interactions are screened on lengths $\mathcal{O}(1/(g\mu)) = \mathcal{O}(\sqrt{\ln(\mu/\Lambda_{\text{QCD}})}/\mu)$. This motivates the conclusion that non-perturbative phenomena such as chiral-symmetry breaking should not occur at sufficiently large μ. Therefore, in QCD with two massless quarks one should expect at least one other phase transition, at a value of μ that we define as μ_1 – a transition characterized by the restoration of chiral symmetry.

The situation in QCD+, with a strange quark, is somewhat more subtle. As has been observed by Alford, Rajagopal, and Wilczek [62] at sufficiently large μ one must reach a phase in which chiral symmetry is broken by a completely different mechanism than that by which it is broken in the QCD vacuum, the mechanism which the authors named color-flavor locking (CFL). We shall discuss the CFL phase of QCD+ in Chapter 9. The mass of the strange quark again crucially affects the phase diagram. It is not known whether this transition to CFL occurs from the chirally symmetric phase $\mu_{\text{CL}} > \mu_1$, or before the chirally symmetric phase even sets in at zero T.[5] Sending m_s to infinity relieves us of this question.

What is the value of μ_1 in QCD, and is it finite? Very little reliable information about the phase transition at μ_1 is available. However, results from several different approaches agree on the conclusion that the value of μ_1 is finite and that $\mu_1 - \mu_0$ is of the order of the typical QCD scale $\Lambda_{\text{QCD}} \approx 200\,\text{MeV} \approx 1\,\text{fm}^{-1}$. For example, equating the quark pressure minus the MIT bag constant to the pressure of nuclear matter yields such an estimate (see, e.g., [64]). The simple model of Section 8.2, which is similar, but neglects the pressure on the hadronic side, gives $\mu_1 = 3\sqrt{\pi}B^{1/4} \approx 1.1\,\text{GeV}$. Here, we should also point out another interesting distinction between QCD and QCD+: the effect of the strange quark in QCD+ is to decrease the value of μ_1 compared with that of QCD. It has even been conjectured that this effect might be sufficient to drive μ_1 below μ_0, which would make normal nuclear matter metastable [60, 61]. Another model that predicts the phase transition at finite μ_1 is the Nambu–Jona Lasinio model, which focuses on the degrees of freedom associated with the spontaneous chiral-symmetry breaking and leads to a similar estimate for μ_1 [53].

---

[5] Perhaps there is no transition at all, as conjectured by Schäfer and Wilczek [63].

What is the order of this phase transition? The MIT bag model predicts that it is a first-order transition since the density, n, of the baryon charge is discontinuous. Unfortunately, analysis of the Nambu–Jona Lasinio model shows that the order of the transition depends on the values of parameters, most notably, on the value of the cutoff. A larger cutoff leads to a second-order transition, a smaller cutoff to a first-order transition [53]. A random-matrix model at $T = 0$ predicts a first-order phase transition [65]. We shall discuss the random-matrix-model predictions in more detail in Chapter 10. Here we shall use more general methods to analyze features of the phase diagram of QCD at finite density *and* temperature.

An additional, qualitative argument for the first-order nature of the chiral phase transition at μ_1 can be drawn from an analogy between QCD and a meta-magnet, such as a crystal of ferrous chloride, $FeCl_2$. At temperatures below the Néel temperature, T_N, and at zero magnetic field, H, such a crystal is antiferromagnetically ordered, i.e. the staggered magnetization, ϕ_{st}, has a nonzero expectation value, $\langle \phi_{st} \rangle \neq 0$. Analogously, $\langle \bar{\psi}\psi \rangle \neq 0$ in QCD below T_c. The magnetic field H is not an ordering field for the staggered magnetization because it couples to a different order parameter (i.e. normal magnetization, ϕ, with $\Delta \mathcal{E} = -H\phi$) and induces nonzero $\langle \phi \rangle$. Similarly, the chemical potential induces a nonzero $\langle \bar{\psi}\gamma_0\psi \rangle$, and the term $\mu \bar{\psi}\gamma_0\psi$ does not introduce explicit breaking of chiral symmetry. At some critical value of H, ferrous chloride undergoes a first-order phase transition, and the staggered magnetization vanishes: $\langle \phi_{st} \rangle = 0$. One could naturally expect that in QCD a similar competition between the low-temperature spontaneous ordering, $\langle \bar{\psi}\psi \rangle \neq 0$, and the ordering $\langle \bar{\psi}\gamma_0\psi \rangle \neq 0$ induced by μ would result in a first-order phase transition. This analogy can be continued into the T–μ plane or the T–H plane in the case of the antiferromagnet. The antiferromagnet has a well-known tricritical point in this plane. Its analog in QCD will be discussed in Section 8.3.5.

Following the arguments of the two preceding paragraphs, we base our subsequent analysis of the phase diagram of QCD with two massless quarks on the following expectations: (i) $\mu_1 \sim \mu_0 + \mathcal{O}(200\,\mathrm{MeV})$ and (ii) the transition is of first order.

8.3.4 Finite T and μ

We shall use two order parameters to analyze the phase diagram of QCD at nonzero T and μ: the chiral condensate $\langle \bar{\psi}\psi \rangle$ (per flavor) given by

$$\langle \bar{\psi}\psi \rangle V = -\frac{1}{N_f}\frac{\partial \Omega}{\partial m} \tag{8.73}$$

and the density of the baryon charge n given by Eq. (8.67). We have already used n to show that there is a singularity at $\mu = \mu_0$ and $T = 0$. It was important for that argument that n is exactly zero for all $\mu < \mu_0$. At nonzero T, however, n is not strictly 0 for any $\mu > 0$. For example, for very small μ and T one finds a very dilute gas of light mesons, nucleons, and antinucleons with calculable density

$$n(T, \mu) \approx \frac{\mu}{T} \left(\frac{2m_N T}{\pi} \right)^{3/2} e^{-m_N/T} \tag{8.74}$$

Since it is nonzero, at finite T the density n cannot be used to predict phase transitions as it can at zero T.

Nevertheless, we can use a continuity argument to deduce that the first-order phase transition at $T = 0$, $\mu = \mu_0$ has to remain a first-order phase transition for sufficiently small T. Therefore, there must be a line emerging from the point $T = 0$, $\mu = \mu_0$. One can think of this transition as boiling the nuclear fluid. The slope of this line can be related to the discontinuities in the entropy density, Δs (or the latent heat per volume $T \Delta s$), and in the baryon density, Δn, across the phase transition line through the generalized Clapeyron–Clausius relation:

$$\frac{dT}{d\mu} = -\frac{\Delta n}{\Delta s} \tag{8.75}$$

This relation follows from the condition that the pressure, temperature, and chemical potential should be the same in the two phases on a phase-co-existence curve and Eq. (8.68). In analogy with ordinary liquid–gas transitions, the gaseous phase has a lower particle density (whence $\Delta n < 0$) and lower entropy density[6] (whence $\Delta s < 0$). Therefore, the slope $dT/d\mu$ must be negative. We further expect that the slope is infinite at $T = 0$ since $s(T = 0) = 0$, and hence $\Delta s(T = 0) = 0$. As there is no symmetry-breaking order parameter that distinguishes the two phases, there is no reason why these two phases cannot be connected analytically. As in a typical liquid–gas transition, it is natural to expect that the first-order phase-transition line terminates at a critical point with the critical exponents of the three-dimensional Ising model.[7] The temperature of this critical point can be estimated from the binding energy per nucleon in cold nuclear matter, $T_0 = \mathcal{O}(10 \, \text{MeV})$. (See Fig. 8.11.) Signatures of this point are seen in heavy-ion collisions at moderate energies (i.e. $\approx 1 \, \text{GeV}$ per nucleon), and the critical properties of this point have been studied through measurements of the yields of nuclear fragments [64, 66]. In particular, the reported critical exponents are in agreement with those of the three-dimensional Ising model [66].

Additional phase transitions that might occur at $T = 0$ would give rise to additional phase-transition lines. One could expect two generic situations. If

---

[6] The entropy per particle is greater in the gaseous phase, but the entropy per volume s is smaller because the particle density is much smaller.

[7] The Ising nature of the universality class follows from the fact that the transition can be modeled as an Ising lattice gas.

there is a breaking of a global symmetry (e.g. translational symmetry in the case of crystallization of nuclear matter), the phase-transition line must separate such a phase from the symmetric phase at higher temperature without any gaps in the line. Otherwise, the transition can terminate at a critical point.

At very high $T \gg \Lambda_{QCD}$, we have a plasma of quarks and gluons with a logarithmically small effective coupling constant, $g(T)$, and we can again calculate the density of the baryon charge n:

$$n(T, \mu) \approx 4 \int \frac{d^3 p}{(2\pi)^3} \left[\exp\left(\frac{|\vec{p}| - \mu/3}{T}\right) + 1 \right]^{-1} - \{\mu \to -\mu\} \qquad (8.76)$$

We expect that the chiral condensate is zero at very high T since the effective coupling is weak, according to asymptotic freedom. Therefore, a phase transition must separate the quark–gluon-plasma phase from the low-temperature phase. This transition has been studied extensively at $\mu = 0$ using a variety of methods. In particular, as we discussed in Chapter 7, lattice calculations have established the value of T_c as approximately 170 MeV [67]. Arguments based on universality suggest that this transition is of second order with the critical exponents of the three-dimensional $SU(2)_L \times SU(2)_R \sim O(4)$ universality class [68]. Results from lattice calculations are consistent with this scenario [69].[8] Here, we assume that this description is true and try to understand what happens to this transition when μ is not zero.

For massless quarks, the low-temperature hadronic phase and the quark–gluon-plasma phase can be distinguished by the expectation value of $\langle \bar{\psi}\psi \rangle$, since this is identically zero in the quark–gluon-plasma phase and nonzero in the hadronic phase with spontaneously broken chiral symmetry. Therefore, when quark masses are strictly zero, a phase transition must separate these two phases, i.e. these phases cannot be connected analytically in the T–μ plane at $m = 0$. Therefore, a line of phase transitions must begin from the point $T = T_c$, $\mu = 0$ and continue into the T–μ plane (see the left-hand side of Fig. 8.11).

As discussed above, restoration of chiral symmetry at $T = 0$ is most likely to proceed via a first-order phase-transition. Therefore, the transition must remain of first order as we continue along a line into the T–μ plane. The slope of this line can again be related to the discontinuity in the baryon charge and the entropy density, Eq. (8.75). Since we expect that both density and entropy will be larger in the quark–gluon-plasma phase, the slope of this line, $dT/d\mu$, should be negative.

This first-order-transition line cannot terminate because the order parameter, $\langle \bar{\psi}\psi \rangle$, is identically zero on one side of the transition. The minimal possibility is

---

[8] A sufficiently light third quark would drive the transition first order [68]. However, lattice calculations also indicate that the strange quark is not sufficiently light for this to occur [70].

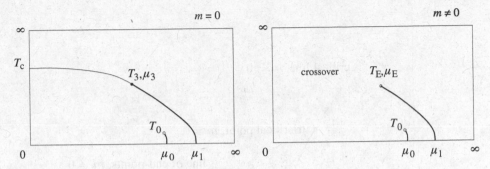

Figure 8.11. Schematic plots of the phase diagram of QCD with two massless (left) and light (right) quarks. The SU(2)$_A$ symmetry of QCD is restored at the second-order phase-transition line begining at $T = T_c$ and $\mu = 0$ (left). This transition is smoothened into a crossover when $m \neq 0$ (right). The first-order phase transition (thick line) remains.

that it merges with the second-order phase-transition line coming from $T = T_c$, $\mu = 0$ (see the left-hand side of Fig. 8.11). The point where the two lines join is a tricritical point. Such a point exists in many physical systems (e.g. for the FeCl$_2$ antiferromagnet), and universal behavior in the vicinity of this point has been studied extensively. In the next section, we review those properties of a tricritical point which follow from universality.

8.3.5 Universal properties of the tricritical point

By analogy with an ordinary (bi)critical point, where two distinct co-existing phases become identical, one can define the tricritical point as a point where three co-existing phases become identical simultaneously. A tricritical point marks an end-point of three-phase co-existence. In order to see this in QCD, it is necessary to consider another dimension in the space of parameters – the quark mass m – see Fig. 8.12. This parameter breaks chiral symmetry explicitly. In such a three-dimensional space of parameters, one can see that there are two surfaces (symmetric with respect to $m \rightarrow -m$ reflection) of first-order phase transitions emanating from the first-order line at $m = 0$. On these surfaces or wings with $m \neq 0$, two phases co-exist: a low-density phase and a high-density phase. There is no symmetry distinguishing these two phases since chiral symmetry is explicitly broken when $m \neq 0$. Therefore, each surface can have an edge that is a line of critical points. These lines, or wing lines, emanate from the tricritical point. The first-order phase-transition line can now be recognized as a line where three phases co-exist: the high-T and -density phase and two low-density and -T phases with opposite signs of m and, hence, also of $\langle \bar{\psi}\psi \rangle$. This line is called, therefore, a triple line.

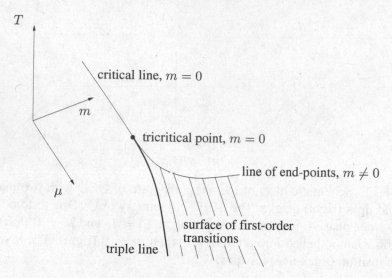

Figure 8.12. Phase transitions in the three-dimensional $T-\mu-m$ space in the vicinity of the tricritical point.

The plane $m = 0$ is a symmetry plane. Chiral symmetry is exact only in this plane, and it is only here that the low- and high-temperature phases must be separated by a transition. One can also view this plane as a first-order phase-transition surface, since $\langle \bar{\psi}\psi \rangle$ has a discontinuity across it. Then, the second-order phase-transition line and the triple line provide a boundary for this surface.

Critical behavior near the tricritical point can now be inferred from universality. The upper critical dimension for this point is 3. Since critical fluctuations are effectively three-dimensional for the second-order phase transition at finite T, we conclude that behavior near this point is described by mean-field exponents with only logarithmic corrections. The effective Landau–Ginsburg theory for the long-wavelength modes, $\phi \sim \langle \bar{\psi}\psi \rangle$, near this point requires a ϕ^6 potential, which has the form (in the symmetry plane $m = 0$)

$$\Omega_{\text{eff}} = \Omega_0(T, \mu) + a(T, \mu)\phi^2 + b(T, \mu)\phi^4 + c(T, \mu)\phi^6 \tag{8.77}$$

with $c > 0$. The ϕ^6 term is necessary in order to create three minima corresponding to the three co-existing phases. This explains why the critical dimensionality is 3, since, for this dimension, the operator ϕ^6 becomes a marginal operator. When $b > 0$, the transition occurs when $a = 0$ and is a second-order transition similar to that seen in a ϕ^4 theory. This corresponds to the second-order line. When $b < 0$ the transition occurs at some positive value of a and is of first order. This is the triple line. When both a and b vanish, we have a tricritical point.

In particular, the following exponents in the symmetry plane $m = 0$ are readily found using mean-field ϕ^6 theory (as noted above, results from renormalization-group studies [71] show that the actual singularities include additional, logarithmic corrections). The discontinuity in the order parameter $\langle\phi\rangle = \langle\bar{\psi}\psi\rangle$ along the triple line as a function of the distance from the critical point μ_3, T_3 (measured as either $T_3 - T$ or $\mu - \mu_3$) behaves like

$$\Delta\langle\bar{\psi}\psi\rangle \sim (\mu - \mu_3)^{1/2} \tag{8.78}$$

The discontinuity in the density, $n = d\Omega_{\text{eff}}/d\mu$, across the triple line behaves like

$$\Delta n \sim \mu - \mu_3. \tag{8.79}$$

The critical behavior along the second-order line is everywhere the same as at the point $\mu = 0$, $T = T_c$ (which is an infrared-attractive fixed point). Therefore, $\langle\bar{\psi}\psi\rangle$ vanishes on the second-order line with three-dimensional O(4) exponents. At the tricritical point, however, the exponent with which $\langle\bar{\psi}\psi\rangle$ vanishes is given by Landau–Ginzburg theory as

$$\langle\bar{\psi}\psi\rangle \sim (T_3 - T)^{1/4} \tag{8.80}$$

When $m \neq 0$, the potential $\Omega_{\text{eff}}(\phi)$ can also contain terms ϕ and ϕ^3 that break $\phi \to -\phi$ symmetry explicitly. (The term ϕ^5 can be absorbed by a shift of ϕ.) The potential $\Omega_{\text{eff}}(\phi)$ still has three minima, and a first-order phase transition can occur when two adjacent minima are equally deep. These transitions form a surface of first-order phase transitions – the wings. The two minima (and an intermediate maximum) can also fuse into a single minimum. This happens on the wing lines at the edge of the surface of first-order phase transitions. The critical behavior along the wing lines is given by the three-dimensional Ising exponents, as is usual at the end-points of first-order liquid–gas-type phase transitions not associated with restoration of a symmetry. In particular, the discontinuity in $\Delta\langle\bar{\psi}\psi\rangle$ and Δn vanishes with exponent $\beta \approx 0.31$. These discontinuities are related to the slope of the wing surface at constant T through a relation similar to Eq. (8.75):

$$\frac{d\mu}{dm} = -\frac{\Delta\langle\bar{\psi}\psi\rangle N_f}{\Delta n} \tag{8.81}$$

There are many other universal properties in the vicinity of a tricritical point that can be derived from the above ϕ^6 Landau–Ginzburg effective potential. One can, for example, show that the $m = 0$ second-order line, the wing lines, and the triple lines approach the triple point with the same tangential direction: The second-order line approaches from one side while the wing lines and the triple line approach from the opposite side. For a more detailed description of the properties of tricritical points, see [71].

8.3.6 Summary and remarks

We reviewed an analysis (qualitative and, in some cases, quantitative) of the salient features of the phase diagram of QCD with two light or massless quark flavors at finite temperature and baryon chemical potential. The most important features of this phase diagram are summarized in Fig. 8.11. The phase diagram can certainly have a much richer structure. The phase transitions shown in Fig. 8.11 are distinguished by the fact that a good order parameter can be associated with each of them. Here the term "good order parameter" implies the existence of some quantity whose expectation value is identically zero in some finite region of parameter space or in one phase and is some function of parameters in the other phase. Two such phases must be separated by a nonanalytic boundary, i.e. a phase-transition. What is crucial here is the identical vanishing of an order parameter or its strict independence of the parameters of the theory. Usually, this is ensured by the existence of some symmetry with respect to which this order parameter transforms nontrivially.

For the chiral phase transition, a good order parameter is the value of $\langle \bar{\psi}\psi \rangle$, which spontaneously breaks the global $SU(2)_L \times SU(2)_R$ chiral symmetry to $SU(2)_V$. Hence, the phase with $\langle \bar{\psi}\psi \rangle = 0$ and the phase with $\langle \bar{\psi}\psi \rangle \neq 0$ cannot be connected without crossing a phase-transition line in the $T-\mu$ plane.

The transition from $n \equiv 0$ to $n \neq 0$ along the $T = 0$ line provides another example of a phase transition associated with a good order parameter. The phases $n \equiv 0$ and $n \neq 0$ cannot be analytically connected; i.e. one must pass through a nonanalytic boundary (phase transition) on passing from one to the other. Since this is a first-order phase transition, continuity requires that there is also a first-order-transition line for some $T \neq 0$. This line can, however, terminate since n is no longer a good order parameter when $T \neq 0$.

Of course, the existence of a good order parameter is a sufficient but not a necessary condition for a phase transition. Other phase-transition lines associated with more subtle phenomena are also possible. In fact, one should expect many subtle surprises in this phase diagram, especially once the finite strange-quark mass is introduced.

The most interesting feature of the phase diagram of Fig. 8.11 is the presence of a tricritical point. Because the critical dimensionality for such a point is equal to 3, critical behavior near this point is given by mean field theory plus logarithmic corrections.

The tricritical point lies in the region expected to be probed by heavy-ion-collision experiments. Discovery of experimental signatures of this point is an exciting prospect for these experiments. Since quark masses are not precisely zero, we should consider a slice of a three-dimensional phase diagram, Fig. 8.12, with $m \neq 0$, shown on the right-hand side in Fig. 8.11. A qualitative difference between the phase diagram for $m \neq 0$ and that for $m = 0$ is the absence of the second-order phase-transition line associated with the restoration of chiral symmetry. This symmetry is explicitly broken for $m \neq 0$. However, continuity from $m = 0$ ensures that the first-order finite-density transition is still present at $m \neq 0$. This transition line is terminated by an ordinary critical point. Criticality at this end-point is not associated with restoration of chiral symmetry, and excitations with the quantum numbers of pions do not become massless there. Criticality at this point is associated with the fact that a correlation length in the channel with the quantum numbers of the sigma meson becomes infinite. (Hence, it is plausible to infer that this point has the critical behavior of the three-dimensional Ising model.) The experimental signatures of the critical point should therefore be associated with this phenomenon.

8.4 The quark–gluon plasma and the energy scales of QCD

Experimental, phenomenological, and theoretical developments point to the possibility that the physics of QCD is characterized by three distinct length scales. At sufficiently short distances, less than $\frac{1}{5}$ fm, calculations and theory confirm asymptotic freedom, which is essential to exposing the underlying gluon and quark degrees of freedom of the theory. At more moderate distances, from $\frac{1}{5}$ fm to $\frac{1}{2}$ fm, for which the running coupling is in the intermediate range, the description of the theory changes qualitatively. On these scales one can argue that semiclassical instanton configurations are breaking the anomalous $U(1)_A$ symmetry and, through properties of their ensemble, are breaking chiral symmetry and are producing the constituent-quark masses. Finally, on greater length scales, phenomenology and lattice methods confirm confinement, the fact that hadronic states are color singlets, and that the confining dynamics comes through thin, discernable but breakable, flux tubes.

One of the major purposes of studying QCD in extreme environments is to understand these phenomena in greater detail. In other words, one of the major purposes of studying QCD in extreme environments is to understand it in ordinary environments.

Scattering experiments in which the quark–gluon plasma is created afford us the chance to look into the workings of QCD with higher spatial and temporal resolutions than can be obtained in ordinary collisions. As emphasized by E. Shuryak [72], the mass scales of the quark–gluon plasma are different and, importantly, smaller than those of the more familiar hadronic phase. The hadronic phase breaks chiral symmetry; the quark–gluon plasma does not. The hadronic phase confines quarks; the quark–gluon plasma does not. The binding mechanism in the hadronic phase is nonperturbative whereas the screening mechanism in the quark–gluon plasma is perturbative. The perturbative screening masses in the quark–gluon plasma are $M^{\text{gluon}} \approx 0.4 \, \text{GeV}$ and

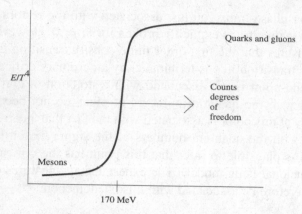

Figure 8.13. The energy density ϵ/T^4 versus T in a typical lattice simulation of QCD.

$M^{\text{quark}} \approx 0.3\,\text{GeV}$ for temperature above but near T_c, at which the plasma first appears. The finer level spacings in the plasma act as a fine-resolution grid to the dynamics in the hadronic phase. A collision that starts in the hadronic phase and ends in the plasma phase is especially sensitive to the levels and dynamics of the hadronic phase because of the relative wealth of open channels in the plasma phase.

This point shows up in several well-known phenomena.

1. The energy density, ϵ.
   Lattice simulations [73] have shown that ϵ/T^4 grows rapidly as T passes through $T_c = 170\,\text{MeV}$. On the low-temperature side of T_c, the energy density is accounted for by the presence of light mesons and on the high-temperature side it is accounted for by quarks and gluons. The number of degrees of freedom grows rapidly as T passes through T_c, and, although interactions are essential for $T \geq T_c$, the Stefan–Boltzmann estimate for the high-temperature energy density does not overshoot lattice-simulation data by more than 10%–20%.

2. The chiral condensate.
   The chiral condensate $\langle \bar{\psi}\psi \rangle$ falls rapidly as T passes through T_c (Fig. 8.14). The hadronic phase is characterized by a large constituent-quark mass, massive baryons and light mesons, whereas the plasma phase has light quarks.

3. The heavy-quark potential.
   When $T < T_c$ the heavy quark potential $V(r)$ rises linearly at large distances before pair production breaks the flux tube which apparently forms between the heavy source and sink of color. However, when $T > T_c$ the

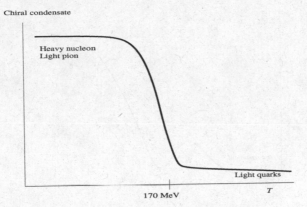

Figure 8.14. $\langle \bar{\psi} \psi \rangle$ versus T in a typical lattice simulation of QCD.

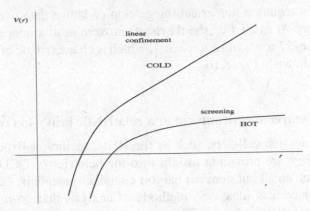

Figure 8.15. The heavy-quark potential above and below the T transition in QCD.

linear component of the potential is replaced by a screening form that saturates as r grows (Fig. 8.15).

4. The velocity of Sound.

Lattice simulations have not yielded the velocity of sound v_s in the two phases, but interesting results are expected. Since $v_s^2 = \partial P / \partial \epsilon$, one expects this quantity to have a dramatic dip in the vicinity of T_c where a change in temperature converts hadronic matter to screened quarks and gluons rather than changing the pressure P. The velocity of sound is an excellent signal for the transition theoretically and should influence the evolution of hadronic matter to the quark–gluon plasma in heavy-ion collisions. Lattice calculations of v_s are demanding because the pressure

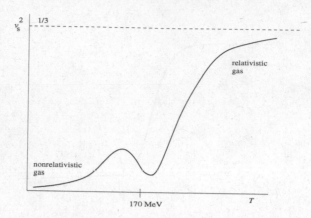

Figure 8.16. The velocity of sound versus T expected in QCD.

estimates require a numerical integration of lattice data over a range of couplings. At small T v_s^2 should rise from zero as in a nonrelativistic gas and at large T v_s^2 should approach $\frac{1}{3}$, which is characteristic of a relativistic gas, as shown in Fig. 8.16.

8.5 The extreme environment at a relativistic heavy-ion collider

Relativistic heavy-ion colliders, such as the RHIC facility at Brookhaven National Laboratory, are providing insight into the behavior of QCD at high energy density and small but nonzero baryon chemical potentials. Let's consider some of the theoretical ideas and methods of analysis that go into the high-energy collisions of large nuclei. This is a vast subject so our discussion will be hardly more than a few snapshots. However, since the goal of heavy-ion collisions is the production and study of the quark–gluon plasma, we should consider the challenges and potential rewards that this experimental subject faces.

RHIC collisions can be analyzed in four stages, which are thought to be characterized by distinct time scales.

First, viewed from the center-of-momentum frame (Fig. 8.17), there are right-moving (positive p_z) and left-moving (negative p_z) nuclei. Each of them can be described by infinite-momentum-frame wavefunctions that give the probability amplitudes of each nucleus to consisting of ensembles of quarks and gluons. The parton model, suitably improved with asymptotic freedom to describe its short distance features, should provide a framework for these initial states.

Second, there is the collision between the constituents in each nucleus. Experiments will shed light on the typical momentum transfers and multiplicities of these underlying collisions.

Third, there is the development of a thermalized plasma of quarks and gluons.

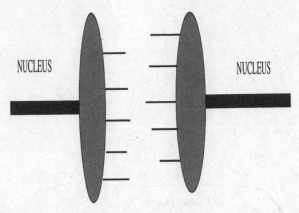

Figure 8.17. Two nuclei described in a parton picture about to collide in the center-of-momentum frame.

Fourth, there is the development of the final states of hadronic debris from the hot soup of colliding and produced constituents.

Consider each stage of the collision in turn.

Begin with the lightcone wavefunctions of the projectiles. That for the right-mover is

$$|\Psi\rangle = \sum_n \int \prod_i^n \frac{d^2 K_i \, dx_i}{x_i} \psi_n(x_1, \vec{K}_1; \ldots; x_n, \vec{K}_n)$$

$$\times \delta^2\left(\sum \vec{K}_i\right) \delta\left(1 - \sum x_i\right)|x_1, \vec{K}_1; \ldots; x_n, \vec{K}_n\rangle \qquad (8.82)$$

Recall from infinite-momentum-frame [74] or lightcone perturbation theory [74] that the wavefunction for a strongly interacting particle is written in a basis of the theory's fundamental constituents, quarks and gluons. In Eq. (8.82) we have a sum over components of the wavefunction of a nucleus, each component consisting of n partons. By the rules of Schrödinger perturbation theory, written using lightcone variables or, equivalently, written in an infinite-momentum frame, the probability amplitude for this component is given by the n-body wavefunction ψ_n whose arguments are the transverse momenta and longitudinal fractions of each parton. The transverse momenta of the constituents add up to that of the nucleus, zero in this case, and the longitudinal fractions add up to unity. Each longitudinal fraction x_i is a positive variable, $0 < x_i < 1$, according to the rules of infinite-momentum-frame perturbation theory. This condition, that each x_i is positive, eliminates vacuum structure from the description of the wavefunction and makes this formulation of the collision particularly physical and direct. In the context of deep inelastic scattering, for example, one views a nucleon in the infinite-momentum frame and probes its structure with a pointlike photon, thereby measuring the longitudinal momentum distributions of the partons in the nucleus, as described by the parton model of Feynman [75] .

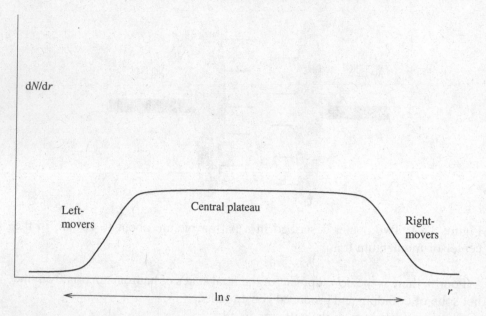

Figure 8.18. The rapidity distribution of partons in a nucleus–nucleus collision.

The bulk features of the wavefunctions of the projectiles are conveniently displayed on a rapidity plot, $r = \frac{1}{2}\ln[(p_0 + p_z)/(p_0 - p_z)]$. The rapidity is particularly handy because it simply translates under a boost along the z axis. One can plot the density of partons of both projectiles on a rapidity plot. Deep inelastic experiments at a fixed Q^2, the four momentum squared of a photon probing the structure of the nuclei, indicate that the plot is essentially flat and boost-invariant. The extent of the rapidity plot is $\ln s$, where s is the usual scattering variable, four times the square of the center-of-momentum energy, describing the collision.

Experiments indicate that rapidity distributions for partons have some simple, statistical features (Fig. 8.18) [75].

1. Short-range correlations in rapidity. The parton distributions lose memory, with a correlation length of 1–2 units, in rapidity. Screening in rapidity is an essential feature. Therefore, features like the quantum numbers of each projectile are limited to the edges of the rapidity plot. Unfortunately, this means that, in very-high-energy heavy-ion collisions, most of the partons are not influenced by the baryon number of the projectiles. The collisions can be characterized by a high energy density and temperature but not high chemical potential μ_B.

2. The "central region" is universal and "vacuum-like." The length of the central region grows like $\ln s$.

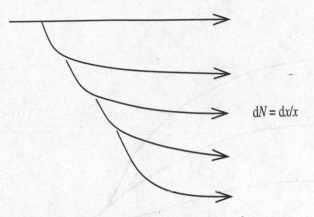

Figure 8.19. A multiperipheral graph in a parton $g\phi^3$ field theory. The graph fills rapidity space uniformly, $dN \sim dx/x \sim d \ln x \sim dr$.

Although first-principles calculations of lightcone wavefunctions are beyond us, models having these features have been known for decades. Multiperipheral graphs of QCD (Fig. 8.19) can populate the rapidity axis uniformly and have suggested that gluons are the most likely parton species in the central region [76].

Asymptotic freedom affects these considerations profoundly. Recall that, in the naive parton model, partons interact only softly with one another, never exchanging high transverse momenta [75]. However, if we consider the structure of the wavefunctions at high spatial resolution where the running coupling is small and perturbative asymptotic freedom applies, then exchanges of high transverse momentum and short-distance additional structure are inevitable [77,78]. Call the gluon-distribution function in momentum space $xG(x, Q^2)$. This is the distribution function that would be resolved by a probe with resolving power $\delta X_\perp^2 \sim 1/Q^2$. Models and theoretical analysis of scattering data suggest that $xG(x, Q^2)$ approaches a constant for small x and fixed Q^2 (Fig. 8.20). However, as Q^2 is increased one discovers that gluons can split perturbatively into three or four gluons of lesser longitudinal fractions x_i. The simplest perturbative Feynman diagrams of QCD apply here and give the probability amplitude of a parton of longitudinal fraction unity and transverse momentum zero consisting of several other partons whose longitudinal fractions sum to unity and whose transverse-momentum distributions are power-law-behaved and given by the features of the low-order pertubation-theory graph relevant for this process. So, for sufficiently small x, $xG(x, Q^2)$ must be an increasing function of Q^2. This is a general feature of the asymptotically free parton model [77] and perturbative evolution equations make this simple physical idea quantitative [78]. For quark-distribution functions that are accessible to deep inelastic electron and neutrino scattering, these ideas have been tested well phenomenologically.

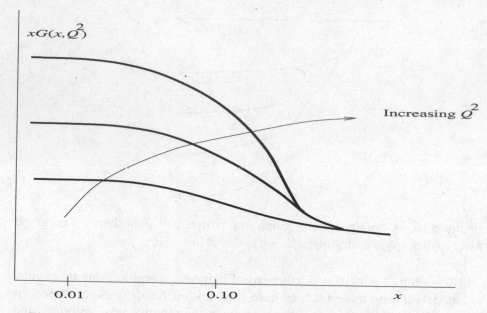

Figure 8.20. The longitudinal momentum distribution for increasing Q^2.

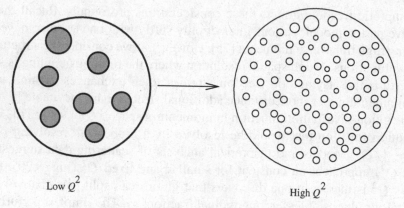

Figure 8.21. Variation in the transverse-plane distribution of partons for increasing Q^2.

If we view the parton density on the plane transverse to the collision axis p_z, then at low Q^2 we see just a few gluons of size $\delta X_\perp^2 \sim 1/Q^2$, but as Q^2 is taken larger we resolve many more gluons of smaller size (Fig. 8.21). This progressive development has led to the idea of "gluon saturation" in the lightcone wavefunction of a nucleus [79].

To understand this idea, consider the wavefunction and imagine that it is coming at you and you have a snapshot of the transverse plane [79]. Call the transverse size of the nucleus R_A. Each parton of transverse momentum Q in the wavefunction occupies an area π/Q^2, as suggested by the uncertainty principle. The parton's cross section is proportional to the square of the running coupling and its geometrical size, $\sigma \sim \alpha_s(Q^2)\pi/Q^2$. However, partons will overlap in the transverse plane when their number N_A becomes comparable to $\pi R_A^2/\sigma$, which is $Q^2 R_A^2/\alpha_s(Q^2)$. One supposes that, when N_A is this large, the overlap stops further growth in the density and "saturation" has been achieved [79]. The transverse momentum of the partons at saturation is $Q_s^2 \sim \alpha_s(Q^2)N_A/R_A^2 \sim A^{1/3}$. It is hoped that the A dependence in this and other similar equations will be testable in heavy-ion collisions. The number of partons at the saturation level is $N_s \sim Q_s^2 R_A^2/\alpha_s(Q_s^2) \sim A$ and this is the number of partons of size scale $1/Q_s$ that could materialize in the quark–gluon plasma. In addition, Q_s^2 sets the scale for the transverse momenta of partons in the central region of the rapidity plot,

$$\frac{1}{\sigma}\frac{dN}{dr\,d^2K_\perp} = \frac{1}{\alpha_s(Q_s^2)}F\left(\frac{K_\perp}{Q_s^2}\right) \tag{8.83}$$

These formulas can be made more precise by considering longitudinal fractions [79]. The discussion here is somewhat simplified, but a useful first guide. When numbers for the various nuclei and the energies available at experimental facilities are substituted here, one finds that $Q_s^2 \sim 1$–$2\,\text{GeV}$. This means that applying perturbative QCD is suspect and competitive nonperturbative effects should be expected. This is potentially the realm of lattice-gauge theory. The relationship of Euclidean lattice calculations and the Minkowski scattering formulation appropriate to heavy-ion collisions must be addressed.

Stages of a heavy-ion collision
The types of physics of interest in the four different stages of a heavy ion collision are thought to be quite distinct.

1. Wavefunctions approach one another.
   As reviewed above, the phase-space parton density is quite large, $\sim 1/\alpha_s(Q_s^2)$. Classical and semi-classical estimates should apply. In particular, the color per volume is large, so the lack of commutivity, $[Q^a, Q^b] = i f^{abc}Q^c$, appears to be a relatively small effect, and the gluon field can be treated classically for many purposes [79].
   At sufficently large resolution Q^2, the partons of that stage fill the transverse plane. The nucleus resembles a "black" absorbing disk. As Q^2 increases further, additional partons are generated at the edges of the disc and cause it to expand. If multiperipheral dynamics is a good guide here,

then the expansion occurs at a logarithmic rate. This becomes the basis for high-energy nulceus–nucleus cross sections to grow at a rate proportional to $\ln^2 s$ and saturate the Froissart bound [80].

2. Interactions.

    The gluon-saturation picture suggests that the characteristic time of inter-actions is quite small, $t_{int} \sim 1/Q_s \sim 0.1$–$0.3$ fm/c and the energy den-sity is very, very large, $\epsilon_{int} \sim Q_s^4/\alpha_s(Q_s^2) \sim 20$ GeV fm^{-3}. Such a large energy density would place the system deep in the quark–gluon-plasma phase. This is good news! Estimates of ϵ_{int} in the past were much more modest, of the order of several GeV fm^{-3}, at most, and estimates of t_{int} were much larger. In this scenario, it is more plausible that thermalization and the creation of a plasma is likely in heavy-ion collisions.

3. Thermalization.

    After the collision the quark–gluon matter expands and thermalizes. The experimental data and models suggest that the time scale for this to occur is $t_{therm} \sim 0.5$–1.0 fm/c.

    RHIC data, especially those showing the spatial distribution of particles produced in non-head-on collisions, strongly suggest that the interactions at early times are strong and the quark–gluon plasma appears very early in the evolution of the final-state [81]. The basic parton–parton interac-tions must be strong and involve high multiplicities. Perturbative pro-cesses alone are not sufficient to explain the data.

4. Hydrodynamic expansion.

    The quark–gluon plasma maintains thermalization and expands until de-coupling sets in at $t_{dec} \sim 10$ fm/c. In the expansion and production of the final-state hadrons, the fastest particles are produced last, as in the "inside–outside" cascade of electron–positron-annihilation processes [82]. The expansion is essentially one-dimensional, along the collision axis, until the latest stages of the development of the final state. A parti-cle's rapidity turns out to be strongly correlated with the spacetime rapid-ity, $\frac{1}{2}\ln[(x_0 + x_3)/(x_0 - x_3)]$, of its point of creation.

Signatures of the quark–gluon plasma

RHIC experiments are complicated and challenging to interpret. Several signatures for the production of the quark–gluon plasma are thought to be the following.

1. Jet quenching.

    The heavy-ion collisions show little or no sign of jets for $p_\perp < 4$ GeV. This is interesting in the light of the fact that, if one scales single-particle

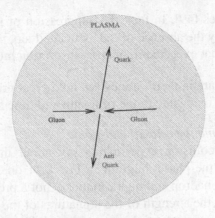

Figure 8.22. Gluons annihilating into quarks within the plasma phase.

hadron spectra from proton–proton data to A–A data, one overshoots the jet data very significantly [81].

It has been argued that this is evidence for the existence of an extended space-time region of the quark–gluon plasma. The idea is that two colliding gluons in the plasma annihilate into an energetic quark–anti-quark pair with the quark and anti-quark having sizable transverse momenta, but their interactions with the plasma medium sap the energy out of the energetic quark and its partner, eliminating ("quenching") those jet-like characteristics (Fig. 8.22). Of course, very-high-energy jets are unaffected by the medium because of the short-range nature of the most prevalent interactions on the rapidity axis, and jets with energies much larger than 4 GeV are seen.

2. Suggestions of deconfinement and restoration of chiral symmetry.
Lepton pairs have long been cited as effective probes into the internal dynamics of the quark–gluon plasma. In particular, if the ρ and ϕ existed in the plasma, then the leptonic decays $\rho \to ee$ and $\phi \to ee$ would easily be seen, as in proton–A collisions. These peaks are missing, however [81]. It is tempting to interpret this experimental result as evidence that the light hadrons have "melted" in the hot plasma.

3. Screening and suppression of J/ψ.
One of the first theoretical suggestions for a signal of the existence of the quark–gluon plasma concerned the heavy quark states J/ψ. It was argued, and backed up with potential model calculations, that quasi-free quark and gluon thermal screening would reduce the attractive forces between the heavy quarks sufficiently to eliminate the binding energies and the J/ψ

states themselves [49]. In fact, the suppression of these states, which is again implied by the absence of the associated lepton pairs, is much more dramatic than that expected in hadron-absorption models.

RHIC experiments are in their infancy but hold great promise for elucidating the dynamics of QCD on a wide range of length and time scales.

Heavy-ion collisions and the critical end-point

What can heavy-ion-collision experiments say about the phase diagram of QCD? On looking at the phase diagram in Fig. 8.11 one observes that, since the transition between hadron and quark matter is not a proper phase transition, but rather a crossover, the strength of the signatures of the quark–gluon plasma should depend on the strength of this transition. Of course, in the ideal limit of $T \to \infty$ the quark–gluon plasma is qualitatively different from the hadron gas, but at the intermediate temperatures probed by heavy-ion colliders the distinction depends on the sharpness of the crossover. The too-well-known fact that hadronic models, if pushed to the extreme, can almost reproduce or at least mimick the signatures of the quark–gluon plasma should not come as a surprise – on the basis of our present-day knowledge of QCD, we must expect that the hadron gas smoothly transforms into the quark–gluon plasma as the temperature rises. There is no good order parameter distinguishing the two states and no phase transition.

Is there a property of the phase diagram that has distinct qualitative, not simply quantitative, features whose signatures cannot be mimicked? The most prominent feature of the phase diagram in Fig. 8.11 is the first-order transition line terminating in a critical point at T_E, μ_E. The end-point is especially attractive since it is the point where thermodynamic properties of QCD are singular. What is the location of this singularity, the coordinates T_E, μ_E? This question has an answer, but we do not know it. If we did, it would appear as one of the fundamental quantitative properties of the QCD phase diagram in this book. Lattice efforts to determine T_E, μ_E will be discussed later.

The strategy for discovering the QCD critical end-point [83, 84] is based on the fact that fluctuations of the sigma field acquire an anomalously large correlation length.[9] The divergence of this correlation length is the reason for the thermodynamic singularity. Such fluctuations induce characteristic correlations between particles that can interact by exchanging sigma quanta. Examples are pions, which are the most convenient particles to observe due to their abundance. By changing the parameters of the collision, notably $\sqrt{s}$, one can vary

---

[9] This correlation length is limited by the finite size and duration of the process to some 3 fm or so [84, 85], which is still very large compared with the correlation length in this channel in the vacuum, and with typical correlation scales at these temperatures.

the location in the phase diagram probed by the heavy-ion-collision experiment. As a function of $\sqrt{s}$, the signatures of the critical point must appear and disappear as the critical point is approached and then passed during such a scan. Measurements of particle ratios allow one to determine the values of the temperature and the chemical potential achieved during the final stages of the collision (before the system breaks up). By matching the two measurements one can determine the location of the critical point on the phase diagram of QCD.

9

Large chemical potentials and color superconductivity

9.1 Color superconductivity and color–flavor locking

Lattice-gauge theory has produced unique insights and predictions for QCD at high temperature, but it has had little to say about QCD in environments rich in baryons. There are several reasons for this impasse. One of them is the fact that the fermion determinant is complex in this environment, so the standard algorithms of the subject fail. This point will be demonstrated later. Other more conventional methods, such as weak-coupling approximations, apparently apply in this environment and suggest that the theory experiences color superconductivity at asymptotically large baryon chemical potentials. There is the hope that the transition to this exotic state of matter actually occurs at a chemical potential of just a few hundred MeV. The subject of color superconductivity produces interesting conjectures for the QCD phase diagram in the $T-\mu_B$ plane where T is small (below a few tens of MeV) and μ_B is very large. In this chapter we will discuss the major points, both strengths and weaknesses, of this approach and refer the reader to the growing literature for more recent, developing events. Once lattice-gauge theory develops an algorithm that applies to this environment, these ideas will be put to stringent tests. However, for much of the rest of this chapter, we must ignore lattice-gauge theory and focus on weak-coupling methods and the BCS theory of superconductivity and its application and extensions to QCD.

We shall see that these weak-coupling nonperturbative methods suggest a fascinating array of phenomena – superconductivity, exotic pairings and condensates, novel spectroscopy, etc.

It is interesting to recall that, in the earliest days of this subject, physicists believed that QCD at large μ_B would be easy to understand. The chemical potential μ_B carries the dimensions of energy, so one argued that the relevant coupling would be $g(\mu_B)$, which would vanish logarithmically as $\mu_B \to \infty$. Therefore, the nonperturbative phenomena, such as the formation of bound states, chiral-symmetry breaking, and confinement itself, would disappear in an environment rich in baryons. At high chemical potential, the argument went, QCD would become a weakly coupled plasma and the phase diagram of the theory would be described by the two-component model, Fig. 8.6.

Physicists born and raised in the era of the BCS explanation of superconductivity knew better [21–24]. If one imagined that the ground state at asymptotic μ_B was a free-quark Fermi sphere, then one knows that even an infinitesimal attractive residual interaction between quarks touches off an instability with respect to a lower-energy ground state consisting of Cooper pairs. Actually BCS superconductivity occurs in a very sophisticated fashion in metals at low temperatures. There the strong Coulomb repulsion between electrons is screened by the nuclei, leaving the weak, but attractive, exchange of low-energy, low-momentum sound waves (phonons) to act between the electrons on the opposite sides of the Fermi surface and lead to the condensation of large, floppy Cooper pairs. These doubly charged pairs condense in the new ground state and further shield it from electromagnetic fields. In the language of field theory, the photon develops a mass in the true ground state through the Meissner effect and there are energy gaps for all the excitations in the superconducting state. The energy gap, Δ, the energy required to break up a Cooper pair, is the source of the system's superconductivity.

Can some of these ideas be applied to QCD? It has been argued that QCD should experience color superconductivity through a BCS mechanism at asymptotic μ_B and it is hoped that this mechanism persists down to reasonable values of the chemical potential, producing color superconductivity inside "typical" neutron stars. One imagines the free-quark model of three massless quarks and three colors at asympototically large μ_B. Sharp Fermi surfaces would appear. A free colored gluon would not propagate easily through such a system. It would experience Debye screening, Landau damping, etc. and single-gluon exchange between quarks on the top of the Fermi surfaces would destabilize them and lead to pairing. Since single-gluon exchange is attractive in the $\bar{3}$ channel, one would expect, perhaps, that $\bar{3}$ diquark condensation would be energetically favorable with respect to a sharp Fermi surface of uncorrelated quarks, and color superconductivity would occur.

This physical picture has been analyzed more critically and there are surprises and subtleties, although color superconductivity is expected with a sustantial mass gap of several tens of MeV. First, one might question the applicability of asymptotic freedom to the dynamics on the surface of the Fermi sphere. The point is that high-energy quarks can scatter through glancing collisions, with small momentum transfers that are not controlled by $g(\mu_B) \ll 1$. One needs to develop, at least, a self-consistent approach to the dynamics here, comparable in quality to discussions of the field-theoretical approach to BCS superconductivity. There one looks for self-consistent superconducting solutions to the coupled Schwinger–Dyson equations for the fermion propagator, the gluon propagator, and the fermion–gluon vertex. We will discuss this program for color superconductivity further below, after we have introduced more of the most elementary features expected in the solution.

If single-vector-gluon exchange is the dominant destabilizing interaction at large chemical potential, then the wavefunction of the ground-state diquark condensate has some simple and striking group-theoretical features, called color–flavor locking [62]. Right-helicity fermions couple with right-helicity fermions through single-gluon exchange and left-helicity fermions couple with left-helicity fermions in such a way that the ground state is a scalar,

$$\langle \psi_{iL}^{a\alpha}(p)\psi_{jL}^{b\beta}(-p)\rangle = -\langle \psi_{iR}^{a\alpha}(p)\psi_{jR}^{b\beta}(-p)\rangle = \Delta(p)\epsilon^{ab}\epsilon^{\alpha\beta\rho}\epsilon_{ij\rho} \qquad (9.1)$$

where α and β are color indices, i and j are flavor indices, and a and b are spinor indices. It is crucial here that there are three flavors and three colors. In fact, the condensate wavefunction carries the indices $\epsilon^{\alpha\beta\rho}\epsilon_{ij\rho} = \delta_a^\alpha\delta_b^\beta - \delta_b^\alpha\delta_a^\beta$ and the color and flavor indices are "locked" together. We see that color and flavor must be rotated together to preserve the condensate. This curious "locking" feature of the ground state is similar to other broken-symmetry states in condensed-matter and high-energy physics. In ^{3}He physics, one has the famous B phase in which spin and orbital rotations are locked together by the condensate. Closer to home, the chiral-symmetry-breaking condensate of QCD at low T and μ_B, the condensate of left-handed quarks and right-handed anti-quarks, locks the $SU(3)_L$ and $SU(3)_R$ flavor rotations.

It is particularly curious that the condensates in Eq. (9.1) lock the flavor chiral groups, $SU(3)_L$ and $SU(3)_R$, together through the intermediary symmetry $SU(3)_{color}$. Apparently, color–flavor locking provides a new mechanism for the dynamical breaking of chiral symmetry. In this picture, chiral symmetry is not restored at large μ_B.

Why should this pattern of condensates be favored over any other? We shall see that, in a small-coupling (large μ) regime, the pattern (9.1) is a solution of a self-consistent equation. In another language, this means that it is at least a *local* minimum of the ground-state energy functional. Whether it is a global minimum has not (yet) been rigorously proven, but a simple physical argument can be given: this pattern involves quarks of all nine color–flavor combinations, providing gaps for all of them. This makes the energy of such a ground state lower than that of other possibilities, which leave some quarks ungapped.

Before considering some of the dynamics here in greater detail, it is interesting to discuss two features of this new ground state. First, note that the condensate carries color; the wavefunction is a $\bar{3}$, so it drives a dynamical Higgs mechanism. A purist would complain that the notation here is breaking the exact, local color symmetry of QCD. In fact, the notation should be understood in the same way as that in which the Higgs mechanism of the standard electroweak model is described: at weak coupling we choose a gauge to do calculations and find the important correlations. This notation is extremely convenient and useful as long as we are considering just weak couplings.

Second, baryon number is broken in the superconducting state as well. Of course, ordinary superconductivity breaks electric charge in the same fashion. This does not mean that the state is violating conservation of charge in a fundamental sense, but rather that the condensate makes the charge quantum number indefinite and, by virtue of Gauss' law, electric charge flows easily into and out of macroscopic volumes.

Third, chiral symmetry is broken in the color–flavor-locked phase, but in a very different fashion than in ordinary hadronic matter. In ordinary hadronic matter, right-handed quark fields and left-handed anti-quark fields are correlated by strong attractive forces, a condensate forms, and a mass gap opens in the fermion propagator. When this state is subjected to an environment having an increasing chemical potential, there is less of a tendency to maintain that mass gap. The reason is that, if there is a mass gap, then, as the fermion states are occupied at nonzero, increasing chemical potential, the new fermions must be placed in states of relatively high energy. Eventually the energy cost is too great and the gap vanishes so that the additional fermions can reside at lower energy. In other words, the attractive forces that favor the formation of a gap are overwhelmed by the energetics of high-energy, additional fermions in states above the gap. So, at some critical μ_B, one would expect restoration of chiral symmetry and the gap to vanish and a sharp Fermi surface to appear in the system. However, this is not the end of the story, because that sharp Fermi surface is unstable with respect to the weakest attraction between the fermions at the top of the Fermi surface, and this new attraction causes its own chiral-symmetry breaking through color–flavor locking. Clearly, even though both phases break chiral symmetry, the mechanisms are totally different (see, however, Section 9.3.4 for another look at this question).

Now on to the development of color superconductivity.

9.2 Calculating the gap at asymptotically large μ

We shall compute the color-superconducting gap at asymptotically large μ. The Green functions describing quark pairing, and the equations which we need to obtain and solve, contain, besides an energy–momentum dependence, a complicated spin, color, and flavor matrix structure. Instead of writing down the equations for such Green functions straightforwardly and only then trying to simplify them in the limit $\mu \to \infty$, we shall choose another strategy. We shall first simplify the Lagrangian and then derive the simplified form of the equation we need from it. In this way we can separate the problem into much more manageable and physically intuitive small steps.

9.2.1 The effective action for quarks near the Fermi surface

The key idea is to identify the fields and their corresponding excitations which will dominate the effects of interest when μ is large. These are, typically, the lowest-energy excitations. In the almost free gas of quarks at large μ, they are the excitations corresponding to adding a quark (or a hole) in close proximity to the Fermi surface, where the kinetic energy of the quark $|\vec{p}|$ is almost cancelled out by the chemical potential μ. In the following we shall identify the fields which are responsible for these excitations and write down the effective action for them.

First consider the full Dirac Lagrangian of quarks in QCD in Euclidean space:

$$\bar{\Psi}[(\partial_\mu + igA_\mu)\gamma_\mu - \mu\gamma_0]\Psi \tag{9.2}$$

The quark fields Ψ carry Dirac, color, and flavor indices, which we have not written out in Eq. (9.2), and the gluon field is a matrix in color indices.

The kinetic term and free spectrum
The free-quark kinetic term (i.e. $A_\mu = 0$) for a mode of a given four momentum p in Euclidean space is

$$(\Psi^\dagger)^i_{\alpha a}[ip_0 + \vec{\alpha} \cdot \vec{p} - \mu]_{\alpha\beta}\Psi^i_{\beta a} \tag{9.3}$$

where, as usual, $\Psi^\dagger = \bar{\Psi}\gamma_0$ and $\alpha_i = i\gamma_0\gamma_i$. First, we see that the Dirac matrix $[ip_0 + \vec{\alpha} \cdot \vec{p} - \mu]_{\alpha\beta}$ does nothing to color a and flavor i indices. Color and flavor just "flow through," and the free action is a sum of actions for quarks of each of $N_c \times N_f$ color and flavor combinations. We shall suppress color and flavor indices in the following.

Since we are dealing with massless quarks, it is convenient to use the chiral representation of the 4×4 Dirac matrices:

$$\alpha_i = \begin{pmatrix} \sigma_i & 0 \\ 0 & -\sigma_i \end{pmatrix} \tag{9.4}$$

where σ_i are the usual 2×2 Pauli matrices. The upper and lower blocks in Eq. (9.4) correspond to spinors of right and left chiralities, respectively. As with color and flavor, the chirality also flows through – the kinetic terms do not mix right and left spinors. We are now left with only $2 \times N_c \times N_f$ actions that involve only two-component spinors. Pick the upper two components of the Dirac spinor (right chirality), so the action becomes

$$(\Psi^\dagger)_\alpha[ip_0 + \vec{\sigma} \cdot \vec{p} - \mu]_{\alpha\beta}\Psi_\beta \tag{9.5}$$

where the indices α and β now run over two values only.

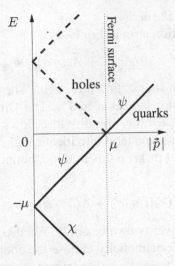

Figure 9.1. The spectrum of free massless quarks at finite μ. The branches with negative energy E correspond to antiparticles, or holes, with energy $|E|$ (dashed lines). The branches corresponding to the fields ψ ($E = |\vec{p}| - \mu$) and χ ($E = -|\vec{p}| - \mu$) are marked correspondingly. The dotted line denotes the Fermi surface $|\vec{p}| = \mu$.

For each vector $\vec{p}$ we can diagonalize the matrix $\vec{\alpha} \cdot \vec{p}$. Denote the eigenvector which satisfies $\vec{\alpha} \cdot \vec{p}\Psi = +|\vec{p}|\Psi$ by ψ. The other eigenvector, corresponding to $\vec{\alpha} \cdot \vec{p}\Psi = -|\vec{p}|\Psi$, will be denoted χ. Both ψ and χ are only single-component fields. The action for the field ψ is

$$\psi^{\dagger}(ip_0 + |\vec{p}| - \mu)\psi \tag{9.6}$$

This action describes an excitation with energy $E = -ip_0$ given by $E_{\psi} = |\vec{p}| - \mu$ (see Fig. 9.1). These are excitations corresponding to a free quark, when $E_{\psi} > 0$, and a free hole, when $E_{\psi} < 0$. The quark excitations can occur only outside of the Fermi surface $|\vec{p}| = \mu$, due to the Pauli principle, while the holes can exist only inside.

Similarly, the dispersion relation for excitations of χ is $E_{\chi} = -|\vec{p}| - \mu$ (see Fig. 9.1). What are these excitations? Their energy is always negative, therefore they describe only antiparticles, with energies $|\vec{p}| + \mu$. These energies are at least μ, because of the penalty that the chemical potential imposes on an extra anti-quark in the system. Thus, if we are interested in the processes which involve energies E much smaller than μ, the effect of such virtual excitations of χ will be suppressed typically by a power of E/μ due to a large energy denominator.[1]

---

[1] We shall also encounter interesting cases in which the effects of the antiparticles appear at leading order, because the contribution of particles vanishes or because the contribution of the antiparticles is additionally enhanced by a power of μ.

The interaction term and the action

Now consider the interaction term in Eq. (9.2):

$$(\Psi^\dagger)^i_{\alpha a}[i(A_0)_{ab} + \vec{A}_{ab} \cdot \vec{\alpha}]_{\alpha\beta} \Psi^i_{\beta b} \tag{9.7}$$

In this case only the flavor index "flows through." The gluon field A_μ is a color $N_c \times N_c$ matrix, with color indices a and b. The Dirac matrix is still block-diagonal in Dirac indices, $\alpha\beta$, i.e. chirality also flows through – see Eq. (9.4). Thus we can split the interaction term into identical $2_{LR} \times N_f$ terms each involving a $2_{\psi\chi} \times N_c$ spinor. Let us keep this in mind, but not write the color indices explicitly:

$$(\Psi^\dagger)_\alpha(k)[iA_0(q) + \vec{A}(q) \cdot \vec{\sigma}]_{\alpha\beta} \Psi_\beta(p) \tag{9.8}$$

The indices α and β run over two values only. We also restored the momentum arguments of the fields. For simplicity choose the coordinate system such that $\vec{p} = (0, 0, |\vec{p}|)$. Then the fields ψ and χ (eigenvectors of $\vec{\sigma} \cdot \vec{p}$) correspond to upper and lower components of the spinor Ψ. For a moment neglect the momentum of the gluon, i.e. set $q = 0$. Then, the momentum of the fermion is not changed, $p = k$, so

$$\Psi^\dagger \vec{A} \cdot \vec{\sigma} \Psi = \psi^\dagger A_3 \psi - \chi^\dagger A_3 \chi + \{\chi^\dagger[A_1 + iA_2]\psi + \text{h.c.}\} \tag{9.9}$$

We see that all terms, except the first one, involve the field χ, which produces contributions typically suppressed by a large energy denominator, as discussed earlier. In the gap equation, diagrams involving the field χ contribute only to the next order in the expansion in powers of E/μ, where E is the typical relevant energy. Therefore, we can neglect these terms for the calculation of the gap to leading order.

Finally, what is the effect of nonzero q? Equation (9.9) is valid when the orientation of the eigenvector $\psi(k)$ is the same as the orientation of the eigenvector $\psi(p)$ in the 2-spinor space. For $k \neq p$ these two eigenvectors are not aligned and additional terms are generated compared with Eq. (9.9). We can estimate the relative magnitude of these terms by calculating the overlap between $\psi(p)$ and $\psi(k)$, which is easiest to do using projectors:

$$\psi(p) = \frac{1 + \vec{v}_p \cdot \vec{\sigma}}{2} \Psi, \qquad \text{where} \qquad \vec{v}_p \equiv \frac{\vec{p}}{|\vec{p}|} \tag{9.10}$$

The vector $\vec{v}_p$ is the velocity of the quark of momentum $\vec{p}$. We can write for the overlap

$$\psi(k)^\dagger \psi(p) = \Psi^\dagger \frac{1 + \vec{v}_k \cdot \vec{\sigma}}{2} \frac{1 + \vec{v}_p \cdot \vec{\sigma}}{2} \Psi$$

$$= \psi(p)\psi(p) + \frac{\vec{v}_k - \vec{v}_p}{2} \Psi^\dagger \cdot \vec{\sigma} \psi(p) \tag{9.11}$$

If $\vec{q}$ is much smaller than μ, and we shall see that these are the gluon momenta that we need in our leading-order calculation, we can expand $\vec{v}_k - \vec{v}_p = \mathcal{O}(|\vec{q}|/|\vec{p}|)$. Since $|\vec{p}| \sim \mu$ for the quark excitations we are focusing on, we conclude that corrections to Eq. (9.9) due to $q \neq 0$ are suppressed by a power of q/μ. Thus, they are negligible in our leading-order calculation.

Now, we can also write $A_3 = \vec{v} \cdot \vec{A}$, where $\vec{v} = \vec{v}_p$. On putting together the kinetic term (9.6) and the interaction term for the fields ψ from Eq. (9.9) as well as the coupling to A_0, we finally obtain the effective Euclidean-space action for the quark excitations near the Fermi surface [86]:

$$S = \int_p \int_k \psi^\dagger(p)[(ip_0 + \epsilon_p)\delta_{pk} + g(iA_0(q) + \vec{v} \cdot \vec{A}(q))]\psi(k) \quad (9.12)$$

In Eq. (9.12) we used momentum modes of the fields, instead of the usual coordinate-space fields, to simplify the fermion kinetic term and to make Feynman-diagram rules easier to derive. We use the following short-hand notations:

$$\epsilon_p \equiv |\vec{p}| - \mu, \qquad \int_p \equiv \int \frac{d^4 p}{(2\pi)^4}, \qquad \delta_{pk} \equiv (2\pi)^4 \delta^4(p - k) \quad (9.13)$$

Note that the interaction with the gluon field has the very simple form of the coupling of a charge density $\psi^\dagger \psi$ to A_0 and current density $\psi^\dagger \vec{v} \psi$ to $\vec{A}$. One should bear in mind that the action (9.12) is applicable to modes of the field ψ describing quark/hole excitations near the Fermi surface, and that the momentum q carried by the gluon should be much smaller than μ in order to ensure that both p and k are in the vicinity of the Fermi surface. We shall see below that such soft gluons give the leading contribution to the gap due to a logarithmic singularity when $|q| \to 0$.

9.2.2 The strategy of the calculation and the meaning of the gap

How do we determine whether the phenomenon of BCS-like fermion pairing/condensation is occurring and how do we quantify the strength of this phenomenon? The general strategy is based on the following observation. If pairing occurs the ground state of the system is modified compared with the "naive" perturbative one. In the Lagrangian or path-integral formulation of the field theory, we can detect it by evaluating certain Green functions, which are *vacuum* expectation values of fields. We look at functions that would be zero in the perturbative vacuum. In our case we shall look at the following two-point function: $\langle \psi(p)\psi(-p) \rangle$. This fermion–fermion propagator is "anomalous" in the sense that it must be zero in the naive vacuum, by virtue of conservation of fermion number, but if the vacuum contains a condensate of fermion pairs, such a fermion-number-violating process becomes possible and the Green function becomes nonzero. In a more common case of spontaneous symmetry breaking, the vacuum expectation value of a scalar field is also such an anomalous (one-point, in this case) Green function.

However, how can the Green function be nonzero if we are able to calculate it only in perturbation theory? The answer is that we must go beyond perturbation theory to evaluate this function. In principle, such a calculation requires resummation of an *infinite* number of Feynman diagrams. The standard method of doing this was introduced in the context of relativistic Lagrangian field theory by Nambu and Jona Lasinio. We assume some trial form of our anomalous Green function and use it in all perturbation-theory calculations. How do we determine whether our particular assumption is correct? We do that by calculating the very same Green function itself. By equating the calculated value of the function to the trial function we obtain a self-consistency equation for the Green function – often called the gap equation, terminology borrowed from BCS. This idea is akin to the mean-field approximation in spin systems.

Obviously, the gap equation always has a trivial solution, corresponding to the naive perturbative value of the Green function. This corresponds, in Hamiltonian language, to the fact that the trivial vacuum, although it is not the most energetically favorable, is still a local extremum of the energy. The nontrivial solution, if and when it exists, signifies that another such extremum exists. Whether this is a more favorable vacuum can be verified by calculating the energy (given by vacuum, or bubble, graphs).

This procedure can, in principle, be carried out to arbitrary order in perturbation theory, but it is seldom practical beyond the leading order. The equation is nonlinear in the Green function already at leading order, and its solution is a function of the coupling constant g, which is often nonanalytic at $g = 0$. This reflects the fact that no *finite* number of Feynman graphs can reproduce this effect. It also means that corrections to the leading-order value of the Green function are not necessarily organized as an expansion in powers of g. However, it is possible to show that they become smaller as $g \to 0$, thus justifying the leading-order result.

From the point of view of the perturbative expansion of the path integral, we can also think of this procedure in the following equivalent way (which is the original view of Nambu and Jona Lasinio). The perturbative expansion is set up by separating out a bilinear, "free" part $\mathcal{L}_0$ from the bare Lagrangian $\mathcal{L}$: $\mathcal{L} = \mathcal{L}_0 + \mathcal{L}_I$. The remaining, "interacting" part of the Lagrangian, $\mathcal{L}_I$, is then used to generate vertices, while $\mathcal{L}_0$ provides propagators. Since the anomalous function is a two-point function, using it in perturbative calculations amounts to adding a certain bilinear term into the Lagrangian $\mathcal{L}_0$. To prevent changing the path integral, one must incorporate the same bilinear into $\mathcal{L}_I$ with the opposite sign – a counterterm. This counterterm generates an additional two-point vertex.

The result of the calculation will not depend on how we separate the Lagrangian into the interacting and free parts. Let us now choose a trial separation, giving a nonzero value for the anomalous Green function. This can be achieved by bringing a term $\Delta_p \psi(p)\psi(-p)$ into $\mathcal{L}_0$. We can now calculate corrections to $\langle \psi(p)\psi(-p) \rangle$ in our modified perturbation theory. If we can find a value for Δ_p that makes the corrections vanish, order by order in perturbation theory, we will have found the exact value of Δ_p.

Now we are almost ready to write the gap equation, but before doing this let's clarify the meaning of the parameter Δ_p. This will not be necessary to derive or solve the equation, but it will help us understand what we are trying to calculate. At this point, we should pay some attention to the color–flavor structure of the condensate $\langle \psi_a^i \psi_b^j \rangle$. As we discussed in the introduction to this section, the pairing occurs between the quarks in the anti-symmetric color (as well as flavor) channel. Thus the fields ψ_a^i and ψ_b^j must always have different colors and different flavors. Denote them $\psi_1 = \psi_a^i$ and $\psi_2 = \psi_b^j$. The bilinear Lagrangian, consisting of the bilinear parts of Eq. (9.12) for fields ψ_1 and ψ_2 with the term $\Delta_p \psi_1(p)\psi_2(-p) + \text{h.c.}$ added, can be written as

$$\begin{pmatrix} \psi_1(p) \\ \psi_2^*(-p) \end{pmatrix}^\dagger \begin{bmatrix} ip_0 + \epsilon_p & -\Delta_p \\ \Delta_p & -ip_0 + \epsilon_p \end{bmatrix} \begin{pmatrix} \psi_1(p) \\ \psi_2^*(-p) \end{pmatrix} \tag{9.14}$$

Thus, the anomalous propagator is given in terms of the function Δ_p as

$$\langle \psi_1(p)\psi_2(-p) \rangle = \frac{\Delta_p}{p_0^2 + \epsilon_p^2 + \Delta_p^2} \tag{9.15}$$

The normal propagator is also modified:

$$\langle \psi_1^\dagger(p)\psi_1(p) \rangle = \langle \psi_2^\dagger(p)\psi_2(p) \rangle = \frac{-ip_0 + \epsilon_p}{p_0^2 + \epsilon_p^2 + \Delta_p^2} \tag{9.16}$$

The poles of this propagator determine dispersion relations for excitations. The energy of propagating excitations, $E = -ip_0$, satisfies $-E^2 + \epsilon_p^2 + \Delta_p^2 = 0$. For simplicity, let us neglect the dependence of Δ_p on p. This is usually (and we shall see it explicitly in our case) a good approximation in the range of values of p_0 or ϵ_p of order Δ_p. Denote the value of Δ_p for momenta p near the Fermi surface by Δ_0. Then we have two branches:

$$E = \pm\sqrt{\epsilon_p^2 + \Delta_0^2} = \pm\sqrt{(|\vec{p}| - \mu)^2 + \Delta_0^2} \tag{9.17}$$

There is a gap between these two branches of the spectrum, equal to $2\Delta_0$ (see Fig. 9.2). It takes a finite energy Δ_0 to excite either a quark or a hole in the system.

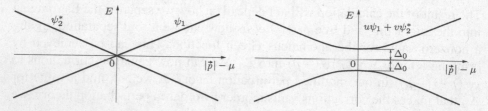

Figure 9.2. The spectrum of the quark and hole excitations near the Fermi surface. The left-hand graph is a zoom-in on the region around $|\vec{p}| = \mu$, $E = 0$ in Fig. 9.1. The spectrum of the field ψ_2^* is inverted due to charge conjugation. On the right-hand side, the spectrum is given by Eq. (9.17). The excitations are now quanta of a field, which is a linear superposition of ψ_1 and ψ_2^*, as indicated.

The excitations of the fields ψ_1 and ψ_2^* mix with each other because of the anomalous propagator (9.15). The fields, corresponding to the energy eigenstates in Eq. (9.17), can be found by diagonalizing the matrix (9.14). For the positive-energy branch it is given by

$$\psi_+ = u_p \psi_1 + v_p \psi_2^*, \qquad \text{with} \qquad \frac{v_p}{u_p} = \tan\phi_p \qquad \text{and } \tan(2\phi_p) = \frac{\Delta}{\epsilon_p}$$

$$(9.18)$$

Near the Fermi surface, $\epsilon_p = 0$, the mixing is maximal, $u_p = v_p$, as it should be because the energy levels cross there (see Fig. 9.2).

9.2.3 The gap equation

The fermion propagator: two flavors
First consider the case of two flavors: $N_f = 2$. The calculation in the case of three flavors is similar, but requires a little more care to set up because of its more complicated color–flavor mixing. As we discussed, the condensate should be anti-symmetric in flavor and color. Such a pairing can be achieved if the anomalous term in the Lagrangian has the following flavor–color form:

$$\Delta_p \epsilon^{ij} \epsilon_{ab3} \psi_a^i(p) \psi_b^i(-p) \tag{9.19}$$

We have arbitrarily chosen the color direction of the condensate to be 3. Let us use letters u and d to denote fields of the two flavors: $u_a = \psi_a^1$ and $d_a = \psi_a^2$, where a is the color index. Then the term (9.19) induces pairing similar to that described by Eq. (9.14) for pairs of fields $u_1 d_2$ and $u_2 d_1$. The propagators in each pair are the same as in Eqs. (9.15) and (9.16), with $\psi_1 \psi_2$ substituted for

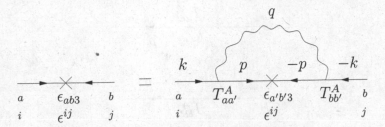

Figure 9.3. The diagrammatic representation of the gap equation. The "flow" of momentum as well as color and flavor indices is indicated.

$u_1 d_2$ or $u_2 d_1$. Therefore, the anomalous Green function can be written as

$$\langle \psi_a^i(-p)\psi_b^j(p)\rangle = \frac{\Delta_p}{p_0^2 + \epsilon_p^2 + \Delta_p^2}\epsilon^{ij}\epsilon_{ab3} \qquad (9.20)$$

The case of three flavors is a little more complicated because the anomalous term has the form

$$\Delta_p \epsilon^{ijk}\epsilon_{abk}\psi_a^i(p)\psi_b^j(-p) \qquad (9.21)$$

and mixes each flavor with two other flavors, e.g. there is a term $u_1 d_2$ as well as $u_1 s_3$, where $s_a = \psi_a^3$. Below we shall discuss a trick that helps us determine the propagator in this case. First consider the simpler case of two flavors.

The equation
Using the Lagrangian (9.12), introducing the bilinear term with a trial function Δ_p as in Eq. (9.19), and applying the self-consistency condition to leading order in the coupling g, we obtain the following gap equation:

$$\Delta_k = -\frac{2}{3}g^2 \int_p \frac{\Delta_p}{p_0^2 + \epsilon_p^2 + \Delta_p^2}D_{\mu\nu}(p-k)v_\mu(p)v_\nu(-p) \qquad (9.22)$$

where the factors $v_\mu(p) \equiv (\mathrm{i}, \vec{v}_p)$ and $v_\nu(-p) = (\mathrm{i}, -\vec{v}_p)$, come from the fermion-gluon vertex $\psi^\dagger v_\mu A_\mu \psi$ in Eq. (9.12) and $D_{\mu\nu}$ is the gluon propagator. Equation (9.22) is graphically represented in Fig. 9.3. The fermion propagator has the form given in Eq. (9.20) and we have removed the identical factors $\epsilon^{ij}\epsilon_{ab3}$ from the right- and left-hand sides of Eq. (9.22). The fact that the flavor and color tensor factors cancel out in Eq. (9.22) is not entirely trivial, and signifies that our choice of the flavor and color structure in Eq. (9.19) was correct – had we chosen a different structure, we might not have been able to satisfy the gap equation. The flow of the flavor indices (see Fig. 9.3) is trivial, because the quark–gluon vertices are "transparent" to flavor. However, color indices require the following algebraic calculation, the result of which is the factor $-\frac{2}{3}$ in Eq. (9.22).

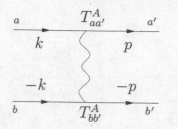

Figure 9.4. The one-gluon exchange between quarks of opposite momenta. Flow of color and momentum is indicated. On glueing the quark lines on the right with an anomalous quark propagator one obtains the diagram in Fig. 9.3

The color indices in the quark–gluon vertex in Fig. 9.3 come from the interaction term in Eq. (9.12), $\psi^\dagger v_\mu A_\mu \psi$, if we recall that A_μ is a matrix in color space, which can be written in the basis of eight color generators T^A as $A_\mu = A_\mu^A T^A$, where the index $A = 1, \ldots, 8$ runs over eight gluon colors. The traceless 3×3 matrices T^A are normalized as

$$\text{tr}(T^A T^B) = \frac{1}{2}\delta^{AB} \tag{9.23}$$

The color factors in the diagram on the right-hand side of the gap equation in Fig. 9.3 then give

$$T^A_{aa'} T^A_{bb'} \epsilon_{a'b'3} = \frac{1}{2}\left(\delta_{ab'}\delta_{a'b} - \frac{1}{N_c}\delta_{aa'}\delta_{bb'}\right)\epsilon_{a'b'3} = -\frac{N_c + 1}{2N_c}\epsilon_{ab3} = -\frac{2}{3}\epsilon_{ab3} \tag{9.24}$$

In Eq. (9.24) we have used the identity[2]

$$T^A_{aa'} T^A_{bb'} = \frac{1}{2}\delta_{ab'}\delta_{a'b} - \frac{1}{2N_c}\delta_{aa'}\delta_{bb'} \tag{9.25}$$

The factor $-\frac{2}{3}$ in Eq. (9.24) carries information about the strength and the sign of the interaction between the two quarks due to the exchange of a gluon. This can be seen by drawing the quark lines in Fig. 9.3 in the way shown in Fig. 9.4. If we assume that Δ_p is positive, the right-hand side of the gap equation (9.22) is manifestly positive – the gluon propagator is positive in Euclidean space and symmetric with respect to $\mu \leftrightarrow \nu$, i.e.

$$-D_{\mu\nu}v_\mu(p)v_\nu(-p) = D_{00} + D_{ij}v_i v_j > 0 \tag{9.26}$$

---

[2] This identity is easy to derive using the fact that any $N_c \times N_c$ matrix can be represented in the basis of matrices T^A, completed with the unit matrix (normalized by a factor $1/\sqrt{2N_c}$). This means that, for an arbitrary matrix X, we can write $X = 2\,\text{tr}(XT^A)\,T^A$, where the summation over $A = 0, 1, \ldots, 8$ also includes the matrix $T^0 = 1/\sqrt{2N_c}$.

where $\vec{v} \equiv \vec{v}_p \equiv \vec{p}/|\vec{p}|$, as before. Like the cancelling out of the color tensors, the consistency of the overall sign in Eq. (9.22) is also an indication that our choice of the trial anomalous term (9.19) was correct. Indeed, suppose for a moment that we had chosen the trial gap term to be *symmetric* in color (this would correspond to a color representation **6** as opposed to $\bar{\mathbf{3}}$). Then, on repeating the algebra in Eq. (9.24) with a symmetric tensor instead of ϵ_{ab3}, we would have found that, although the tensor structure is preserved, the factor is now positive, $(N_c - 1)/(2N_c) = +\frac{1}{3}$. This is because the members of a pair of quarks in a symmetric color representation repel each other. Also note that the signs of the electric and the magnetic terms in Eq. (9.26) being the same reflects the fact that the quarks which pair, according to Eq. (9.19), are moving in opposite directions, and thus, as in electrodynamics, if they attract electrically, they also attract magnetically.

Pointlike interaction
For a moment imagine that the gluon does not propagate, i.e. it provides a pointlike interaction, both in space and in time. In other words, consider a propagator that is independent of the 4-momentum $p - k$: $D_{\mu\nu} = \delta_{\mu\nu}/\Lambda^2$, where Λ is some constant mass parameter. Then Eq. (9.22) takes the form

$$\Delta_k = \frac{4}{3}\frac{g^2}{\Lambda^2} \int_p \frac{\Delta_p}{p_0^2 + \epsilon_p^2 + \Delta_p^2} \tag{9.27}$$

Since the right-hand side does not depend on k now, we conclude that $\Delta_k \equiv \Delta$ is constant. The integral over p can be done. We see that, when $\Delta \ll \mu$, the integrand is appreciably nonzero only for p_0 and ϵ_p of order Δ, and that the integral diverges logarithmically when $\Delta \to 0$. This justfies our use of the effective action (9.12) as long as we limit ourselves to logarithmic precision. Within this precision, we can write the integral in the following way, by using the fact that it is dominated by a thin shell around the Fermi surface:

$$\int_p \equiv \int \frac{d^4 p}{(2\pi)^4} = \frac{\mu^2}{2\pi^2} \int \frac{dp_0 \, d\epsilon_p}{2\pi} \frac{d\Omega_p}{4\pi} \tag{9.28}$$

We find, to logarithmic precision,

$$\Delta = \frac{\mu^2}{2\pi^2}\frac{4}{3}\frac{g^2}{\Lambda^2} \int dp_0 \frac{\Delta}{2\sqrt{p_0^2 + \Delta^2}} = \frac{\mu^2}{2\pi^2}\frac{4}{3}\frac{g^2}{\Lambda^2} \Delta \ln\left(\frac{\mu}{\Delta}\right) \tag{9.29}$$

where we cut off the intergal in the ultraviolet at a scale of order μ, since at this scale our approximate action (9.12) breaks down. The nontrivial solution to the gap equation (9.29) is thus

$$\Delta = \mu \exp\left(-\frac{3\pi^2 \Lambda^2}{2\mu^2}\frac{1}{g^2}\right) \tag{9.30}$$

with exponential precision. This result – an exponentially small value of the gap $\exp(-\text{constant}/g^2)$ is familiar from BCS theory, in which the role of an approximately pointlike interaction is played by phonon exchange.

We should note two features of the solution (9.30). First, the gap is non-analytic in the coupling g, so it is not expandable in powers of g. Second, the gap is exponentially small at small coupling $\Delta \ll \mu$ (we assume that $\mu < \Lambda$). Both features will remain qualitatively true when we include the effect of the non-locality of gluon exchange.

Long-range magnetic-gluon exchange

In QCD, we have to take into account the momentum $q = p - k$ dependence of the propagator $D_{\mu\nu}$, i.e. the long range and the temporal retardation of the interaction. This can be done to logarithmic precision in Eq. (9.22). If we take the tree-level propagator $D_{\mu\nu}(q) \sim 1/q^2$ and specialize to $|\vec{k}| \approx |\vec{p}| \approx \mu$, we can write

$$D_{\mu\nu}(p - q) \sim \frac{1}{(p_0 - k_0)^2 + 2\mu^2(1 - \cos\theta)} \tag{9.31}$$

where θ is the angle between $\vec{p}$ and $\vec{k}$. The integral over this angle in Eq. (9.22) is logarithmically divergent at $\theta = 0$ when $p_0 = k_0$ and is cut off at $\theta \sim q_0/\mu$. This means that small-angle scattering of the quarks dominates, and it will saturate the result to logarithmic precision. The right-hand side of the gap equation now has an additional logarithmic divergence, a collinear one, apart from the usual BCS logarithmic divergence in Eq. (9.29). This will turn out to change the coupling-constant dependence of the gap. To calculate this dependence correctly, one needs to go beyond the free-gluon propagator. The reason is that the cutoff $\theta \sim q_0/\mu$ may become irrelevant if q_0 is too small, and we must take into account the effect of the gluon polarization tensor $\Pi_{\mu\nu}$ in the medium of quarks. Although the polarization tensor is of order g^2, its effect on small-angle scattering can become dominant if q_0 is sufficiently small, typically when $q_0^2 < \Pi_{\mu\nu}$. For example, Debye screening will provide the cutoff for the scattering angle $\theta_{\min} \sim |\vec{q}|_{\min}/\mu \sim g$, which becomes relevant when $q_0 < g\mu$.

We shall need to consider the gluon propagator only in the regime $q_0 \ll |\vec{q}|$, for, as we shall see, the logarithmically dominant contribution to the gap equation is accumulated in that regime. In this case, the propagator, with the effect of the polarization included and to the precision we require, can be written in Coulomb gauge as[3]

$$D_{00}(q) = \frac{1}{|\vec{q}|^2 + g^2\mu^2} \tag{9.32}$$

The actual coefficient of the Debye mass term $g^2\mu^2$ is not unity as in Eq. (9.32)

---

[3] We shall show how to derive this result and discuss the issue of gauge invariance below.

but its precise value does not affect the result in the logarithmic approximation. To the same precision,

$$D_{ij}(q) = \frac{\delta_{ij} - \hat{q}_i \hat{q}_j}{|\vec{q}|^2 + g^2 \mu^2 \frac{q_0}{|\vec{q}|}} \tag{9.33}$$

where $\hat{q}_i \equiv \vec{q}/|\vec{q}|$. The last term in the denominator in Eq. (9.33), whose actual numerical coefficient is also unimportant, describes Landau damping of the spacelike ($|\vec{q}| \gg q_0$) gluon modes. Note that it causes screening of the slowly *varying* magnetic fields and vanishes for static fields ($q_0 = 0$) – this is dynamical screening. It cuts off fields at the wave vector $|\vec{q}| < (g^2 \mu^2 q_0)^{1/3}$. For $q_0 \ll g\mu$ this screening scale is much lower than $g\mu$. Thus the collinear logarithmic divergence of the magnetic interaction is cut off much closer to the singularity than is the electric, Coulomb, interaction. In other words, the leading contribution to the collinear logarithmic singularity is from the magnetic gluons, as long as $q_0 \ll g\mu$. Retaining only the contribution of the magnetic gluons, one arrives at

$$\Delta_k = \frac{2g^2}{3} \int_p \frac{\Delta_p}{p_0^2 + \epsilon_p^2 + \Delta_p^2} \frac{1}{|\vec{q}|^2 + g^2 \mu^2 \frac{q_0}{|\vec{q}|}} \tag{9.34}$$

The electric gluons contribute significantly only when $q_0 > g\mu$. We shall return to the contribution of electric gluons later and see that their effect is indeed subleading in the limit $g \to 0$.

The integral in Eq. (9.34) is over the energy p_0, magnitude of momentum $|\vec{p}|$, or $\epsilon_p = |\vec{p}| - \mu$, and direction of momentum Ω_p and can be approximated as in Eq. (9.28). Taking the integral over ϵ_p first, using the method of residues:

$$\Delta_k = \frac{2g^2}{3} \frac{\mu^2}{2\pi^2} \int dp_0 \frac{d\Omega_p}{4\pi} \frac{\Delta_p}{2\sqrt{p_0^2 + \Delta_p^2}} \frac{1}{|\vec{q}|^2 + g^2 \mu^2 \frac{q_0}{|\vec{q}|}} \tag{9.35}$$

The gap on the right-hand side, which was initially a function both of p_0 and of $|\vec{p}|$ ($\epsilon_p = |\vec{p}| - \mu$) is now a function of p_0 alone, since, as a result of integration over ϵ_p we have put p on the mass shell: $p_0^2 + \epsilon_p^2 + \Delta_p^2 = 0$, i.e. $|\vec{p}|$ is a function of p_0 in Eq. (9.35). The integral over the angle Ω_p can also be taken, in logarithmic approximation, giving

$$\Delta_k = \frac{2g^2}{3} \frac{\mu^2}{2\pi^2} \int dp_0 \frac{\Delta_p}{2\sqrt{p_0^2 + \Delta_p^2}} \frac{1}{2\mu^2} \ln\left(\frac{\mu}{(g^2 \mu^2 q_0)^{1/3}}\right) \tag{9.36}$$

The argument of the collinear logarithm in Eq. (9.36) is equal to $1/\theta_{min}$, where $\theta_{min} \sim |\vec{q}|_{min}/\mu \sim (g^2 \mu^2 q_0)^{1/3}/\mu$ is the scattering angle at which magnetic-gluon exchange is cut off by Landau damping. The integral over p_0 adds one more logarithmic divergence, which has the same origin as in the usual BCS case. One can divide the interval spanned by $|p_0|$ into four regions, by order of magnitude: $(0, \Delta_0)$, $(\Delta_0, |k_0|)$, $(|k_0|, g\mu)$, and $(g\mu, \mu)$. In the first region, the integral does not contribute to the BCS logarithm (the singularity is regularized by Δ in the denominator). In the second and third regions, we can replace $|q_0| = |p_0 - k_0|$ by the largest of the momenta: $|k_0|$ in the second region and $|p_0|$ in the third, i.e. to logarithmic precision we have $|q_0| = \max(|p_0|, |k_0|)$. To the same precision we can also neglect Δ compared with p_0 in $\sqrt{p_0^2 + \Delta_p^2} = |p_0|$. The last region $(g\mu, \mu)$ gives a parametrically small contribution, as we shall see shortly, and we can ignore it in the leading logarithmic approximation. Within the same precision we can also neglect the power of g under the logarithm in Eq. (9.36):

$$\Delta_k = \frac{g^2}{6\pi^2}\left[\frac{1}{3}\int_{\Delta_0}^{k_0} dp_0 \frac{\Delta_p}{p_0}\ln\left(\frac{\mu}{k_0}\right) + \frac{1}{3}\int_{k_0}^{\mu} dp_0 \frac{\Delta_p}{p_0}\ln\left(\frac{\mu}{p_0}\right)\right] \quad (9.37)$$

It is convenient to introduce logarithmic variables:

$$x = \ln\left(\frac{\mu}{k_0}\right), \qquad y = \ln\left(\frac{\mu}{p_0}\right), \qquad x_0 = \ln\left(\frac{\mu}{\Delta_0}\right),$$

where $\Delta_0 \equiv \Delta_{p_0 = \Delta_0}$

$$(9.38)$$

The variables x and y measure the "order of magnitude" of k_0 and p_0 on the scale between Δ_0 and μ. The value Δ_0 is the value of the actual gap near the Fermi sphere (see Fig. 9.2). As we shall see at the end of this calculation, the gap is exponentially smaller than μ when $g \to 0$, i.e. the total number of e-foldings between Δ_0 and μ is proportional to $1/g$ in this limit (see Eq. (9.45)). This fact should make it clear why the contribution of the scale interval $(g\mu, \mu)$ is small. It spans only $\ln(1/g)$ e-foldings, which is parametrically a negligible number compared with $1/g$ (see also Fig. 9.9). Then

$$\Delta(x) = \frac{g^2}{6\pi^2}\left(\frac{1}{3}\int_x^{x_0} dy\,\Delta(y)x + \frac{1}{3}\int_0^x dy\,\Delta(y)y\right) \quad (9.39)$$

where we introduced $\Delta(x) \equiv \Delta_k$ with x from Eq. (9.38).

This integral equation can be transformed into a differential equation with boundary conditions and solved. First note that Eq. (9.39) implies that

$$\Delta(0) = 0 \quad (9.40)$$

– i.e. the pairing strength, or gap, vanishes far from the Fermi surface, at the scale of order μ. Taking the derivative of Eq. (9.39) gives

$$\Delta'(x) = \frac{g^2}{18\pi^2}\int_x^{x_0} dy\,\Delta(y) \quad (9.41)$$

This equation implies another boundary condition:

$$\Delta'(x_0) = 0 \tag{9.42}$$

– the gap does not vary with energy near the Fermi surface. Now, on differentiating once again, we find

$$\Delta''(x) = -\frac{g^2}{18\pi^2}\Delta(x) \tag{9.43}$$

which has a two-parameter solution,

$$\Delta(x) = A \sin\left(\frac{g}{3\sqrt{2\pi}}x + \phi\right) \tag{9.44}$$

The boundary conditions (9.40) and (9.42) should be used here. Equation (9.40) means that $\phi = 0$. The condition (9.42) means that the gap (9.44) reaches its maximum at $x = x_0$, i.e. $A = \Delta(x_0)$.

It may seem that we cannot fix the amplitude A because the differential equation we are solving, Eq. (9.43), as well as the boundary conditions (9.40) and (9.42) are homogeneous. However, Eq. (9.42) means that the argument of the sine, $gx/(3\sqrt{2\pi})$, is equal to $\pi/2$ at $x = x_0$, i.e.

$$x_0 = \frac{3\pi^2}{\sqrt{2}g} \tag{9.45}$$

However, the value of x_0 is the number of e-foldings between the scales of the gap Δ_0 and μ, Eq. (9.38), i.e. [87–89]

$$\Delta_0 = \mu e^{-x_0} = \mu \exp\left(-\frac{3\pi^2}{\sqrt{2}g}\right) \tag{9.46}$$

Since $\Delta_0 = \Delta(x_0) = A$, the energy dependence of the gap reads finally

$$\Delta_p = \Delta_0 \sin\left[\frac{g}{3\sqrt{2\pi}}\ln\left(\frac{\mu}{|p_0|}\right)\right] \tag{9.47}$$

which is sketched in Fig. 9.5.

We see that the appearance of the collinear logarithm in addition to the usual BCS one changed the power of $1/g$ in the exponent of the gap from $1/g^2$ in Eq. (9.30) to $1/g$ in Eq. (9.46). Another notable consequence of the nonlocality of the interaction is the nonnegligible energy dependence of the gap in Eq. (9.47).

Electric gluons and higher-order corrections
What is the effect of the electric gluons described by the propagator (9.32)? As we discussed, the collinear divergence in the Coulomb propagator (9.32) is regulated by Debye screening at the scale $|\vec{q}| \sim g\mu$, i.e. electric gluons contribute

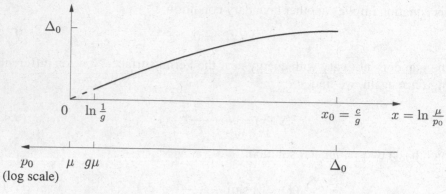

Figure 9.5. The energy dependence of the gap in the leading logarithmic approximation, Eq. (9.47). Note also that the energy dependence in the tiny interval $(g\mu, \mu)$ should be corrected for the effects of the electric gluons, if the
terms of order $\ln(1/g)$ are to be retained together with terms of order $1/g$ as in
Eq. (9.48).

to the collinear logarithm only when q_0 is in the interval $(g\mu, \mu)$. This means
that at least p_0 or k_0 must lie in this interval. In terms of the logarithmic variable x, Eq. (9.38), the size of the corresponding interval is $\ln(1/g)$, compared
with the total scale interval $x_0 = 1/g$. One can therefore expect the effect of the
electric gluons to (a) modify the function $\Delta(x)$ significantly in the small interval
$(0, \ln(1/g))$ and (b) shift the value of x_0 in Eq. (9.45) by a correction of order
$\ln(1/g)$. We leave the detailed calculation as an exercise to the reader. We note
only that the corrections of order $\ln(1/g)$ from the magnetic gluons, which we
neglected, must also be included at this level of precision. The result is [87]

$$x_0 = \frac{3\pi^2}{\sqrt{2}g} - 5\ln\left(\frac{1}{g}\right) \tag{9.48}$$

where we neglected terms parametrically smaller than $\ln(1/g)$. Terms $\ln(1/g)$
and higher produce pre-exponential multiplicative corrections to the value of the
gap at the Fermi surface:

$$\Delta_0 = \mu e^{-x_0} = \mu g^{-5} \exp\left(-\frac{3\pi^2}{\sqrt{2}g}\right) \tag{9.49}$$

We also note that, since electric gluons provide extra attraction, they increase
the gap, as is implied by the sign of the correction in Eqs. (9.48) and (9.49). Of
course, Eq. (9.49) is good only up to an unknown constant factor.

The estimates of higher-order corrections to the gap equation indicate that
neither the coefficient of $1/g$ nor that of the logarithm in Eq. (9.48) is modified

by such corrections [90]. Higher-order corrections contribute only additional sub-leading terms, which are finite as $g \to 0$, to Eq. (9.48). Consider, for example, corrections that can be absorbed into the renormalization of the coupling g. These corrections amount to a replacement $g \to g + O(g^3 \ln(\mu/\Delta))$. On performing such a shift of the coupling in Eq. (9.48) and accounting for the fact that the large renormalization logarithm $\ln(\mu/\Delta)$ is of order $1/g$, we see that x_0 is modified by at most $O(g^0)$.

The precise value of the next-order, constant, term in Eq. (9.48) is unknown at the time of writing. This is especially unfortunate, because a constant term in Eq. (9.48) translates into a constant *factor* in Eq. (9.49) that determines the absolute size of the gap.

The case of three flavors

As we have already mentioned, the case of three flavors is a little more involved because the anomalous, or pairing, term (9.21) in the Lagrangian pairs each quark flavor with two others. In order to obtain the anomalous quark propagator one can use the following trick. Instead of ψ_a^i, use a different basis for the $N_f \times N_c = 9$ quark fields, ψ^A, such that

$$\psi_a^i = \sqrt{2} \sum_{A=0}^{8} (T^A)_{ia} \psi^A, \qquad \text{i.e.} \qquad \psi^A = \sqrt{2} \sum_{ia} (T^A)_{ai} \psi_a^i \qquad (9.50)$$

The term (9.21) is diagonal in the new basis:

$$\begin{aligned}
\epsilon^{ijk} \epsilon_{abk} \Delta_p \psi_a^i(p) \psi_b^j(-p) &= 2[\text{tr}\, T^A \, \text{tr}\, T^B - \text{tr}(T^A T^B)] \Delta_p \psi^A(p) \psi^B(-p) \\
&= (3\delta^{A0}\delta^{B0} - \delta^{AB}) \Delta_p \psi^A(p) \psi^B(-p) \\
&\equiv -\Delta_p^A \psi^A(p) \psi^A(-p)
\end{aligned} \qquad (9.51)$$

where in the last term we introduced

$$\Delta^A \equiv \begin{cases} \Delta & \text{for} \quad A = 1, \ldots, 8 \\ -2\Delta & \text{for} \quad A = 0 \end{cases} \qquad (9.52)$$

Since the normal term in the Lagrangian remains diagonal in the new basis,

$$\psi^{\dagger i}_a \psi_a^i = 2\,\text{tr}(T^A T^B \psi^{\dagger A} \psi^B) = \psi^{\dagger A} \psi^A \qquad (9.53)$$

we can now easily find the propagator (both the normal and the anomalous one):

$$\langle \psi^{\dagger A}(p) \psi^B(p) \rangle = \delta^{AB} \frac{-ip_0 + \epsilon_p}{p_0^2 + \epsilon_p^2 + (\Delta_p^A)^2}$$

$$\langle \psi^A(p) \psi^B(-p) \rangle = \delta^{AB} \frac{\Delta^A}{p_0^2 + \epsilon_p^2 + (\Delta_p^A)^2} \qquad (9.54)$$

We see that all nine quarks ψ^A acquire gaps. These gaps are not all equal, however. Eight of the quarks acquire the same gap Δ but one has the gap 2Δ. This fact creates a complication when we try to close the gap equation, as we shall see soon. Rewrite the anomalous propagator in Eq. (9.54) in the original basis of ψ_a^i:

$$
\begin{aligned}
\langle \psi_a^i \psi_b^j \rangle &= 2 \sum_{A=0}^{8} T_{ia}^A T_{jb}^B \frac{\Delta^A}{p_0^2 + \epsilon_p^2 + (\Delta_p^A)^2} \\
&= -\epsilon^{ijk} \epsilon_{abc} \frac{\Delta}{p_0^2 + \epsilon_p^2 + \Delta_p^2} \\
&\quad - \frac{1}{3} \delta_{ia} \delta_{jb} \left(\frac{2\Delta}{p_0^2 + \epsilon_p^2 + 4\Delta_p^2} - \frac{2\Delta}{p_0^2 + \epsilon_p^2 + \Delta_p^2} \right)
\end{aligned}
\tag{9.55}
$$

The last term, proportional to $\delta_{ia} \delta_{jb}$, corrects for the fact that, in the first term, proportional to $\epsilon^{ijk} \epsilon_{abc}$, we have put all denominators equal. We see that, although the anomalous term in the Lagrangian (9.21) is proportional to $\epsilon^{ijk} \epsilon_{abc}$, the anomalous propagator (9.55) is not.

After we have substituted the propagator into the gap equation in Fig. 9.3, the right-hand side of this equation will not be proportional to $\epsilon^{ijk} \epsilon_{abc}$. The first term in Eq. (9.55), sandwiched by the generators T^A from the gluon vertices, will give the term $\epsilon^{ijk} \epsilon_{abc}$, as in the two-flavor case (simply because it is anti-symmetric in color indices, see Eq. (9.24)). However, the second term, $\delta_{ia} \delta_{jb}$, will produce a mixture of terms symmetric and antisymmetric with respect to $a \leftrightarrow b$. This means that already at the one-loop level the anomalous term is a mixture of an anti-symmetric term (9.21) and a symmetric term, $\delta_a^i \delta_b^j + (a \leftrightarrow b)$. In other words, the quarks pair in a mixture of color $\bar{3}$ and color **6** states.

To account for this we should set up two equations for two gaps. However, to the leading logarithmic precision with which we are working, this is not needed. Indeed, the last term in the propagator (9.55), which is responsible for the color-symmetric admixture in the gap equation, does not produce the BCS logarithmic divergence $\ln(\mu/\Delta)$. Thus the symmetric (color **6**) pair condensate is suppressed by a factor $1/\ln(\mu/\Delta) \sim g$, and its effect can be neglected to the order with which we are working at large μ. In this case the gap equation closes and we obtain the same equation for Δ as in the two-flavor case.

Debye screening and Landau damping
Here we shall derive the expression for the gluon propagator $D_{\mu\nu}$ which we needed for the calculation of the gap. We can use the same effective action (9.12)

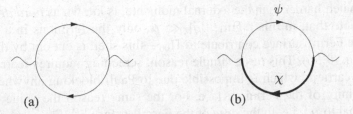

Figure 9.6. Quark loop diagrams contributing to the gluon polarization tensor $\Pi_{\mu\nu}$. Thin lines denote propagation of the field ψ, while the thick line is the propagator of the field χ – see Fig. 9.1. When the momentum running in the loop is near the Fermi surface, and the energy is small, the χ field is far off shell and its propagator can be effectively shrunk to a point, producing a local contribution to $\Pi_{\mu\nu}$. Only diagram (a) contributes to Π_{00}, see Eq. (9.56), while Π_{ij} receives contributions both from (a) and from (b), see Eqs. (9.59) and (9.63).

to calculate the one-loop polarization operator:

$$\Pi_{00}(q) = \mathcal{N}g^2 \int_p \frac{1}{ip_0 + \epsilon_p} \frac{1}{i(p_0 + q_0) + \epsilon_{p+q}}$$

$$= \frac{\mathcal{N}g^2\mu^2}{2\pi^2} \int \frac{d\Omega}{4\pi} d\epsilon_p \frac{\theta(\epsilon_p) - \theta(\epsilon_{p+q})}{iq_0 + \epsilon_{p+q} - \epsilon_p} = \frac{\mathcal{N}g^2\mu^2}{2\pi^2} \int \frac{d\Omega}{4\pi} \frac{\vec{v}\cdot\vec{q}}{iq_0 + \vec{v}\cdot\vec{q}}$$

$$= \frac{\mathcal{N}g^2\mu^2}{2\pi^2} f\left(\frac{q_0}{|\vec{q}|}\right) = m_D^2 f\left(\frac{q_0}{|\vec{q}|}\right) \tag{9.56}$$

where the number $\mathcal{N} = 2N_f N_c$ counts the two chiralities, N_f flavors, and N_c colors of quarks running in the loop in Fig. 9.6(a), and the Debye mass is given by

$$m_D^2 = \frac{\mathcal{N}g^2\mu^2}{2\pi^2} = N_f N_c \frac{g^2\mu^2}{\pi^2} \tag{9.57}$$

The dimensionless function f of the ratio $q_0/|\vec{q}|$ is given by

$$f(x) = 1 + \frac{ix}{2} \ln\left(\frac{x+i}{x-i}\right) = 1 + \pi x + O(x^2) \tag{9.58}$$

The difference $\theta(\epsilon_p) - \theta(\epsilon_{p+q})$ appears in Eq. (9.56) because the integral is zero when both poles are on the same side of the real axis. It has a simple meaning as the difference $n_{p+q} - n_p$ of two occupation numbers, step-functions at zero temperature. We also used the fact that $|\vec{q}| \ll |\vec{p}| \sim \mu$ to neglect $|\vec{q}|/\mu$ in $\epsilon_{p+q} - \epsilon_p = \vec{v}\cdot\vec{q}$. In other words we are considering the response of a very dense medium, in which the typical Fermi momenta of particles are large (hard) with respect to a very slow (soft) variation of the external field A_0. Such a regime, and the corresponding approximation, when the momenta of particles inside the loop are much harder than the external momenta, is known as *hard dense loops* (HDL). Note that, in this regime, $|\vec{q}| \ll \mu$, only the fermions in a thin shell around the Fermi surface contribute to Π_{00} – this shell is cut out by the θ functions in Eq. (9.56). This has a simple reason: screening requires rearrangement of charge carriers, which is impossible due to Pauli blocking anywhere except in the vicinity of the Fermi surface. For the same reason the value of Π_{00} is proportional to μ^2, i.e. to the *area* of the Fermi surface.

We neglected the modification to the fermion propagator due to the presence of the gap (9.16). This does not change the result within the precision we are after. The effect of Δ is to "smear" the θ functions in Eq. (9.56) into distributions n_p with a typical width of the crossover near $|\vec{p}| = \mu$ from 0 to 1 of order Δ. In the regime in which the photon propagator is employed in our calculation of the gap, $|\vec{q}| \gg \Delta$, the width of this crossover is negligible.

For the spacelike components of the polarization operator, the contribution of the near-surface modes described by the effective action (9.12) is, similarly,

$$
\begin{aligned}
\tilde{\Pi}_{ij} &= \mathcal{N} \frac{g^2 \mu^2}{2\pi^2} \int \frac{d\Omega}{4\pi} \frac{\vec{v} \cdot \vec{q}}{iq_0 + \vec{v} \cdot \vec{q}} v_i v_j \\
&= m_D^2 \left[\left(-\frac{1}{6} + \frac{1}{2} f + \frac{1}{2} x^2 f \right) \delta_{ij} + \left(\frac{1}{2} - \frac{1}{2} f - \frac{3}{2} x^2 f \right) \hat{q}_i \hat{q}_j \right] \\
x &\equiv \frac{q_0}{|\vec{q}|}
\end{aligned} \tag{9.59}
$$

The correct value of the polarization operator, $\Pi_{\mu\nu}$, cannot be equal to $\tilde{\Pi}_{ij}$ since the transversality of the polarization operator $q_\mu \Pi_{\mu\nu} = 0$ requires that

$$
\Pi_{ij} = -\hat{q}_i \hat{q}_j a^2 \Pi_{00} + G(\delta_{ij} - \hat{q}_i \hat{q}_j) \tag{9.60}
$$

where the function G is arbitrary, not constrained by the requirement of transversality. In other words, only the spatially transverse part of Π_{ij} is independent; the longitudinal part is related to Π_{00}. Using Eqs. (9.59) and (9.56), we find that $\Pi = \tilde{\Pi}$ does not satisfy Eq. (9.60), but, if

$$
\Pi_{ij} = \tilde{\Pi}_{ij} + \frac{m_D^2}{3} \delta_{ij} \tag{9.61}
$$

then Eq. (9.60) is satisfied. The function G is given by

$$
G = m_D^2 \left(-\frac{1}{2} + \frac{1}{2} f + \frac{1}{2} x^2 f \right) \tag{9.62}
$$

Why are we missing the term $(m_D^2/3)\delta_{ij}$ in Eq. (9.59)? As we shall see, this term comes from the contribution of the modes in the bulk of the Fermi sphere. These modes are already integrated out when we use the effective action (9.12) for near-surface modes. The effect of these modes is local, i.e. momentum-independent, and can by accounted for by a counterterm in the effective action: $(m_D^2/6)A_i A_i$. This counterterm comes from the diagram in Fig. 9.6(b), where one of the two fermion lines describes the propagator of the field χ. The vertices of this diagram arise from the last terms in the curly brackets in Eq. (9.9). As we have seen, the excitations created by the field χ carry large negative energy, $-|\vec{p}| - \mu$, i.e. describe antiparticles that can be created virtually for a very short time if the available energy is small compared with μ. Thus, on the scale of the effective theory the thick line in Fig. 9.6(b) can be effectively shrunk to a point, creating a local counterterm. The diagram can be evaluated easily using the method of residues to do the integral over p_0 first:

$$2\mathcal{N}g^2 \int_p \frac{1}{ip_0 + (|\vec{p}| - \mu)} \frac{1}{ip_0 + (|\vec{p}| + \mu)} (\delta_{ij} - v_i v_j)$$

$$= 2\mathcal{N}g^2 \int_{|\vec{p}|<\mu} \frac{d^3 p}{(2\pi)^3} \frac{1}{2|\vec{p}|} (\delta_{ij} - v_i v_j) = \frac{m_D^2}{3} \delta_{ij} \qquad (9.63)$$

The overall factor of two accounts for the possibility of either line in the diagram being thick. The last integral is restricted to momenta inside the Fermi sphere – only for those momenta will the two propagator poles be on opposite sides of the real axis. The denominator $2|\vec{p}|$ is the difference between the energies of the particle and antiparticle states at the same momentum $\vec{p}$. The factor $\delta_{ij} - v_i v_j$ accounts for the fact that, according to Eq. (9.9), only two components of the gluon field (A_1 and A_2 – orthogonal to $\vec{v} = (0, 0, 1)$) can flip a positive-energy quark into a negative-energy quark.

Finally, to obtain the gluon propagator $D_{\mu\nu}$, we need to invert the matrix:

$$q^2 \delta_{\mu\nu} - q_\mu q_\nu + \Pi_{\mu\nu} + \frac{1}{\xi} \bar{q}_\mu \bar{q}_\nu \qquad (9.64)$$

where the last term is the gauge-fixing term and $\bar{q}_\mu = (0, \vec{q})$. We choose the Coulomb gauge.[4] On inverting Eq. (9.64) and setting $\xi \to 0$, we obtain

$$D_{00} = \frac{1}{\vec{q}^2 + \Pi_{00}}; \qquad D_{ij} = (\delta_{ij} - \hat{q}_i \hat{q}_j) \frac{1}{q^2 + G} \qquad (9.65)$$

In the regime relevant for our calculation of the gap, $q_0 \ll |\vec{q}|$, the function Π_{00} tends to a constant, the square of the Debye mass (9.56). The electrostatic

---

[4] To the order in which we are working the gap is gauge-invariant. This is because, in our calculation, both quarks p and k in Fig. 9.3 are on the mass shell. A gauge-dependent term $q_\mu q_\nu$ in the propagator will produce a factor $iq_0 + \vec{q} \cdot \vec{v}$ in the vertices, which vanishes, since it is (up to $O(|\vec{q}|/\mu)$) equal to $(ik_0 + \epsilon_k) - (ip_0 + \epsilon_p)$, which is zero when p and k are on the mass shell. Note that only the gauge invariance of the value of the gap on the mass shell is ensured.

gluon fields are screened on the length scale of $1/m_D$, according to Eq. (9.65). The function G is given by, expanding Eq. (9.62),

$$G(q) = \frac{\pi}{2}m_D^2 \frac{q_0}{|\vec{q}|} + O\left(\frac{q_0^2}{|\vec{q}|^2}\right) \tag{9.66}$$

Note that, if we rotate back to Minkowski space, $q_0 \to iq_0$, this term becomes imaginary, as it should since it is the Landau *damping* term.

Since Landau damping provides screening only for the dynamical magnetic fields, we might worry that the effect of screening due to the Meissner effect is significant, since it screens even static fields. We neglected this effect when we replaced the full propagator, including the anomalous one, by the free-fermion propagator in the calculation of Π_{ij}. The contribution of the Meissner effect, however, is sub-leading, as are many other contributions that we neglected in our leading-order calculation. The reason is that the integral over q_0 in the gap equation is (logarithmically) dominated by the region where $q_0 \gg \Delta$. Thus, dynamical screening operates on the scale of $|\vec{q}| \sim (q^0 g^2 \mu^2)^{1/3} \gg (\Delta g^2 \mu^2)^{1/3}$, which is much larger than the scale $g\Delta$ on which Meissner screening becomes significant. Thus the momentum region where the Meissner effect is significant does not contribute to the collinear logarithm and can affect the result only in the next order in $1/\ln(\mu/\Delta) \sim g$.

9.3 Lowest excitations of the CFL phase

Our discussion in the previous section was concerned with the ground state of QCD at large baryochemical potential μ. We saw that, in this regime, the state in which the quarks simply fill all available levels inside the Fermi sphere is unstable. An arbitrarily small coupling to gluons creates a more energetically favorable state, which we can characterize by the presence of an anomalous Green function $\langle \psi \psi \rangle$. We also saw that the spectrum of quarks acquires a gap of order Δ. In this section we shall address the following question: what are the lowest-lying excitations of such a ground state?

Motivation

One reason why we need to understand the properties of the exitations is because some potentially observable consequences of the color superconductivity depend on the spectrum of excitations. For example, the rate of cooling of a neutron star depends on the specific heat, which, in turn, is a measure of the number of excitable degrees of freedom at a given temperature. At small temperatures, the lowest-lying excitations dominate the specific heat.

There is also another, not completely unrelated, conceptual reason why we want to understand the excitations. The reader must already have noted that most of our discussion of the ground state has been phrased in terms of quantities that are *not* gauge-invariant, and thus unobservable. This is not an uncommon situation for a gauge theory. In perturbation theory, the language based on quarks and gluons (or leptons and vector bosons, in electroweak theory), is appropriate in the situations in which the effective gauge coupling is small, as it is in our case. The reason for this is that, as long as the coupling is small, the actual physically observable (i.e. gauge-invariant) states have the same properties as the elementary excitations of the quark and gluon fields, e.g. insofar as energy–momentum relations and conserved charges are concerned. However, to make a conceptually sensible physical prediction, we have to phrase our results in terms of the gauge-invariant quantities.[5]

The situation we are facing is very similar to the Higgs phenomenon in the electroweak theory. In this case, we cannot, strictly speaking, make a sensible prediction that involves the value of the expectation value v_H of the Higgs field, because it is not a gauge-invariant, and thus not an observable, quantity. An example of a quantity that can be observed/measured is the mass of electroweak bosons m_W. This mass is proportional to the Higgs expectation value to leading order in perturbation theory (which is the main purpose of the Higgs mechanism). At higher orders this relation becomes gauge-dependent. Thus, we can still use the Higgs expectation value as a parameter in the theory to leading order, but not to all orders. To avoid this limitation we find other quantities, which are observable, and can also be expressed through the Higgs expectation value, in a gauge-dependent way, of course. The key is that, if we write the relation between such two observable quantities, eliminating the Higgs expectation value, the gauge dependence will also disappear from such a relation, to all orders in perturbation theory. In electroweak theory, an example of such an observable is the Fermi coupling constant G_F (or the lifetime of the muon). The relationship between G_F and v_H is not gauge-invariant to all orders, but the relationship between m_W and G_F is.

Symmetries and Goldstone excitations

Let us begin with the considerations of symmetry. We need to begin by comparing the symmetries of the microscopic, i.e. QCD, Lagrangian with the symmetries of the ground state. If we find that a *global* symmetry of a Lagrangian is broken in the ground state, the Goldstone theorem will tell us that there are massless particles, associated with the broken-symmetry generators. If the symmetry is only approximate, the masses of the corresponding "almost" Goldstone

---

[5] The lattice theory also makes this point, in its own way. According to Elitzur, an attempt to measure a non-gauge-invariant quantity on the lattice must give zero identically. For example, we cannot measure the anomalous Green function $\langle \psi \psi \rangle$. Of course, this is not a limitation of the lattice theory; rather it is a demonstration that the physical requirement of gauge invariance of observed phenomena is built into the foundation of the theory.

particles will be small. These Goldstone particles will then be low-energy excitations.

As we discussed, the Lagrangian symmetry $SU(3)_L \times SU(3)_R \times SU(3)_{color}$ breaks to $SU(3)_{L+R+color}$. The color $SU(3)_{color}$ does not count as a global symmetry and we thus have $16(= 8 + 8)$ generators of symmetry in the Lagrangian, but only eight in the ground state. Thus, we can expect eight ($= 16 - 8$) Goldstone particles. We can also guess that these eight particles fill a multiplet under the remaining $SU(3)$ symmetry group. In addition, the global $U(1)_B$ symmetry, which is responsible for conservation of baryon charge, is also broken in the ground state, which gives us one Goldstone particle.

The $U(1)_A$ symmetry requires special consideration. The Cooper-pair condensate breaks this symmetry also. However, due to the anomaly, this symmetry of the Lagrangian is not a symmetry of the quantum theory at all. In other words, the vacuum degeneracy corresponding to the rotation of the angle parametrizing $U(1)_A$ – which is also known as the θ angle – is lifted. However, it turns out that, at large μ_B, the effect of the anomaly on the symmetry of the ground state becomes weak, and the degeneracy of the ground state with respect to $U(1)_A$ is restored. In other words, the dependence on the angle θ disappears. This happens because the instanton configurations responsible for the θ dependence are suppressed [91]. These configurations of the gauge field carry long-range color electric fields, which are screened by the Fermi sea of quarks. We can therefore consider $U(1)_A$ as an approximate symmetry (and parametrically so, in the sense that, at $\mu_B \to \infty$, this symmetry becomes exact). This means that we should expect in total ten Goldstone bosons, one octet and two singlets.[6]

9.3.1 The effective Lagrangian for Goldstone modes

Our next step is to determine the effective Lagrangian which describes the Goldstone excitations which we have predicted. Our guiding principle is again based on symmetry: any symmetry of the microscopic Lagrangian has to be correctly reproduced in the effective Lagrangian. If we limit ourselves to the terms of lowest order in momenta, the above symmetry considerations allow a small number of terms. These considerations leave only a finite number of undetermined constants in the effective Lagrangian.

---

[6] It is important to recognize that the $U(1)_B$ boson is exactly massless whereas the $U(1)_A$ boson has a small but nonzero mass. This mass is actually calculable [92]. This difference between the bosons becomes even more significant if one considers global properties, such as vortices. A vortex of $U(1)_B$ phase, similarly to an Abrikosov vortex in liquid helium, carries only the energy associated with the length of the vortex. However, in the $U(1)_A$ case, there is additional energy associated with a two-dimensional surface attached to the vortex. Properties of such nontopological domain walls are also calculable at high μ [92].

We recall that the ground state of the CFL phase is characterized by the following diquark condensates:

$$X^{ia} \sim \epsilon^{ijk}\epsilon^{abc}\langle \psi_L^{bj}\psi_L^{ck}\rangle^* \qquad \text{and} \qquad Y^{ia} \sim \epsilon^{ijk}\epsilon^{abc}\langle \psi_R^{bj}\psi_R^{ck}\rangle^* \quad (9.67)$$

where the complex conjugation was added for convenience so that X and Y transform under $SU(3)_c \times SU(3)_L \times SU(3)_R$ as (3,3,1) and (3,1,3), respectively:

$$X \to U_L X U_c^T \qquad \text{and} \qquad Y \to U_R Y U_c^T \quad (9.68)$$

Note that left quarks pair with left quarks and right with right, and the orientations of these condensates X and Y are free for us to choose. The low-energy excitations in the CFL phase are given by the slow rotations of the phases of X and Y. Therefore, we can factor out the norm of the condensates and consider unitary matrices X and Y. Together they give us $9 + 9 = 18$ degrees of freedom. Eight of them are eaten by the gluons through the Higgs mechanism, and the surviving ten become the low-energy excitations we have anticipated.

For simplicity, we shall at first ignore the overall $U(1)$ phases of X and Y and assume that $\det X = \det Y = 1$. The Lagrangian should be symmetric under the $SU(3)_c \times SU(3)_L \times SU(3)_R$ rotations (9.68). This condition fixes the Lagrangian to the leading (second) order in derivatives [93]:

$$\mathcal{L}_{\text{eff}} = \frac{f_\pi^2}{2}\,\text{tr}[(X^\dagger\,\partial_0 X)^2 + (Y^\dagger\,\partial_0 Y)^2] + \text{spatial gradients} \quad (9.69)$$

The spatial gradients enter in a similar way but with a different constant instead of f_π. The cross term $\text{tr}[(X^\dagger\,\partial_0 X)(Y^\dagger\,\partial_0 Y)]$ is allowed by the symmetries, but, as we shall see in Section 9.3.2, it is suppressed in weak coupling (i.e. at large μ). Roughly speaking, to leading order in g^2, left and right quarks decouple from each other.

So far we have ignored the fact that $SU(3)_c$ is a *local* symmetry. To take this into account, we must replace the derivatives in Eq. (9.69) by covariant derivatives,

$$\begin{aligned}
\mathcal{L}_{\text{eff}} &= \frac{f_\pi^2}{2}\,\text{tr}[(X\,\partial_0 X^\dagger - gA_0)^2 + (Y\,\partial_0 Y^\dagger - gA_0)^2] + \cdots \\
&= \frac{f_\pi^2}{4}\,\text{tr}[(X\,\partial_0 X^\dagger - Y\,\partial_0 Y^\dagger)^2 + (X\,\partial_0 X^\dagger + Y\,\partial_0 Y^\dagger - 2gA_0)^2] + \cdots
\end{aligned} \quad (9.70)$$

The second term is responsible for the Higgs effect: the vector-like fluctuations, i.e. fluctuations with equal changes in X and Y, $\delta X = \delta Y$, become longitudinal components of the gluon A_μ. The gluons acquire a mass of order $O(gf_\pi)$ (electric and magnetic masses are, in general, different). We shall see that $f_\pi \sim \mu$, and thus the gluon mass is much larger than the momentum scales p that we are considering ($p < 2\Delta \ll g\mu$), so gluons decouple from the low-energy theory. The axial-like fluctuations of the phases, $\delta X = -\delta Y$, can be written as fluctuations of the phases of a new unitary matrix,

$$\Sigma = XY^\dagger \tag{9.71}$$

and the effective Lagrangian takes the form (in Euclidean space)

$$\mathcal{L}_{\text{eff}} = \frac{f_\pi^2}{4} \text{tr}\big(\partial_0 \Sigma \, \partial_0 \Sigma^\dagger + v_\pi^2 \, \partial_i \Sigma \, \partial_i \Sigma^\dagger\big) \tag{9.72}$$

which is the usual Lagrangian of the nonlinear sigma model except that the speed of the mesons, v_π, can be different from the speed of light. The matrix Σ is a singlet in color, transforms under $\text{SU}(3)_\text{L} \times \text{SU}(3)_\text{R}$ as

$$\Sigma \to U_\text{L} \Sigma U_\text{R}^\dagger \tag{9.73}$$

and describes the meson octet.

What happens if we take into account the $\text{U}(1)$ phases of X and Y? Then it is possible to add into the chiral Lagrangian a term proportional to

$$[\text{tr}(X^\dagger \, \partial_0 X)]^2 + [\text{tr}(Y^\dagger \, \partial_0 Y)]^2 \tag{9.74}$$

It is possible to add a cross term, $[\text{tr}(X^\dagger \, \partial_0 X)][\text{tr}(Y^\dagger \, \partial_0 Y)]$ but it is suppressed in weak coupling for the same reason that the cross term was omitted from Eq. (9.69). We shall make the consequences of the term (9.74) clear by splitting X and Y into $\text{SU}(3)$ and $\text{U}(1)$ parts,

$$X = \tilde{X} e^{2i\theta + 2i\phi} \quad \text{and} \quad Y = \tilde{Y} e^{-2i\theta + 2i\phi} \tag{9.75}$$

where $\tilde{X}$ and $\tilde{Y}$ are $\text{SU}(3)$ matrices, and the angle ϕ is the variable conjugate to the baryon charge, normalized to 1 for a single quark. The normalization of the $\text{U}(1)_\text{A}$ phase θ is fixed analogously. Consequently, the field Σ defined in Eq. (9.71) now has the form

$$\Sigma = \tilde{\Sigma} \, e^{4i\theta}, \qquad \tilde{\Sigma} = \tilde{X} \tilde{Y}^\dagger \tag{9.76}$$

In terms of $\tilde{\Sigma}$, ϕ, and θ, the lowest-order chiral Lagrangian consistent with the symmetries has the form

$$\mathcal{L}_{\text{eff}} = \frac{f_\pi^2}{4} \text{tr}\big(\partial_0 \tilde{\Sigma} \, \partial_0 \tilde{\Sigma}^\dagger + v_\pi^2 \, \partial_i \tilde{\Sigma} \, \partial_i \tilde{\Sigma}^\dagger\big) + 12 f_{\eta'}^2 \big[(\partial_0 \theta)^2 + v_{\eta'}^2 (\partial_i \theta)^2\big]$$
$$+ 12 f_\text{H}^2 \big[(\partial_0 \phi)^2 + v_\text{H}^2 (\partial_i \phi)^2\big] \tag{9.77}$$

The difference between the decay constants of η' and H mesons from f_π arises from the additional allowed term (9.74), whereas the difference between $f_{\eta'}$ and f_H arises from the cross term $[\text{tr}(X^\dagger \partial_0 X)][\text{tr}(Y^\dagger \partial_0 Y)]$, and, therefore, is small at weak coupling. We shall use $f_{\eta'} = f_H$ in the rest of the book.

The elementary meson fields π^A, η', and H are defined as

$$\tilde{\Sigma} = \exp\left(i\frac{\lambda^A \pi^A}{f_\pi}\right), \qquad \theta = \frac{\eta'}{\sqrt{24}\,f_{\eta'}}, \qquad \phi = \frac{H}{\sqrt{24}\,f_H} \qquad (9.78)$$

where λ^A ($A = 1, \ldots, 8$) are Gell-Mann matrices normalized so that $\text{tr}(\lambda^A \lambda^B) = 2\delta^{AB}$.

9.3.2 Decay constants of the Goldstone bosons

Let us now show that the decay constants of the Goldstone bosons, f_π, $f_{\eta'}$, and f_H, can all be computed in the high-density regime and are all proportional to the chemical potential μ. We shall demonstrate the method for f_H, whose calculation is the simplest. Let us imagine that the symmetry $U(1)_B$ generated by conserved baryon charge is not a global symmetry, but instead a local one. So we introduce into the theory a gauge field A_μ coupled to the baryon current. Owing to the Higgs mechanism, the gauge bosons acquire a finite mass, which can be computed both in the microscopic theory, in which it is expressed in terms of the chemical potential, and in the effective theory, in which it is proportional to f_H. Matching the two results determines f_H in terms of μ.

In order to compute f_H we have to deal only with the part of the effective chiral Lagrangian (9.77) that contains the phase ϕ. Since ϕ is the phase of the $U(1)_B$ rotation, which is now a local symmetry, one should replace the derivatives by covariant ones,

$$\mathcal{L} = 12 f_H^2 \left[(\partial_0 \phi + e A_0)^2 + v_H^2 (\partial_i \phi + e A_i)^2\right] \qquad (9.79)$$

where e is a small auxiliary coupling constant. From Eq. (9.79) we find that the gauge bosons acquire finite mass terms, which are different for A_0 and A_i,

$$m_{A_0}^2 = 24 e^2 f_H^2 \qquad (9.80)$$
$$m_{A_i}^2 = 24 v_H^2 e^2 f_H^2 \qquad (9.81)$$

On the other hand, the masses in Eqs. (9.80) and (9.81) can be computed from the microscopic theory, in which $m_{A_0}^2$ has the meaning of the Debye mass of the electric A_μ field, while $m_{A_i}^2$ is the Meissner mass of the magnetic components of A_μ. To compute these masses, it is most convenient to use the low-energy effective theory containing only fermion modes near the Fermi surface that we derived in Section 9.2.1.

Figure 9.7. The leading-order diagrams contributing to decay constants.

To leading order, the only contribution to $m^2_{A_0}$ is from the sum of two one-loop diagrams shown in Fig. 9.7, which, for zero external momentum, is equal to

$$2\frac{\mu^2}{2\pi^2} \int \frac{dp_0\,d\epsilon_p}{(2\pi)} \sum_{A=1}^{9} \left(-\frac{(ip_0+\epsilon_p)^2}{(p_0^2+\epsilon_p^2+\Delta_A^2)^2} + \frac{\Delta_A^2}{(p_0^2+\epsilon_p^2+\Delta_A^2)^2} \right) \quad (9.82)$$

where we use the formula for the phase space near the Fermi surface (9.28). The overall factor of two in Eq. (9.82) comes from summation over left- and right-handed quarks in the internal loop. We find that the two diagrams in Fig. 9.7 give equal contributions to the squared Debye mass, which is given by

$$m^2_{A_0} = 18e^2\frac{\mu^2}{2\pi^2} \quad (9.83)$$

This value is the same as the Debye mass that A_μ would have in the absence of the superconductivity – see Eq. (9.57). The origin of this coincidence can be made clear by the following argument. What we have computed is simply the 00 component of the polarization operator of A_μ, Π_{00}. Since A_0 is coupled to the baryon charge $n = \psi^\dagger\psi$, Π_{00} has the following interpretation in linear-response theory: the baryon charge density generated by an external uniform field A_0 is given by

$$n = \Pi_{00}(0)A_0 \quad (9.84)$$

However, the uniform A_0 field is simply a shift of the chemical potential, or the Fermi energy. Therefore, Π_{00} is equal to $e^2\,\partial n/\partial\mu$, i.e. the density of states near the Fermi surface, which is exactly the right-hand side of Eq. (9.83). It is easy to see that $\partial n/\partial\mu$ is the same in the normal phase and superconducting phase, provided that the gap in the latter is small.

On comparing Eq. (9.83) with the result from the effective theory, Eq. (9.80), one finds the decay constant f_H,

$$f_H^2 = \frac{3}{4}\frac{\mu^2}{2\pi^2} \quad (9.85)$$

An important remark is in order here. Equation (9.85) tells us that f_H depends only on the chemical potential μ, not on the gap Δ. This might seem to contradict the fact that, at $\Delta = 0$, the U(1)$_B$ symmetry would be restored and no

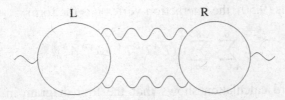

L   R

Figure 9.8. A higher-order diagram contributing to the decay constants. One loop contains a left-handed quark; the other has a right-handed quark.

Goldstone boson is expected in the theory. The explanation is that, as Δ decreases, the domain of applicability of the effective Lagrangian $p < 2\Delta$ shrinks and disappears at $\Delta = 0$. Therefore, the persistence of f_H does not contradict the restoration of $U(1)_B$ symmetry at $\Delta = 0$.

To find the Meissner mass, we have to evaluate the same two diagrams of Fig. 9.7 and add the "counterterm" mass given by the diagram in Fig. 9.6(b). In the case of the Meissner mass the vertices are attached to A_i and A_j. Instead of being equal, the two diagrams in Fig. 9.7 now cancel each other out, since the vertex factor of the first diagram is $v_i v_j$, whereas that in the second diagram is $v_i(-v_j)$. Therefore, the Meissner mass in the superconducting phase is equal to the "counterterm" Meissner mass (9.63), which (squared) is equal to $m_{A_0}^2/3$. On comparing this with Eqs. (9.81) and (9.80), one finds that the velocity of the H boson is given by

$$v_H = \frac{m_{A_i}}{m_{A_0}} = \frac{1}{\sqrt{3}} \tag{9.86}$$

i.e. it is equal to the speed of sound in relativistic fluids. This result is not completely surprising, since the $U(1)_B$ phase ϕ of the condensate is the variable conjugate to the baryon density $\psi^\dagger \psi$, whose fluctuations give rise to sound waves. Therefore, H quanta can be considered to be phonons. However, they are not hydrodynamic phonons since they exist outside the hydrodynamic regime (as in our case at zero temperature where the mean free path diverges).

One can repeat the same calculation for the η' meson and see that, to leading order, all diagrams remain the same. Thus we find that $f_{\eta'} = f_H$, and the η' meson also propagates with the speed of sound. There is, however, no symmetry that requires the decay constants of H and η' to be equal; in fact, they are not equal in higher orders of perturbation theory. For example, the diagram drawn in Fig. 9.8 gives contributions of opposite signs to f_H and $f_{\eta'}$.

The computation of f_π and v_π is completely analogous to that of f_H. We could introduce a fictitious field coupled to the flavor currents, but we can also make use of the existing coupling to the gluons to compute f_π. The only additional complication is that the interaction vertex now has a nontrivial structure. In our color–flavor basis (9.50), the interaction vertex has the form

$$2 \sum_{A,C=0}^{8} \sum_{B=1}^{8} \mathrm{tr}(T^A T^B T^C) \, \psi^{\dagger A} A^B \psi^C \tag{9.87}$$

A straightforward calculation shows that the first diagram in Fig. 9.7 gives $(\frac{3}{4})g^2\mu^2/(2\pi^2)$ and the second diagram contributes $-[(3+4\ln 2)/18]g^2\mu^2/(2\pi^2)$ to the Debye mass. Taking into account the factor of two arising from left- and right-handed quarks, the result reads

$$m_{A_0}^2 = \frac{21 - 8\ln 2}{18} \frac{g^2\mu^2}{2\pi^2} \tag{9.88}$$

The mass in Eq. (9.88) is not equal to the Debye mass in the normal phase, in contrast to the previous case of the fictitious U(1) boson coupled to the baryon current.

Notice that, if the gluons are coupled to the left-handed quarks alone, the squared Debye mass would be halved, which is exactly what one obtains by throwing away Y from the Lagrangian (9.70). However, if we add the XY cross term to the Lagrangian, this will no longer be true. Therefore, this cross term, though it is not forbidden by the symmetry, has a small coefficient in weak coupling. This can be seen also from the fact that the coupling of the left- and right-handed flavor currents is zero to leading order. In higher orders of perturbation theory it receives contributions from diagrams like those in Fig. 9.8, where one vertex corresponds to the left-handed flavor current and the other to the right-handed one.

In the effective theory, the gluon mass is given by $m_{A_0}^2 = g^2 f_\pi^2$. Therefore, one can determine the decay constants of the mesons in the pseudo-scalar octet [94],

$$f_\pi^2 = \frac{21 - 8\ln 2}{18} \frac{\mu^2}{2\pi^2} \tag{9.89}$$

The ratio $f_\pi^2/f_{\eta'}^2 = f_\pi^2/f_H^2 = 2(21 - 8\ln 2)/27 \approx 1.14$ is quite close to unity. This fact is reminiscent of the OZI rule [95].

The Meissner mass is equal to

$$m_{A_i}^2 = 2\frac{g^2\mu^2}{2\pi^2}\left(\frac{1}{2} - \frac{1}{4} - \frac{3+4\ln 2}{54}\right) = \frac{21 - 8\ln 2}{54}\frac{g^2\mu^2}{2\pi^2} \tag{9.90}$$

where the first term in the parentheses in the intermediate expression comes from the bare Meissner mass, while the last two terms come from the two diagrams in Fig. 9.7. Again, in contrast to the case of U(1)$_B$, the contributions of the two diagrams do not cancel each other out. It is somewhat surprising that the squared Meissner mass is a third of the squared Debye mass, as it is for the U(1)$_B$ case.

Since the ratio of these two masses is the velocity of the Goldstone bosons, we find that the velocities of *all* Goldstone modes in our theory equal the speed of sound $1/\sqrt{3}$ to leading order of perturbation theory.[7]

9.3.3 Meson masses

We now turn on finite small quark masses and compute the resulting masses of the Goldstone bosons. By introducing bare-quark masses into the microscopic theory, we break the $SU(3)_L \times SU(3)_R$ symmetry. Less symmetry means more freedom: more terms are allowed in $\mathcal{L}_{eff}$. However, constraints on the form of these terms can still be imposed if we note that the bare-quark mass term

$$\Delta\mathcal{L} = \Psi_L^\dagger M \Psi_R + \text{h.c.} \tag{9.91}$$

with M being the 3×3 mass matrix, could be made invariant under $SU(3)_L \times SU(3)_R$ if the matrix M is not passive under this symmetry but also transforms together with $\Psi_{L,R}$ in the following way:

$$\Psi_L \to U_L \Psi_L, \qquad \Psi_R \to U_R \Psi_R \qquad \text{and} \qquad M \to U_L M U_R^\dagger \tag{9.92}$$

Any term in the effective Lagrangian, written as a function of Σ and M, must respect the extended symmetry (9.73) and (9.92). At large μ the $U(1)_A$ symmetry is effectively restored, which imposes an additional constraint on the possible form of mass terms in the chiral Lagrangian. Under the $U(1)_A$ transformation, the microscopic degrees of freedom transform as

$$\Psi_L \to e^{i\zeta} \Psi_L, \qquad \Psi_R \to e^{-i\zeta} \Psi_R \tag{9.93}$$

where ζ is an arbitrary pure phase. Equation (9.93) implies the following transformation law of X, Y, and Σ:

$$X \to e^{-2i\zeta} X, \qquad Y \to e^{2i\zeta} Y \qquad \text{and} \qquad \Sigma \to e^{-4i\zeta} \Sigma \tag{9.94}$$

The quark-mass term (9.91) is invariant under the $U(1)_A$ symmetry if one requires that M transforms as

$$M \to e^{2i\zeta} M \tag{9.95}$$

under $U(1)_A$. Thus, any mass term in the effective Lagrangian must be invariant both under the $SU(3)_L \times SU(3)_R$ symmetry and under the $U(1)_A$ symmetry extended by the transformation of M.

Since the bare-quark masses are assumed to be small, we want to construct the mass terms of lowest possible order in M. To first order, the only candidate is the

---

[7] Similar results, $v = v_F/\sqrt{3}$, have been found by Bogolyubov and Anderson for phase waves in BCS superconductors and by Leggett for spin waves in superfluid ^{3}He [96].

mass term of the chiral Lagrangian at zero chemical potential, $\text{tr}(M^\dagger \Sigma) + \text{h.c.}$ This term is invariant under $SU(3)_L \times SU(3)_R$; however, it is not invariant under the $U(1)_A$ symmetry. Therefore, we must consider terms of higher order in M. At order M^2 we find that the following term is allowed by the symmetries:[8]

$$\Delta\mathcal{L}_{\text{eff}} = -c \det M \, \text{tr}(M^{-1}\Sigma) + \text{h.c.} \qquad (9.96)$$

There are two other possible terms allowed by the symmetries:[9]

$$\{\text{tr}[(M\Sigma^\dagger)^2] + [\text{tr}(M\Sigma^\dagger)]^2\} \det\Sigma + \text{h.c.} \qquad \text{and} \qquad \text{tr}(M\Sigma^\dagger M^\dagger\Sigma) \qquad (9.97)$$

However, only the coefficient of the term (9.96) is nonzero to leading order in the expansion in powers of g, or $1/\ln(\mu/\Delta)$, as we shall see below.

Meson spectroscopy in CFL: inverse mass ordering

The term (9.96) gives rise to an interesting pattern of meson masses in the CFL phase.

To write the (mass)^2 matrix for the Goldstone bosons we expand the field Σ using 3×3 Gell-Mann matrices λ^a:

$$\Sigma = \exp\left(\frac{i\pi^a\lambda^a}{f_\pi}\right) = 1 + \frac{i\pi^a\lambda^a}{f_\pi} - \frac{\pi^a\pi^b\lambda^a\lambda^b}{2f_\pi^2} + \cdots \qquad (9.98)$$

where $a = 1,\ldots,9$, $\text{tr}(\lambda^a\lambda^b) = 2\delta^{ab}$, $\lambda^9 = \sqrt{2/3}$ and $\pi^9 = \eta' f_\pi/f_{\eta'}$. Since the kinetic term is conventionally normalized we can read off the (mass)^2 matrix from the mass term (9.96) (except for the trivial rescaling of the η' field):

$$\frac{c}{f_\pi^2} \det M \, \text{tr}(M^{-1}\lambda^a\lambda^b) \, \pi^a\pi^b = \frac{C}{2} \det M \, \text{tr}(M^{-1}\lambda^a\lambda^b) \, \pi^a\pi^b \qquad (9.99)$$

where

$$C = \frac{2c}{f_\pi^2} = \frac{108}{21 - 8\ln 2} \frac{\Delta_0^2}{\mu^2} \qquad (9.100)$$

which comes from Eq. (9.89), which we have derived, and Eq. (9.114), which we shall derive below.

---

[8] One way of looking at this term is to realize that, under $SU(3)_L$, both M and Σ transform as fundamental 3-plets. We require two powers of M and one of Σ in order to satisfy the $U(1)_A$ neutrality. We can construct an $SU(3)_L$ singlet out of a product of M, M, and Σ if we antisymmetrize with respect to the first index of each of these matrices (the index on which $SU(3)_L$ acts). By antisymmetrizing also with respect to the second index we obtain an $SU(3)_L \times SU(3)_R$ singlet, $\epsilon_{a'b'c'}\epsilon_{abc}M_{aa'}M_{bb'}\Sigma_{cc'}$, which coincides with Eq. (9.96) up to a factor of two.

[9] Taking a minus sign in the curly brackets in the first term would give the term in Eq. (9.96).

With the quark-mass matrix $M = \mathrm{diag}(m_\mathrm{u}, m_\mathrm{d}, m_\mathrm{s})$, the 9×9 meson-mass matrix (9.99) decomposes into a diagonal 6×6 matrix and a nondiagonal 3×3 matrix. The former gives rise to the mass formulas for $\pi^\pm$, $\mathrm{K}^\pm$, K^0, and $\bar{\mathrm{K}}^0$:

$$m^2_{\pi^\pm} = C(m_\mathrm{u} + m_\mathrm{d})m_\mathrm{s}$$
$$m^2_{\mathrm{K}^\pm} = C(m_\mathrm{u} + m_\mathrm{s})m_\mathrm{d}$$
$$m^2_{\mathrm{K}^0, \bar{\mathrm{K}}^0} = C(m_\mathrm{d} + m_\mathrm{s})m_\mathrm{u} \qquad (9.101)$$

The remaining 3×3 matrix corresponding to the π^0, η, and η' mesons is more complicated. Unlike in the QCD vacuum, in the CFL phase the π^0, η, and η' excitations mix strongly. The mixing pattern is easy to understand if we neglect the small difference between f_π and $f_{\eta'}$. Then the mass matrix (9.99) has the simplest form in the basis $\bar{\mathrm{u}}\mathrm{u}$, $\bar{\mathrm{d}}\mathrm{d}$, $\bar{\mathrm{s}}\mathrm{s}$, which is related to π^0, η, η' by the quark-model relations $\pi^0 = (\bar{\mathrm{u}}\mathrm{u} - \bar{\mathrm{d}}\mathrm{d})/\sqrt{2}$, $\eta = (\bar{\mathrm{u}}\mathrm{u} + \bar{\mathrm{d}}\mathrm{d} - 2\bar{\mathrm{s}}\mathrm{s})/\sqrt{6}$, and $\eta' = (\bar{\mathrm{u}}\mathrm{u} + \bar{\mathrm{d}}\mathrm{d} + \bar{\mathrm{s}}\mathrm{s})/\sqrt{3}$. In this basis $\lambda_{\bar{\mathrm{u}}\mathrm{u}} = \mathrm{diag}(1, 0, 0)$, $\lambda_{\bar{\mathrm{d}}\mathrm{d}} = \mathrm{diag}(0, 1, 0)$, $\lambda_{\bar{\mathrm{s}}\mathrm{s}} = \mathrm{diag}(0, 0, 1)$, and the matrix in this basis becomes diagonal. The mixing among π^0, η, and η' is thus ideal, and the masses of the mixed states are

$$m^2_{\bar{\mathrm{u}}\mathrm{u}} \cong 2Cm_\mathrm{d}m_\mathrm{s}, \qquad m^2_{\bar{\mathrm{d}}\mathrm{d}} \cong 2Cm_\mathrm{u}m_\mathrm{s}, \qquad m^2_{\bar{\mathrm{s}}\mathrm{s}} \cong 2Cm_\mathrm{u}m_\mathrm{d} \quad (9.102)$$

The masses of $\bar{\mathrm{u}}\mathrm{u}$ and $\bar{\mathrm{d}}\mathrm{d}$ mesons are of the same order of magnitude as those of $\pi^\pm$ and K mesons. The $\bar{\mathrm{u}}\mathrm{u}$ meson is the heaviest and is approximately $\sqrt{2}$ times heavier than $\mathrm{K}^\pm$, while the $\bar{\mathrm{d}}\mathrm{d}$ is $\sqrt{2}$ times heavier than neutral kaons. The $\bar{\mathrm{s}}\mathrm{s}$, in contrast, is particularly light. In reality, $f_\pi \neq f_{\eta'}$, so the mixing is not completely ideal.

The reader will notice that the mass ordering in the CFL phase is completely reversed compared with what one sees in the QCD vacuum. While this is quite surprising, it can easily be explained. The key point is that the mesons, which are the fluctuations of Σ in Eq. (9.71), should be thought of as bound states of a triplet anti-diquark X and an anti-triplet diquark $Y^\dagger$, i.e. as $\bar{\mathrm{q}}\bar{\mathrm{q}}\mathrm{qq}$ states, rather than $\bar{\mathrm{q}}\mathrm{q}$ states. Consider, for example, the lightest meson in the CFL phase, which in the normal quark model would be labeled as the $\bar{\mathrm{s}}\mathrm{s}$ state. Now the s quark should be replaced by the $\bar{\mathrm{u}}\bar{\mathrm{d}}$ anti-diquark, and the $\bar{\mathrm{s}}$ anti-quark should be replaced by the ud diquark, so this meson is represented as $\bar{\mathrm{u}}\mathrm{d}\mathrm{u}\mathrm{d}$. Since such a meson does not contain the strange quark, it is not surprising that its mass does not depend on m_s, as in Eq. (9.102). For all other mesons, one can write the quark structure as well, by making the replacements $\mathrm{u} \to \bar{\mathrm{d}}\bar{\mathrm{s}}$, $\mathrm{d} \to \bar{\mathrm{u}}\bar{\mathrm{s}}$, and $\mathrm{s} \to \bar{\mathrm{u}}\bar{\mathrm{d}}$. Since one replaces the heaviest quark by the lightest anti-diquark and vice versa, the inverse mass ordering can be expected. It is important to note, however, that the mesons in the CFL phase have the same quantum numbers (up to mixing) as do the mesons in the QCD vacuum.

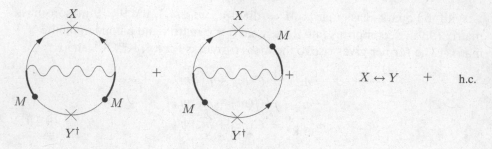

Figure 9.9. Lowest-order diagrams contributing to the M and Σ dependence of the vacuum energy in the CFL phase. Arrows indicate the direction of flow of the fermion (baryon) charge. This direction is reversed only at the X and Y vertices.

Calculation of the coefficient of the mass term

In order to determine the coefficient of the mass term we can use the same strategy as we did in determining the decay constants. We need to choose a quantity that can be calculated in two independent ways, using the effective Lagrangian and the microscopic theory (QCD), and match the results. In the present case it is convenient to calculate the shift of the vacuum energy due to the quark mass.

The mass-dependent term in the (Euclidean) effective Lagrangian (9.96) is equal to the shift of the energy of the ground state as a function of the quark-mass matrix M and of the given orientation of the slow field Σ. In order to calculate such a shift in the microscopic theory, we need to identify quark–gluon-vacuum bubble diagrams that depend on M and Σ. The diagrams we are interested in must carry two M vertices as well as one X vertex and one $Y^\dagger$ vertex, since $\Sigma = XY^\dagger$. The diagrams of this type, which are of lowest order in g, are shown in Fig. 9.9.

The gluon line is needed in this diagram for the following reason. Let us go clockwise around the quark loop in Fig. 9.9 starting from the $Y^\dagger$ vertex. This vertex sits on the propagator of the low-energy field ψ_R in our notation of Section 9.2.1. The insertion of M has two effects on the quark field. It flips the chirality of the field, R $\rightarrow$ L, according to Eq. (9.91). However, there is another effect. The left-chirality field which we obtain is not ψ_L, but χ_L – the field whose excitations describe antiparticles.[10] On the other hand, the insertion of X, which

---

[10] This can be seen using the representation of the $\vec{\alpha}$ Dirac matrices (9.4). The fields ψ_L and ψ_R satisfy the eigenvalue equation $\vec{\alpha} \cdot \vec{p}\,\psi = |\vec{p}|\psi$, which gives them the necessary dispersion relation $E = |\vec{p}| - \mu$. The fields χ satisfy $\vec{\alpha} \cdot \vec{p}\,\chi = -|\vec{p}|\chi$. If we chose the z axis along the vector $\vec{p}$, the components of the Dirac 4-spinor are $\Psi \equiv (\Psi_L; \Psi_R) = (\psi_L, \chi_L; \chi_R, \psi_R)$. Thus we have a mass term $\psi_L^\dagger \Psi_R = \psi_L^\dagger \chi_R + \cdots$.

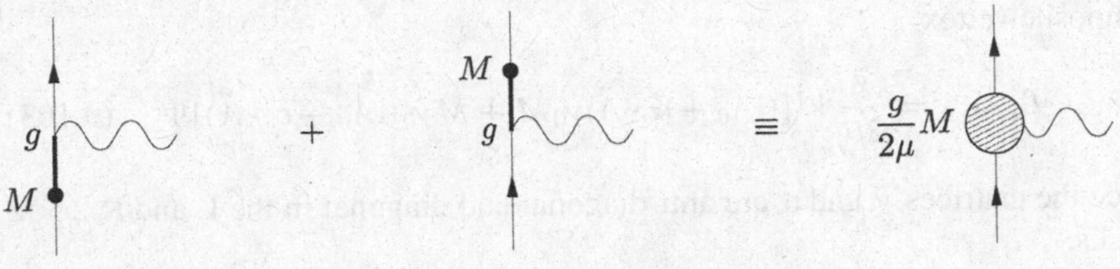

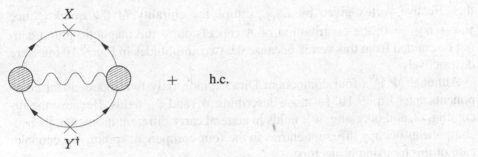

Figure 9.10. Diagrams that combine into the composite-vertex, simplifying diagrams in Fig. 9.9.

Figure 9.11. The same diagrams as in Fig. 9.9, but using the composite vertex from Fig. 9.10.

we need to make, has to be on a ψ_L line, since X describes the color–flavor orientation of the anomalous term $\psi_L \psi_L$. The insertion of the gluon vertex is necessary in order to provide the flip $\chi_L \to \psi_L$.

The diagrams in Fig. 9.9 can be grouped into a simpler diagram, which has a composite vertex – a sequence of an M and a gluon insertion, as illustrated in Fig. 9.10.

With the composite vertex defined in Fig. 9.10, the sum of the diagrams in Fig. 9.9 can be replaced by the diagrams in Fig. 9.11. We could also have arrived at this diagram by integrating out the χ fields in the path integral and using the resulting effective theory for the ψ fields.

It turns out that the diagram in Fig. 9.11 is easy to evaluate despite its being a two-loop diagram. Let us first determine what expression corresponds to the composite vertex in Fig. 9.10. Both of the diagrams which contribute to this vertex contain an antiparticle (i.e. χ) propagator, represented by a thick line. If the external fermion lines are on shell, $E \equiv ip_0 = |\vec{p}| - \mu$ (and we shall use this vertex only for on-shell external quark lines), this internal antiquark propagator is off shell by $E - (-|\vec{p}| - \mu) = 2|\vec{p}|$, which we can approximate, within our precision, i.e. up to terms of order E/μ, as 2μ. By using four-component spinors, using Eq. (9.7) for the quark–gluon vertex, and writing the quark-mass term (9.91) as $\Psi^\dagger \gamma_0 M \Psi$, we obtain for the sum of the two contributioins to the composite vertex

$$\mathcal{L}_{\psi_L\psi_R A} = \frac{g}{2\mu}\Psi^\dagger[(iA_0 + \vec{\alpha}\cdot\vec{A})\gamma_0 M + M\gamma_0(iA_0 + \vec{\alpha}\cdot\vec{A})]\Psi \qquad (9.103)$$

Since the matrices γ and α are anti-diagonal and diagonal in the L and R 2×2 blocks,

$$\gamma_0 = \begin{pmatrix} 0 & 1 \\ 1 & 0 \end{pmatrix} \qquad \text{and} \qquad \alpha = \begin{pmatrix} \sigma & 0 \\ 0 & -\sigma \end{pmatrix} \qquad (9.104)$$

the effective vertex given by $\mathcal{L}_{\psi_L\psi_R A}$ flips the chirality of the quark. Since $\gamma_0\vec{\alpha} + \vec{\alpha}\gamma_0 = 0$, the contribution of $\vec{A}$ cancels out – the magnetic gluon cannot be emitted from this vertex because the two amplitudes in Fig. 9.10 interfere destructively.

Although Ψ is a four-component Dirac spinor, only two independent components enter Eq. (9.103) – those describing ψ_L and ψ_R fields. Because the incoming, Ψ, and outgoing, $\Psi^\dagger$, fields in general carry different momenta, their ψ components occupy different entries in the four-component spinor. We can take care of this by using projectors:

$$P_{\vec{p}}^\pm = \frac{1 \pm \vec{\alpha}\cdot\vec{v}_{\vec{p}}}{2} \qquad (9.105)$$

Thus, we can finally write for the composite $\psi_L\psi_R A$ vertex, using momentum components of the fields,

$$\frac{ig}{\mu}\Psi^\dagger P_{\vec{k}}^+ \gamma_0 P_{\vec{p}}^+ M A_0 \Psi = \frac{ig}{\mu}\Psi^\dagger \gamma_0 P_{\vec{k}}^- P_{\vec{p}}^+ M A_0 \Psi \qquad (9.106)$$

Another simplifying feature of this vertex is that it vanishes if the gluon does not carry any momentum, i.e. $\vec{p} = \vec{k}$, since $P_{\vec{p}}^- P_{\vec{p}}^+ = 0$. This can be understood as a consequence of conservation of angular momentum: the emitted electric gluon carries away no angular momentum, while the spin of the quark has to flip if its momentum is unchanged, since its chirality flips.

Now we can write the expression corresponding to the diagram in Fig. 9.11:

$$\Delta E_{\text{vac}} = -\frac{1}{2}\left(\frac{ig}{\mu}\right)^2 \int_p \int_k \frac{\Delta_p}{p_0^2 + \epsilon_p^2 + \Delta_p^2} \frac{\Delta_k}{k_0^2 + \epsilon_k^2 + \Delta_k^2} \frac{1}{|\vec{p} - \vec{k}|^2}$$

$$\times \frac{1}{2}\text{tr}(P_k^- P_p^+) \sum_{A=1}^{8} \text{tr}\,\text{tr}(MT^A X M T^A Y^\dagger) + \text{h.c.} \qquad (9.107)$$

We have collected all the color- and flavor-carrying matrices under the double (color and flavor) trace at the end of the expression. The first factor of $\frac{1}{2}$ is due to the symmetry of the diagram in Fig. 9.11. The factor of $\frac{1}{2}$ in front of tr($P_k^- P_p^+$) accounts for the fact that the diagram shown in Fig. 9.11 includes only the $\psi_R \to \psi_L$ part of the vertex (9.103) – the other part, $\psi_L \to \psi_R$, is in the complex-conjugate diagram denoted by h.c. Evaluating the trace over Dirac indices gives

$$\mathrm{tr}\,(P_k^- P_p^+) = 1 - \vec{v}_{\vec{p}} \cdot \vec{v}_{\vec{k}} = 1 - \cos\theta \tag{9.108}$$

and we see that the dependence on the angle θ between $\vec{p}$ and $\vec{k}$ cancels out between this trace and the Coulomb propagator:

$$\frac{1}{|\vec{p} - \vec{k}|^2} = \frac{1}{2\mu^2(1 - \vec{v}_{\vec{p}} \cdot \vec{v}_{\vec{k}})} \tag{9.109}$$

As a result, the integrals over p and k factorize:

$$\Delta E_{\mathrm{vac}} = \frac{1}{2}\left(\frac{g}{\mu}\right)^2 \left(\int_p \frac{\Delta_p}{p_0^2 + \epsilon_p^2 + \Delta_p^2}\right)^2 \frac{1}{4\mu^2} \sum_{A=1}^{8} \mathrm{tr}\,\mathrm{tr}(MT^A X M T^A Y^\dagger) + \text{h.c.} \tag{9.110}$$

The integral can be evaluated within our (logarithmic, or $\mathcal{O}(g)$) precision:

$$\int_p \frac{\Delta_p}{p_0^2 + \epsilon_p^2 + \Delta_p^2} \approx \frac{\mu^2}{2\pi^2} \int_0^\infty \frac{dp_0\, \Delta(p_0)}{\sqrt{p_0^2 + \Delta^2}}$$

$$\approx \frac{\mu^2}{2\pi^2} \int_{\Delta_0}^\mu \frac{dp_0}{p_0} \Delta_0 \sin\left[\frac{g}{3\sqrt{2}\pi} \ln\left(\frac{\mu}{p_0}\right)\right]$$

$$= \frac{\mu^2}{2\pi^2} \Delta_0 \frac{3\sqrt{2}\pi}{g} \tag{9.111}$$

where we took into account the energy dependence of Δ from Eq. (9.47) and also Eq. (9.46) for Δ_0 (in particular, the fact that, for $p_0 = \Delta_0$, the argument of the sine in Eq. (9.111) is equal to $\pi/2$). Note that this integral receives its value from the large interval of energies between Δ and μ, in close similarity to our calculation of the gap. The logarithmic enhancement of the integral is clear from the factor $1/g$ in the result which, according to Eq. (9.46), is equal to $\ln(\mu/p_0)$ up to a numerical factor.

Finally, let us evaluate the color and flavor traces in Eq. (9.110):

$$\sum_{A=1}^{8} \mathrm{tr}\,\mathrm{tr}(MT^A X M T^A Y^\dagger)$$

$$= \sum_{A=1}^{8} M^{ii'} T_{aa'}^A X_c^k \epsilon^{ijk} \epsilon_{abc} M^{jj'} T_{bb'}^A (Y^\dagger)_{c'}^{k'} \epsilon^{i'j'k'} \epsilon_{a'b'c'}$$

$$= -\tfrac{2}{3} M^{ii'} X_c^k \epsilon^{ijk} \epsilon_{a'b'c} M^{jj'} (Y^\dagger)_{c'}^{k'} \epsilon^{i'j'k'} \epsilon_{a'b'c'}$$

$$= -\tfrac{4}{3} M^{ii'} M^{jj'} (XY^\dagger)^{kk'} \epsilon^{ijk} \epsilon^{i'j'k'} = -\tfrac{8}{3}\det M\,\mathrm{tr}(M^{-1}\Sigma) \tag{9.112}$$

where we began by applying Eq. (9.24) to the underlined factors. Note that we have obtained the term which matches the effective Lagrangian (9.96). The other two terms in Eq. (9.97) can appear only in a higher order in g.

On putting together Eqs. (9.110), (9.111), and (9.112) we obtain

$$\Delta E_{\text{vac}} = -\frac{3\Delta_0^2}{2\pi^2} \det M \operatorname{tr}(M^{-1}\Sigma) + \text{h.c.} \tag{9.113}$$

By comparing this with the dependence of the vacuum energy on M and Σ in the effective theory given by Eq. (9.96), we determine the value of the coefficient c [94,97]:

$$c = \frac{3\Delta_0^2}{2\pi^2} \tag{9.114}$$

This completes our calculation in this section. We can now insert the constant c into the effective Lagrangian Eq. (9.96), and obtain the expressions for the meson masses (9.101) and (9.102), using the constant $C = 2c/f_\pi^2$ as in Eq. (9.100).

To add to our discussion of the spectrum in Section 9.3.3, since $C \sim \Delta^2/\mu^2$, the masses become smaller and smaller as μ increases. The values of the masses depend on the value of the gap Δ, which, unfortunately, we know only up to a factor.

9.3.4 Continuity of quark and nuclear matter?

It is amusing to notice that, insofar as the quantum numbers of the particles are concerned, the spectrum of the lowest-lying pseudo-scalar excitations is the same as one would expect in superfluid nuclear, or rather hypernuclear, matter. The breaking of chiral symmetry produces the usual pseudo-scalar octet, while the breaking of the singlet $U(1)_B$, which is superfluidity, is responsible for the scalar singlet. In hypernuclear matter, the candidate for such a state is the H dibaryon – a bound state of $\Lambda\Lambda$, which, similarly to the H boson in the CFL phase, has the quark content udsuds. For the Goldstone-type excitations the similarity is a consequence of the similar patterns of symmetry breaking in the ground state, $SU(3)_L \times SU(3)_R \times U(1)_B \to SU(3)_V$, and the Goldstone theorem.

It is also interesting that the fermion excitations are also similar. Since the baryon charge is not a good quantum number in the CFL phase, the baryons should be characterized by their flavor quantum numbers under the unbroken $SU(3)_V$ symmetry (neglecting quark masses). The fermion excitations in the CFL phase are, on the perturbative level, quarks. There are nine of them, counting color and flavor combinations. It is easy to see that they fall into an octet and a singlet. For the same reason eight of the gaps are equal whereas one is larger in Eq. (9.52). The lowest-lying baryons also fill out an octet (a heavier singlet might exist, but, if it does, is certainly unstable due to decay into an octet baryon and a pseudo-scalar).

Furthermore, the vector excitations in the CFL phase – the gluons, which acquire mass by a Meissner–Higgs-type mechanism, also occupy an octet that is similar to the octet of the ρ, K^*, etc. in the hadronic phase.

All of these observations suggest that the transition to hypernuclear matter is the only transition necessary as the chemical potential increases at $T = 0$. In other words, in QCD with three light quarks at $T = 0$, there might only be the transition to hypernuclear matter, with the breaking of the $U(1)_B$ symmetry and the existence of superfluidity. The symmetry of the ground state is the same at asymptotically high μ and no additional phase transitions are necessary in order to connect to this asymptotic regime, as pointed out in [63].

Since the operator of the electric charge is one of the $SU(3)_V$ generators (the charges of u, d, and s sum to zero), the charge content of the octets is the same in the CFL phase and in hypernuclear matter. However, how can the quarks appear as integer-charge excitations? This is possible because, in addition to flavor, quarks carry color. In the CFL phase neither color nor flavor is a good quantum number of an excitation, but their difference is, since simultaneous equal color and flavor rotations preserve the ground-state condensate (9.1). There are eight such transformations, and one of them is the electric charge of the CFL phase. The electric charges in the octet are thus given by differences of the eigenvalues of the actual electric charge $Q_{\text{flavor}} = \text{diag}(\frac{2}{3}, -\frac{1}{3}, -\frac{1}{3})$ and the color $Q_{\text{color}} = \text{diag}(\frac{2}{3}, -\frac{1}{3}, -\frac{1}{3})$, which are always integral.

The modified electric-charge generator $Q_{\text{flavor}} - Q_{\text{color}}$ is not broken, and, therefore, if a gauge field is coupled to this charge it will remain massless, as the photon should be. However, since the corresponding transformation, involving both color and flavor rotations, is different from the transformation in the vacuum, the corresponding *modified* photon field is a mixture of the usual photon and the field coupled to the $\text{diag}(\frac{2}{3}, -\frac{1}{3}, -\frac{1}{3})$ color generator. The admixture of the latter is of the order of the ratio of electromagnetic and strong couplings e/g (this is the analog of the Weinberg angle in electroweak theory). This also means that the modified photon couples slightly more weakly (by a relative amount of order $(e/g)^2$) in the CFL phase. It is also interesting that, since all charged excitations which can be excited by a passing photon are gapped, a chunk of CFL matter would be transparent, like a diamond.[11]

---

[11] In a realistic environment, e.g. in the interior of a cold compact star, one would need to be sure that no electrons are present in the CFL matter. This is the case, due to the condition of electric neutrality [98], provided that condensation of charged kaons does not occur (see [99] and Section 9.4).

9.4 Comments and some further developments

In this chapter we concentrated on finding the properties of the QCD ground state and the lowest excitations using first-principles perturbative calculations. The success of such an approach is due to the asymptotic freedom of QCD which makes the expansion in powers of $g(\mu)$ systematically meaningful at large μ. We have seen that such calculations are nontrivial even at lowest order and the dependence on the coupling constant g is far from analytic. This is due to the nonperturbative nature of the phenomenon of BCS pairing.

How far in μ does one have to go to achieve reasonable precision in such calculations? A straightforward analysis [89, 100] suggests that this value of μ may need to be as large as 10^8 MeV. This means that, although they are systematically correct, the results obtained using first-principles perturbative calculations need not be practically reliable at values of μ of the order of a few hundred MeV – typical values expected to pertain to the interiors of neutron stars. The same region of moderate μ is also of great theoretical interest – the phase transitions are expected to occur there. Although perturbative calculations cannot address these issues, these first-principles results establish the existence of effects that one might expect to persist, in qualitatively similar ways, at moderate μ.

Phenomenologically motivated models of strong interactions, such as Nambu–Jona Lasinio-type theories with interactions modeled by one-gluon or instanton effects [23, 24], allow one to study these phenomena in the region of interesting μ. In particular, phase transitions can be studied as a function of μ as well as of T [101].

Another parameter, which we kept small in our calculations, is the value, or values, of the quark masses. Our calculation in Section 9.3.3 is based on an expansion in m_q. This expansion may be a good guide for realistic up- and down-quark masses, but we must be concerned about higher-order corrections in the expansion for the strange-quark mass which is, physically, not small. The most important such corrections to the meson masses have been analysed in [97, 99]. It turns out that these additional mass terms lower the masses of the kaons (K^0 and K^+ – kaons with an antistrange quark) further in the CFL phase, thus making condensation of these mesons possible once $m_s \sim m_{u,d}^{1/3} \Delta^{2/3}$. One can turn this condition around to determine the value of μ at which such condensation occurs in the phase diagram of QCD [102].

In addition, if the value of m_s is large enough, the pairing of strange with up and down quarks is not energetically favorable, because of the mismatch in the Fermi surfaces. Again, analysis based on models such as the Nambu–Jona Lasinio-type models in [103] is necessary in order to address this regime. The corresponding *unlocking* transition from the CFL phase to the two-flavor pairing phase occurs at a value $m_s \sim \Delta^{1/2} \mu^{1/2}$. For the physical value of m_s, such a condition determines the boundary of the CFL phase in the QCD phase diagram.

The mismatch of the Fermi surfaces of the paired quarks could also trigger the following interesting phenomenon, as pointed out in [104]. It has been known, at least on a theoretical level since the work of Larkin, Ovchinnikov, Fulde, and Ferrell [105], that, in the BCS theory of pairing, a mismatch of the electron up- and down-spin Fermi surfaces does not lead immediately to an unpairing. Rather, there is a critical value of the mismatch $\delta\mu$ at which there occurs a transition to a pairing with nonzero total momentum of a pair – essentially, because each of the pairing fermions sits close to its own Fermi sphere. The resulting variable-phase condensate breaks translational invariance. It has also been known that it is more energetically favorable to condense several such waves breaking all (continuous) translational symmetries. The resulting state has all the properties of a crystal. In the context of QCD such a phase has been termed a *crystalline color superconductor*. The study of such a crystal in QCD is still going on, but preliminary analysis indicates that the face-centered cubic crystal is energetically preferred [106].

Lattice-field theory as a means of addressing nonperturbative problems from first principles would be indispensible for obtaining reliable information about all such phenomena at intermediate μ and realistic m_s. Unfortunately, as we have already pointed out, this regime is still off bounds to a direct Monte Carlo simulation, for the reasons we discuss in detail in Section 10.1.

The application of these results to the physics of compact stars is a developing subject. Such an appliction would require, besides reaching into the regime of relevant intermediate μ and realistic m_s, also the inclusion of electromagnetism, which plays an essential role in the energetics of the QCD phases (see, e.g., [107]). A growing number of potential manifestations of color-superconducting phenomena is being investigated. They include the mass–radius relation, spin-down effects of gravitational r-mode instabilities, glitches, and cooling rates. The interested reader should find the recent reviews on the subject a valuable source of information. Further references include [108, 109].

10

Effective Lagrangians and models of QCD at nonzero chemical potential

10.1 QCD at finite μ and the sign problem

As discussed briefly in earlier chapters, the QCD partition function at nonzero baryon chemical potential cannot be evaluated by standard Monte Carlo simulation methods. This is so because the integrand of the Euclidean path integral is not positive for all configurations of the fields if the color gauge group is SU(3). This problem is known as the *sign problem*. Later in this book we shall look at the state of the art of developing new approaches and algorithms that might provide some limited simulation information on QCD at nonzero μ_B and T.

The *sign problem* is one of the most pressing bottlenecks in computational physics. In addition to limiting our knowledge of QCD, it occurs in condensed-matter physics and has undermined simulation studies of correlated electron systems and high-T_c superconductors.

The grand canonical partition function is formally, when it is written as a path integral,

$$Z \equiv e^{-\Omega(T,\mu)/T} = \int \mathcal{D}A \, \mathcal{D}\bar{\psi} \, \mathcal{D}\psi \, \exp(-S_E) \tag{10.1}$$

The Euclidean action, S_E, is given by

$$S_E = \int_0^{1/T} dx_0 \int d^3x \left(\frac{1}{2g^2} \text{tr}(F_{\mu\nu}F_{\mu\nu}) - \sum_{f=1}^{N_f} \bar{\psi}_f(\partial\!\!\!/ + i A\!\!\!/ + m_f - \mu\gamma_0)\psi_f \right) \tag{10.2}$$

where $N_f = 2$ is the number of flavors, $N_c = 3$ is the number of colors, and $m_f = m$ is the quark mass. The Euclidean matrices γ_μ are Hermitian. On integrating over the fermion fields we can also write

$$Z = \int \mathcal{D}A \exp\left(-\frac{1}{2g^2} \text{tr}(F_{\mu\nu}F_{\mu\nu})\right) \det(D\!\!\!/ + m_f - \mu\gamma_0) \tag{10.3}$$

The first factor, the exponent, is manifestly positive. The problem arises because of the second factor, the determinant of the Dirac operator. This determinant becomes complex when $\mu \neq 0$.

Let us first check that this determinant is positive when $\mu = 0$. A note on terminology: we shall use the term "matrix" and "operator" interchangeably when referring to the Dirac operator. In a discretized version of the theory, e.g. on a lattice, or in a random-matrix approximation, it is a finite, albeit large, dimensional matrix. In a continuum theory it is an operator acting on Dirac 4-spinor functions.

Consider the case $m = 0$ first. The matrix $\slashed{D}$ is anti-Hermitian; therefore, if $m = 0$ and $\mu = 0$, the Dirac determinant is at least real. In fact, it is positive because of the vectorlike nature of the Dirac fermion. It is easy to see this explicitly in the Weyl (chiral) representation of the gamma-matrices. In this representation all four gamma-matrices are block diagonal in 2×2 blocks:

$$\gamma_0 = \begin{pmatrix} 0 & 1 \\ 1 & 0 \end{pmatrix}, \qquad \gamma = \begin{pmatrix} 0 & i\sigma \\ -i\sigma & 0 \end{pmatrix} \tag{10.4}$$

All four gamma-matrices are anti-Hermitian (we are working in Euclidean space) and the upper block is the complex conjugate of the lower block. Therefore the matrix $\slashed{D}$ has the form

$$\slashed{D} = \begin{pmatrix} 0 & iX \\ iX^\dagger & 0 \end{pmatrix}, \qquad iX = D_0 + i\sigma \cdot D. \tag{10.5}$$

– the upper block is the anti-Hermitian conjugate of the lower block. The determinant of such a matrix is manifestly positive: $\det \slashed{D} = \det(XX^\dagger) > 0$.

The mass enters on the diagonal in Eq. (10.5) and does not spoil the argument above: $\det(\slashed{D} + m) = \det(XX^\dagger + m^2)$.

Now consider $\mu \neq 0$. Then

$$\slashed{D} + m - \mu\gamma_0 = \begin{pmatrix} m & iX - \mu \\ iX^\dagger - \mu & m \end{pmatrix} \tag{10.6}$$

and

$$\det(\slashed{D} + m - \mu\gamma_0) = \det[(X + i\mu)(X^\dagger + i\mu) + m^2] \tag{10.7}$$

is not positive definite. It would be if we formally chose μ to be imaginary [110]. However, for real μ, the physical situation, the determinant is in general complex. This means that we cannot straightforwardly apply Monte Carlo methods to Eq. (10.3). [1]

---

[1] The problem could be called the "complex-phase" problem, but, at least in principle, one could reduce such a problem to a sign problem by combining the gauge-field configurations into pairs such that the phase of the determinant cancels out explicitly. This leaves an indefinite sign for the integrand.

One natural step would be to investigate the partition function (10.3) in the quenched approximation. That is, use the Monte Carlo weight given by the pure gluon action and set the determinant to unity. One could then argue that the limit $N_f \to 0$ is smooth and we could obtain results that are in qualitative agreement with real QCD, i.e. with $N_f = 2$ or 3. Though such an approach is very fruitful for finite-temperature QCD, it fails when $\mu \neq 0$.

The essence of the problem is that there are many theories that have the same $N_f \to 0$ limit. We need to know which one of these theories the quenched case approximates. The answer to this question is crucial, since the different theories have qualitatively different quark contents and physical properties. It turns out that the quenched case approximates a "wrong" theory. It also turns out that this "wrong" theory also has physical significance (a theory of finite isospin density) and we shall discuss it at the end of this chapter.

A random-matrix model of the QCD partition function becomes very useful when we want to understand the failure of the quenched approximation. We shall introduce this model and see what it actually predicts for real QCD *with* fermions. We shall see that these predictions agree with the physical expectations outlined in Chapter 8. Owing to the simplicity of the model, we shall be able to carry out the explicit analytic calculations needed to investigate the nature of the quenched limit.

10.2 The random-matrix model of QCD

Random-matrix theory provides an effective description of those degrees of freedom in QCD which are responsible for the spontaneous breaking of chiral symmetry. In this respect, it is similar to Landau–Ginzburg effective theory. Random-matrix theory is based on the observation that spontaneous chiral-symmetry breaking is related to the density of small eigenvalues of the Dirac operator ($\lambda \ll \Lambda_{QCD}$). This relationship is expressed quantitatively by the Banks–Casher formula (5.117), $\langle \bar{\psi} \psi \rangle = \pi \rho_{ev}(0)$. Here, $\rho_{ev}(0)$ is the density of small (but nonzero) eigenvalues (per unit λ and per unit four-volume, V_4) of the Euclidean Dirac operator in the thermodynamic limit $V_4 \to \infty$. The dynamics of these eigenvalues can be described using a random matrix (of infinite size) in place of the Dirac operator. This approximation can be shown to give exact results in a certain limit – the mesoscopic limit [111, 112].

The random-matrix approximation consists of replacing the matrix X in the expression (10.5) for the Dirac matrix by a completely random matrix. It is surprizing that many properties of the small eigenvalues are correctly reproduced in this case. This means that small eigenvalues are very insensitive to the precise form of correlations in the matrix X introduced by its dependence on fluctuating gauge fields A_μ. We shall not attempt to prove this, but rather take a pragmatic approach and see what happens if we do make such an approximation. The virtue of the random-matrix model is that it is solvable analytically.

10.2.1 A random-matrix-model description of the QCD phase diagram

We write

$$Z_{RM} = \int \mathcal{D}X \exp\left(-\frac{N}{\sigma^2} \, \mathrm{tr}(XX^\dagger)\right) \det{}^{N_f}(D+m) \qquad (10.8)$$

where D is the $2N \times 2N$ matrix approximating the Dirac operator $\slashed{D} - \mu\gamma_0$:

$$D = \begin{pmatrix} 0 & iX - iC \\ iX^\dagger - iC & 0 \end{pmatrix} \qquad (10.9)$$

The random matrix X has dimensions $N \times N$. The total dimension of D is $2N$. This is the number of small eigenvalues, which is proportional to V_4. In QCD we expect N to be approximately equal to the typical number of instantons (or anti-instantons) in V_4; therefore, $N/V_4 \approx n_{\mathrm{inst}} \approx 0.5\,\mathrm{fm}^{-4}$ [113]. The matrix C is deterministic and describes the effects of temperature and chemical potential [114]. In the simplest (and original) $T \neq 0$, $\mu = 0$ model [115], the choice $C = \pi T$ describes the effect of the smallest Matsubara frequency. As noted in [116], it is possible to simulate the effects of the eigenvalue correlations induced by the pairing of instantons and anti-instantons into molecules by choosing a more general form for the diagonal matrix C with elements C_k, which are (increasing) functions of T. In the $T = 0$, $\mu \neq 0$ model of [65], $C = i\mu$ describes the effect of the chemical potential. Here we consider the more general case $T \neq 0$, $\mu \neq 0$. Although we do not know the detailed dependences of the elements of C on T and μ, we understand that T primarily affects the real (i.e. Hermitian) part of C and μ affects the imaginary (i.e. anti-Hermitian) part. We shall adopt the following approximate form for this dependence: $C_k = a\pi T + ib\mu$ for one half of the eigenvalues and $C_k = -a\pi T + ib\mu$ for the other half with a and b dimensionless parameters.[2] This form accounts for the fact that there are two smallest Matsubara frequencies,[3] $+\pi T$ and $-\pi T$. Such a linear *Ansatz* for C is certainly very naive, but it probably would be pointless to refine it while accepting the other crude approximations we have already made. This form reflects our understanding of the properties of C sufficiently well.

The chiral condensate is calculated as

$$\langle \bar{\psi}\psi \rangle = \frac{1}{N_f V_4} \frac{\partial \ln Z_{RM}}{\partial m} \qquad (10.10)$$

---

[2] These parameters are intended to reflect the degree of overlap and correlation between instantons and anti-instantons. One can therefore predict that a and b are smaller than 1. We shall estimate the values of a and b below.

[3] Alternatively, this form preserves the relation $\langle \gamma_0 D \rangle = 0$ at $\mu = 0$.

Current algebra fixes the value of $\langle \bar{\psi}\psi \rangle_0 \approx 2\,\text{fm}^{-3}$ at $T = \mu = m = 0$. The only dimensional parameter remaining in the partition function, Z_{RM}, is the variance of the random matrix, σ. Thus,

$$\langle \bar{\psi}\psi \rangle_0 = \text{constant} \times \frac{N}{V_4\sigma} \approx \text{constant} \times \frac{n_{\text{inst}}}{\sigma} \qquad (10.11)$$

The dimensionless constant will be found below and is equal to 2. This fixes the value of $\sigma \approx 0.5\,\text{fm}^{-1} \approx 100\,\text{MeV}$. It is convenient to use σ as a unit of mass in the model and also absorb the coefficients πa and b into T and μ. In other words, we measure m in units of σ, T in units of $\sigma/(\pi a)$, and μ in units of σ/b.

The $N \to \infty$ (i.e. thermodynamic) limit of the partition function, Eq. (10.8), can be found in the following way. Let's first sketch the strategy of the calculation and then carry out the details below. First, write the determinant in Eq. (10.8) as an integral over Grassmann N vectors (quarks). Second, perform the now Gaussian integration over X. As a result, one is left with an integral over Grassmann N vectors, which appear in the exponent in a quadrilinear form. Third, to make this integration Gaussian, introduce integrations over auxiliary $N_f \times N_f$ matrices ϕ. Finally, on performing the Gaussian integral over Grassmann N vectors, one arrives at Eq. (10.16). The limit $N \to \infty$ in Eq. (10.16) is given by the saddle point of the integrand.

Now let's perform the above steps explicitly. We must introduce Grassmann variables that carry two indices. The index $i = 1, \ldots, N$ – the same as the index numbering the rows and columns of X. This index corresponds to color as well as space indices of the original Dirac matrix (from the instanton point of view, it counts instanton zero modes). To account for the power of the determinant, there should be N_f such Grassmann variables, and we label them with the index $a = 1, \ldots, N_f$. These are simply quark-flavor indices. We need two such double-indexed variables, since the matrix D is $2N \times 2N$. We denote them Ψ_L and Ψ_R. The subscripts L and R refer to chirality. Concluding this step, we write ($\sigma = 1$ in our units now)

$$Z_{\text{RM}} = \int \mathcal{D}X \exp[-N\,\text{tr}(XX^\dagger)]$$

$$\times \int \mathcal{D}\Psi \exp\left(\Psi_L^{\dagger a}(iX - iC)\Psi_L^a + \Psi_R^{\dagger a}(iX^\dagger - iC)\Psi_R^a\right)$$

$$\times \exp\left(\Psi_L^{\dagger a}m\Psi_R^a + \Psi_R^{\dagger a}m\Psi_L^a\right) \qquad (10.12)$$

where we suppressed "color/space" indices i of Ψ and X to focus on flavor indices a.

The integration over matrix elements of X can be done now using this formula, which is valid for arbitrary matrices A and B:

$$\int \mathcal{D}X \exp[-\text{tr}(XX^\dagger)]\exp[\text{tr}(AX) + \text{tr}(BX^\dagger)] = \exp[\text{tr}(AB)] \qquad (10.13)$$

which is easy to derive by writing $X = \operatorname{Re} X + i \operatorname{Im} X$ and integrating over real and imaginary parts of the matrix elements X using standard Gaussian integration rules. Thus in Eq. (10.12) the terms linear in X collapse into a four-fermion term:

$$Z_{RM} = \int \mathcal{D}\Psi \exp\left(\frac{1}{N}(\Psi_L^{\dagger a}\Psi_R^b)(\Psi_R^{\dagger b}\Psi_L^a)\right)$$

$$\times \exp\left(-i\Psi_L^{\dagger a}C\Psi_L^a - i\Psi_R^{\dagger a}C\Psi_R^a\right)$$

$$\times \exp\left(\Psi_L^{\dagger a}m\Psi_R^a + \Psi_R^{\dagger a}m\Psi_L^a\right) \tag{10.14}$$

Note that, in order to write the four-fermion term in Eq. (10.14), we had to permutate some Ψ variables, which affects the signs of the terms above.

Now we shall break the four-fermion term into bilinears, but using the $N_f \times N_f$ matrices such as $\Psi_L^{\dagger a}\Psi_R^b$, instead of the $N \times N$ matrices such as $\Psi_L^a\Psi_L^{\dagger a}$. This requires a new auxiliary complex matrix ϕ^{ab}:

$$Z_{RM} = \int \mathcal{D}\phi \exp[-N \operatorname{tr}(\phi\phi^{\dagger})] \int \mathcal{D}\Psi \exp[\operatorname{tr}(\Psi_L^{\dagger}\phi\Psi_R + \Psi_R^{\dagger}\phi^{\dagger}\Psi_L)]$$

$$\times \exp(-i\Psi_L^{\dagger}C\Psi_L - i\Psi_R^{\dagger}C\Psi_R)$$

$$\times \exp(\Psi_L^{\dagger}m\Psi_R + \Psi_R^{\dagger}m\Psi_L) \tag{10.15}$$

In Eq. (10.15) we suppressed flavor indices a and b. Finally, on integrating over Ψ we obtain, taking into account the fact that C has two kinds of diagonal elements, $\pm T - i\mu$,

$$Z_{RM} = \int \mathcal{D}\phi \exp[-N \operatorname{tr}(\phi\phi^{\dagger})]$$

$$\times \det^{N/2}\begin{pmatrix} \phi + m & \mu + iT \\ \mu + iT & \phi^{\dagger} + m \end{pmatrix} \det^{N/2}\begin{pmatrix} \phi + m & \mu - iT \\ \mu - iT & \phi^{\dagger} + m \end{pmatrix}$$

$$= \int \mathcal{D}\phi \exp[-N\Omega(\phi)] \tag{10.16}$$

where

$$\Omega(\phi) = \operatorname{tr}\left(\phi\phi^{\dagger} - \frac{1}{2}\ln\{[(\phi + m)(\phi^{\dagger} + m) - (\mu + iT)^2]\right.$$

$$\left. \times [(\phi + m)(\phi^{\dagger} + m) - (\mu - iT)^2]\}\right) \tag{10.17}$$

The integration in Eq. (10.16) is performed over $2 \times N_f \times N_f$ variables, which are the real and imaginary parts of the elements of the complex matrix ϕ. In the limit $N \to \infty$ this integral is determined by a saddle point of the integrand or, alternatively, the minimum of $\Omega(\phi)$:

$$\lim_{N \to \infty} \frac{1}{N} \ln Z_{RM} = -\min_{\phi} \Omega(\phi) \tag{10.18}$$

The function $\Omega(\phi)$ is an effective potential for the degrees of freedom describing the dynamics of the chiral phase transition. One can see that the value of ϕ at the minimum, i.e. the equilibrium value $\langle \phi \rangle$, gives us the value of the chiral condensate, Eq (10.10):

$$\langle \bar{\psi}\psi \rangle = \frac{1}{N_f V_4} \frac{N}{\sigma} 2 \operatorname{Re} \operatorname{tr} \langle \phi \rangle \tag{10.19}$$

(cf. Eq. (10.11)).

For real m, it is reasonable to expect that the minimum occurs when ϕ is a real matrix proportional to the unit matrix. With this assumption, we need to find only one real parameter, ϕ, which minimizes the potential:

$$\Omega = N_f \left(\phi^2 - \frac{1}{2} \ln\{[(\phi + m)^2 - (\mu + iT)^2][(\phi + m)^2 - (\mu - iT)^2]\} \right) \tag{10.20}$$

This is not a simple ϕ^6 potential, but it has remarkably similar properties. In particular, the condition $\partial\Omega/\partial\phi = 0$ gives a fifth-order polynomial equation in ϕ.

The symmetry plane $m = 0$

Let us first consider the symmetry plane $m = 0$. Then the equation $\partial\Omega/\partial\phi = 0$ always has one trivial root, $\phi = 0$. The remaining four roots are the solutions of a quartic equation, which has the form of a quadratic equation in ϕ^2:

$$\phi^4 - 2(\mu^2 - T^2 + \tfrac{1}{2})\phi^2 + (\mu^2 + T^2)^2 + \mu^2 - T^2 = 0 \tag{10.21}$$

Above the second-order line, i.e. in the high temperature phase, Eq. (10.21) does not have real roots (since $\phi^2 < 0$ for each pair of roots). This corresponds to the fact that the potential Ω has only one minimum at $\phi = 0$ (i.e. the trivial root). On the second-order line, a pair of roots of Eq. (10.21) becomes zero, i.e. the potential is $\Omega \sim \phi^4$ near the origin. This means that, on the second-order line,

$$(\mu^2 + T^2)^2 + \mu^2 - T^2 = 0 \tag{10.22}$$

The second-order line ends when the remaining pair of roots also becomes zero, i.e. the potential becomes $\Omega \sim \phi^6$ near the origin. This happens when

$$\mu^2 - T^2 + \tfrac{1}{2} = 0 \tag{10.23}$$

on the second-order line. The condition Eq. (10.23), together with Eq. (10.22), determines the location of the tricritical point in the $T - \mu$ plane:

$$T_3 = \frac{1}{2}\sqrt{\sqrt{2} + 1} \approx 0.776 \quad \text{and} \quad \mu_3 = \frac{1}{2}\sqrt{\sqrt{2} - 1} \approx 0.322 \tag{10.24}$$

The equation for the triple line is obtained from the requirement that the depth of the minima in Ω at ϕ given by the pair of solutions of Eq. (10.21) (furthest from the origin) should coincide with the depth at the origin $\phi = 0$. The equation for the triple line is therefore

$$\mu^2 - T^2 + \frac{1}{2} + \frac{1}{2}\sqrt{1 - 16\mu^2 T^2} - \frac{1}{2}\ln\left(\frac{1 + \sqrt{1 - 16\mu^2 T^2}}{2}\right)$$

$$+ \ln(\mu^2 + T^2) = 0 \tag{10.25}$$

In particular, when $T = 0$, we obtain the elementary equation $\mu^2 + 1 + \ln\mu^2 = 0$ whose solution is $\mu \approx 0.528$ [65]. This is the value of μ_1. On setting $\mu = 0$ in the equation for the second-order line, Eq. (10.22), we find $T_c = 1$ [115].

We recall that the units of T and μ depend on the unknown dimensionless parameters a and b. However, these unknown factors cancel out from the ratios

$$\frac{T_3}{T_c} \approx 0.78; \qquad \frac{\mu_3}{\mu_1} \approx 0.61 \tag{10.26}$$

Taking $T_c = 160\,\text{MeV}$ and $\mu_1 N_c = 1200\,\text{MeV}$, we find that $T_3 \approx 120\,\text{MeV}$ and $\mu_3 N_c \approx 700\,\text{MeV}$.[4]

Note that the second-order line, Eq. (10.22), marks the location of the points on the phase diagram where the symmetric minimum $\phi = 0$ disappears, i.e. turns into a maximum. On continuing this line below point T_3, we obtain the location of *spinodal* points (see Fig. 10.1). In the region between this line of spinodal points and the first-order phase-transition line, the chirally symmetric phase $\phi = 0$ can exist as a metastable state. Such a state can be reached by supercooling, and it is unstable against the nucleation of bubbles of the broken phase $\phi \neq 0$. A similar line with equation $4T\mu = 1$, the superheating line, and the supercooling line, Eq. (10.22), bound the region around the first-order phase-transition line where the potential $\Omega(\phi)$ has three minima. All these lines meet at the tricritical point as shown in Fig. 10.1.

Away from the $m = 0$ plane
Away from the symmetry plane $m = 0$, the expressions for the wing surfaces and the wing lines become rather lengthy. The principle, however, remains simple. The minima of the potential, Ω, satisfy the equation $\partial\Omega/\partial\phi = 0$ and are given by (three out of five) roots of a fifth-order polynomial. On the wing surface, the depth $\Omega(\phi)$ in a pair of adjacent minima is the same. On the wing line, the two adjacent minima fuse into one. In terms of the roots of the polynomial, three roots coincide (two minima and one maximum). In other words, the potential is $\Omega \sim (\phi - \langle\phi\rangle)^4$ on the wing line and near the minimum.

---

[4] We remind the reader that we denote by μ the chemical potential per quark, therefore $\mu_B = \mu N_c$. With these values for T_c and μ_1, the parameters a and b have the values $a \approx 0.2$ and $b \approx 0.13$.

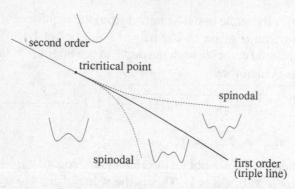

Figure 10.1. The phase diagram in the symmetry plane ($m = 0$) near the tricritical point with the corresponding shapes of the effective potential Ω indicated
schematically. The dashed (spinodal) lines bound the region where both phases
(chirally symmetric and broken) can exist, one as a stable phase and one as a
metastable phase. Within this region the transition from one phase to another
occurs through nucleation (e.g. one phase boils), whereas outside, provided that
the system has been superheated/supercooled carefully, it would occur via a process known as spinodal decomposition.

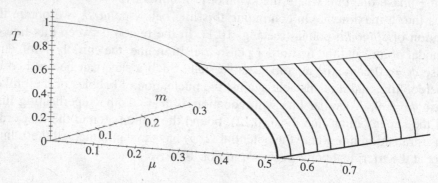

Figure 10.2. The phase diagram of QCD with two light flavors of mass m as calculated from
the random-matrix model. The almost-parallel curves on the wing surface are cross sections of
this surface with $m =$ constant planes. The units of m are $\sigma \approx 100\,\text{MeV}$, those of T are $T_c \approx$
$160\,\text{MeV}$, and those of μ are $\mu_1 N_c/0.53 \approx 2300\,\text{MeV}$, with the choices of T_c and μ_1 from the
text.

The resulting phase diagram is plotted in Fig. 10.2. One can see, as
expected from mean field theory near the tricritical point, that the wing
lines together with the triple line approach the tricritical point with the
same slope as the second-order line but from the other side. The critical exponents near the tricritical point given by the random-matrix model
can also be seen to coincide, as expected, with the mean-field exponents of Eqn. (8.78), (8.79), and (8.80). The corresponding dependences

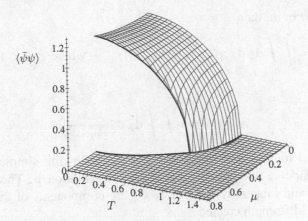

Figure 10.3. The chiral condensate $\langle \bar{\psi} \psi \rangle$ (in units of $\langle \bar{\psi} \psi \rangle_0 \approx 2 \, \text{fm}^{-3}$) as a function of T and μ in the random-matrix model. The units of T and μ are as in Fig. 10.2.

of $\langle \bar{\psi} \psi \rangle$ on T and μ in the random matrix model are shown in Fig. 10.3.

10.2.2 Chiral-symmetry breaking, Lee–Yang zeros, and an electrostatic analogy

As observed by Lee and Yang [117], nonanalytic behavior of a thermodynamic quantity, including, in the present case, the discontinuity in the value of an order parameter in the thermodynamic limit, is caused by the coalescence of zeros of the partition function to form a boundary crossing the relevant parameter axis. In the case of QCD, the signature of spontaneous chiral-symmetry breaking is the discontinuity in $\langle \bar{\psi} \psi \rangle$ as m is varied along the real axis and crosses $m = 0$. This discontinuity is equal to $2\pi \rho(0)$, where $\rho(0)$ is the density of the zeros on the imaginary-m axis near $m = 0$ in the thermodynamic limit.

Indeed, in finite volume, when the number of degrees of freedom is finite, the partition function (10.3) is an analytic function of its parameters. The parameter we are interested in is the quark mass m, and it is easy to see that the partition function is a polynomial in m. It has a set of discrete zeros. The derivative of the logarithm of Z with respect to m is the order parameter, the chiral condensate:

$$\langle \bar{\psi} \psi \rangle = \frac{\partial Z}{\partial m} \tag{10.27}$$

(To simplify equations that follow we chose a different normalization for $\langle \bar{\psi} \psi \rangle$ compared with Eq. (10.19).) If $\rho(m)$ is the distribution of the zeros in the complex plane of m, then

$$\ln Z(m) = \int\int dx\,dy\,\rho(x,y)\ln(m-z), \qquad \text{where } z = x + iy \quad (10.28)$$

and

$$\langle\bar\psi\psi\rangle(m) = \int\int dx\,dy\,\frac{\rho(x,y)}{m-z} \qquad (10.29)$$

It is helpful to look at this equation using the following simple electrostatic analogy. Consider a two-dimensional charge distribution ρ. Then Eq. (10.29) says that real and imaginary parts of $\langle\bar\psi\psi\rangle$ are components of the 2-vector of the electric field strength created by ρ:

$$\vec E = (\text{Re}\,\langle\bar\psi\psi\rangle, -\text{Im}\,\langle\bar\psi\psi\rangle) \qquad (10.30)$$

This analogy makes it easy to invert the relation (10.29):

$$\rho = \frac{1}{2\pi}\vec\nabla\vec E = \frac{1}{\pi}\frac{\partial}{\partial z^*}\langle\bar\psi\psi\rangle \qquad (10.31)$$

where $\partial/\partial z^* \equiv (\partial/\partial x + i\,\partial/\partial y)/2$. From Eq. (10.31) we see that ρ vanishes if the function $\langle\bar\psi\psi\rangle$ is holomorphic (depends only on $x + iy$). The function $\rho(z)$ is a measure of the nonanalyticity of $\langle\bar\psi\psi\rangle(z)$ at a given point z. If the function $\langle\bar\psi\psi\rangle(z)$ has a cut, the zeros form a one-dimensional distribution in the limit $N \to \infty$. The discontinuity across the cut along the imaginary axis at $m = 0$ is the signature of spontaneous symmetry breaking, and the size of this discontinuity is related to the *linear* density of zeros at the origin, $\rho(0)$:

$$\langle\bar\psi\psi\rangle(+0) - \langle\bar\psi\psi\rangle(-0) = 2\pi\rho(0) \qquad (10.32)$$

Note how this relation resembles the Banks–Casher relation (5.117). The difference is that here ρ is the density of partition-function zeros rather than the density of the eigenvalues as in Eq. (5.117). In fact, when the chemical potential is nonzero, the Dirac operator is not even Hermitian, and its determinant is no longer real. As a result, when $\mu \neq 0$, the density of Dirac eigenvalues can be defined straightforwardly only in quenched QCD, i.e. when the contribution of the (complex) fermion determinant to the measure is approximated by unity.

Because the random-matrix model is solvable analytically, we can determine the locations of the cuts of the function $\langle\bar\psi\psi\rangle$, and therefore learn how the zeros of the partition function in the complex-m plane evolve with changes in temperature and chemical potential. A few typical cases are illustrated in Fig. 10.4. The results in this figure are actual locations of the zeros of the partition function (10.8) at finite N. It illustrates that the zeros do indeed line up in the limit $N \to \infty$. At zero T and μ, the zeros form a line, which (in the

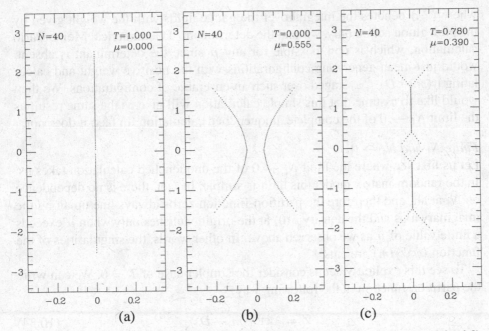

Figure 10.4. Zeros of the partition function of a finite-sized N-random-matrix model (10.8) in the complex-m plane calculated numerically for various values of T and μ. The calculation has been done for $N_f = 1$, but the $N \to \infty$ limit is independent of N_f. The density of points is proportional to the strength of the cut (discontinuity in $\langle \bar{\psi}\psi \rangle$) in the $N \to \infty$ limit.

$N \to \infty$ limit) means a cut on the Riemann sheet of the function $\langle \bar{\psi}\psi \rangle$ along the imaginary axis. Raising the temperature pushes the zeros away from the origin along the imaginary axis until the density at the origin vanishes (continuously), the cut breaks in two, as in Fig. 10.4(a), and chiral symmetry is restored (cf. Eq. (10.32)). The chemical potential pushes the zeros away from the origin in the direction of the real axis until the cut splits in two, as illustrated in Fig. 10.4(b). Note that the density $\rho(0)$ is finite just before the split. Therefore, the transition is of first order. Near the tricritical point, the split in the direction of the real axis (due to the chemical potential) occurs at the same time as the density $\rho(0)$ vanishes (due to the effects of the temperature). This is illustrated in Fig. 10.4(c).

10.2.3 The quenched limit of QCD and the random-matrix model

Why does the quenched approximation fail at $\mu \neq 0$? In the quenched approximation we calculate the order parameter $\langle \bar{\psi}\psi \rangle$ as the average over configurations with a weight from which the determinant is thrown out:

$$\langle \bar{\psi}\psi \rangle_{\text{quench}}(m) = \langle \text{tr}(m - D)^{-1} \rangle \tag{10.33}$$

where $\langle \ldots \rangle$ denotes the integral over the gauge fields with the weight given by the pure gluon action only, without the determinant. The numerical Monte Carlo calculation, which is now possible for any μ since the determinant is absent, would just mean generating configurations with the positive weight and calculating $\text{tr}(m - D)^{-1}$ averaged over such an ensemble of configurations. We then would like to assume that this simple calculation will give us the same result as the limit $N_f \to 0$ of the complete, unquenched calculation. In fact, it does not.

Finite N_f and $N_f \to 0$

Let us first see where the limit $N_f \to 0$ of the unquenched calculation takes us. In the random-matrix model this limit is simple. In fact, there is no dependence on N_f at all, and therefore the partition-function zeros always line up along the imaginary axis and the density $\rho(0)$ at the origin vanishes only when μ exceeds a finite value of μ as we observed above. In other words, the singularities of the function $\langle \bar{\psi}\psi \rangle(m)$ are cuts.

To see this explicitly, let us consider the simpler case of $T = 0$. We can write the partition function in the following way:

$$Z = \langle \det^{N_f}(m - D) \rangle \tag{10.34}$$

using the same notations as in Eq. (10.33). In the random-matrix approximation the $\langle \cdot \rangle$ denotes integration over X and it can be performed using the same sequence of transformation and auxiliary variables as in Eq. (10.16). The value of $\langle \bar{\psi}\psi \rangle(m)$ in the $N \to \infty$ limit is determined by Eq. (10.19), which, in our current normalization becomes simply $\langle \bar{\psi}\psi \rangle = \phi$, with ϕ given by the saddle point of the effective potential:

$$\Omega(\phi) = \text{tr}\{\phi\phi^\dagger - \ln[(\phi + m)(\phi^\dagger + m) - \mu^2]\} \tag{10.35}$$

(We remind the reader that the real and imaginary parts of ϕ should be considered as independent *complex* variables.) Assuming that the saddle point is achieved when ϕ is a diagonal matrix, we find the following equation for ϕ:

$$\phi + m = \phi[(\phi + m)^2 - \mu^2] \tag{10.36}$$

The solution of this cubic equation is an analytic function of m, which has three Riemann sheets, connected by cuts. The correct Riemann sheet is selected by the condition that $\phi \to 1/z$ when $z \to \infty$, which follows from $\int dx\,dy\,\rho = 1$ and Gauss' theorem in the electrostatic analogy of Section 10.2.2. The cuts evolve with μ as shown schematically in Fig. 10.5.

The quenched result

Now, what are the singularities of the function $\langle \bar{\psi}\psi \rangle_{\text{quench}}(z)$ in the quenched approximation? From Eq. (10.33) the value of $\langle \bar{\psi}\psi \rangle_{\text{quench}}(z)$ can be related to

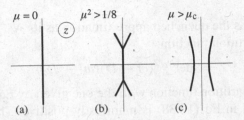

Figure 10.5. The evolution of cuts of the function $\langle \bar{\psi}\psi \rangle(z)$ with increasing μ. $\mu_c = 0.53$.

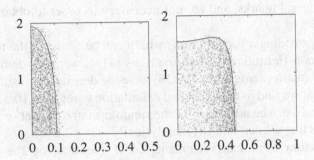

Figure 10.6. Eigenvalues of 20 random 100×100 matrices on the complex-z plane at two values of μ^2, 0.06 (left) and 0.40 (right). The line is the boundary of the $\rho_{ev} \neq 0$ region calculated analytically.

the averaged distribution of the eigenvalues ρ_{ev} of the matrix D:

$$\langle \bar{\psi}\psi \rangle_{\text{quench}}(m) = \int \int dx \, dy \, \frac{\rho_{ev}(x, y)}{m - z}, \qquad z = x + iy \qquad (10.37)$$

This distribution is easy to obtain numerically and, in the random-matrix model, it has the form shown in Fig. 10.6. One sees that, as soon as μ is different from zero, the eigenvalues spread into a two-dimensional distribution. This is not a surprise, since the matrix D loses its anti-Hermiticity properties. However, on comparing Fig. 10.6 with Fig. 10.5 we conclude that singularities of the function $\langle \bar{\psi}\psi \rangle_{\text{quench}}(m)$ are qualitatively different from line-like (cut) singularities of $\langle \bar{\psi}\psi \rangle(m)$ at any finite N_f. In particular, the *linear* density of eigenvalues vanishes immediately when μ differs from zero. In other words, chiral simmetry is restored at $\mu = 0$ in the quenched approximation, instead of at finite $\mu = \mu_c$. The limit $N_f \to 0$ is not the same as the quenched approximation at finite μ.

What does this mean? It means that the quenched approximation does not approximate, even qualitatively, the behavior of the full theory. This can be further understood if we find a theory that has the quenched approximation as its $N_f \to 0$ limit.

Conjugate quarks

The theory which has the quenched approximation as its $N_f \to 0$ limit is given by the following partition function:

$$Z = \langle \det^{N_f/2}(m - D)(m^* - D^\dagger) \rangle \tag{10.38}$$

On comparing this partition function with the one given by Eq. (10.34), we note that the determinant in Eq. (10.38) is manifestly positive. This is achieved by introducing a set of *conjugate quarks* – fermion fields whose Dirac matrix is the Hermitian conjugate to that of normal quarks. For m real, this means that the sign of the chemical potential is reversed, and for $\mu = 0$ conjugate quarks are identical to normal quarks, and are not necessary in order to obtain the correct quenched limit.

Instead of providing a formal proof, which can be found in the mathematical literature on non-Hermitian random matrices [118], we shall demonstrate that the partition-function zeros in Eq. (10.38) have the density $\rho(z)$ which coincides with $\rho_{ev}(z)$ that we find in the quenched calculation using Eq. (10.33) and which we see in Fig. 10.6. The advantage of the random-matrix model (over full QCD calculation) is that we can find $\rho(z)$ analytically.

The integration over matrix elements of X (denoted by the brackets) in Eq. (10.38) is similar to the previous calculation we have done in Eqs. (10.34) and (10.16). However, instead of one auxiliary $N_f \times N_f$ matrix field ϕ we have to introduce four $N_f/2 \times N_f/2$ matrices, which we denote a, b, c, and d:

$$Z = \int \mathcal{D}a\,\mathcal{D}b\,\mathcal{D}c\,\mathcal{D}d \, \det^N \begin{pmatrix} z+a & \mu & 0 & id \\ \mu & z+a^\dagger & ic & 0 \\ 0 & id^\dagger & z^*+b^\dagger & \mu \\ ic^\dagger & 0 & \mu & z^*+b \end{pmatrix}$$
$$\times \exp[-N(|a|^2 + |b|^2 + |c|^2 + |d|^2)] \tag{10.39}$$

On differentiating with respect to m, we find that the value of the condensate $\langle \bar{\psi}\psi \rangle$ can be found by taking the value of a at the saddle point of the integrand in Eq. (10.39).

The set of solutions of the saddle-point equation is richer in this case. There is a solution with $c = d = 0$. In this case the conjugate quarks decouple and we simply obtain the same holomorphic function $\langle \bar{\psi}\psi \rangle = a = b$ as before. However, there is another solution in which the condensates c and d are not zero! Then the function $\phi = a$ is not holomorphic and therefore $\rho \neq 0$. This saddle point dominates the integral at small z for $0 < \mu < 1$.

The condensates c and d are bilinears of the type $\langle \bar{\psi}\chi \rangle$, mixing original ψ and conjugate χ quarks. These condensates do not break the original chiral symmetry but break a spurious (replica-type) symmetry involving both original and conjugate quarks. We shall encounter similar condensates carrying baryon number when we discuss the SU(2) model of QCD *with* quarks in Section 10.3. In the quenched theory, as in SU(2) QCD, the original chiral symmetry is always restored at $\mu > 0$. The spurious symmetry is spontaneously broken for $\mu < 1$ and is restored for $\mu > 1$.

The boundary of the $\rho_{ev} \neq 0$ region is given by

$$y^2 = (\mu^2 - x^2)^{-2}[4\mu^4(1 - \mu^2) - (1 + 4\mu^2 - 8\mu^4)x^2 - 4\mu^2 x^4] \quad (10.40)$$

It is plotted in Fig. 10.6 for comparison with numerical data. The baryonic condensates c and d inside of the "blob" are given by

$$|c|^2 = |d|^2 = \frac{\mu^2}{\mu^2 - x^2} - \mu^2 - \frac{x^2}{4(\mu^2 - x^2)^2} - \frac{y^2}{4} \quad (10.41)$$

On the boundary, Eq. (10.40), they vanish and the two solutions (holomorphic and nonholomorphic) match. In the outer region $c = d = 0$ and a is the solution of the cubic equation (10.36). Inside the "blob" the value of a is given by

$$\langle \bar{\psi}\psi \rangle = a = \frac{1}{2}\frac{x}{\mu^2 - x^2} - x - \frac{iy}{2} \quad (10.42)$$

and the density of the partition-function zeros (and of the eigenvalues in the quenched case) (10.31) is

$$\rho = \rho_{ev} = \frac{1}{4\pi}\left(\frac{x^2 + \mu^2}{(\mu^2 - x^2)^2} - 1\right) \quad (10.43)$$

To appreciate the nontriviality of this result one should notice that expression (10.33) appears to depend only on m, not on m^*! The limit $N_f \to 0$ must be taken with great care.

10.3 Two-color QCD and effective Lagrangians

Effective Lagrangians are used in high-energy physics to extract aspects of the low-energy solution of the theory by emphasizing symmetries and their realizations, through the Goldstone mechanism and current algebra. The approach is akin to spin-wave analyses of condensed-matter systems. Although its reach is limited to the lowest momenta and energies in the system, it can produce nontrivial results that are strongly constrained by symmetries and their realizations. Since many such results are nonperturbative, they are difficult to derive from other approaches that begin with the microscopic Lagrangian and its interacting quarks and gluons. For energies and momenta less than or of the same order as the pion mass, the effective-Lagrangian method, and chiral perturbation theory in particular, may be invaluable.

Two-color QCD has some peculiar properties, of which some present advantages and others present disadvantages. Since its representations are pseudo-real, as we will discuss in detail below, its quark–anti-quark meson spectrum is identical to its two-quark "baryon" spectrum. Therefore, its lowest-lying "baryon" has the same mass as its pion and vanishes in the chiral limit. When it is placed into an environment with a chemical potential, it will experience a transition to a diquark condensate at a critical quark chemical potential of $\mu_c = m_\pi/2$ because at this point the sum of the lowest energy levels of the two quarks, -2μ, cancels out the energy of the lightest "baryon" and triggers condensation. This critical chemical potential vanishes when the bare mass of the theory's quarks is taken to zero. This is qualitatively different from QCD, in which quark–anti-quark mesons and three-quark baryons are distinct, both through dynamics and through quantum numbers. In particular, a small chemical potential should hardly affect the theory's meson spectroscopy since mesons carry no baryon number. The nucleons are the lightest, colorless, physical fermion states with the lowest physical baryon number and, when μ exceeds the gap in this channel, $3\mu_c = M_N$, a transition should occur. At very high chemical potential one expects the BCS mechanism to prevail and $\bar{3}$ Cooper pairs to condense off a sharp Fermi surface and form a color superconductor. The mass of the nucleon is large in the chiral limit of the model where $m_\pi \to 0$, so the separation of the two energy scales, chiral symmetry on the one hand and the formation of a color superconductor on the other, should be clear.

We shall see that, in the SU(2) version of QCD, one predicts a superfluid phase of diquark condensation at $\mu_c = m_\pi/2$. In the diquark phase, the baryon number is spontaneously broken, so one expects a single Goldstone boson to appear. Since the diquarks are meson-like and colorless, the phase resembles a superfluid rather than the color superconductor expected in SU(3) QCD. Note that the Goldstone bosons of baryon-number breaking are exactly massless even for a theory with nonvanishing quark masses. In sharp contrast, the color-diquark phase of the SU(3) model is similar to the Higgs phase of the electroweak model because in both cases the condensate breaks a global part of the gauge symmetry. The diquarks of the SU(3) theory transform nontrivially under the gauge group, like the Higgs particle of electroweak physics and the Cooper pairs of electronic superconductivity, and, when they condense, one expects the formation of a vacuum with color screening and no exact Goldstone excitations. The energetics of the superfluid phase of the SU(2) model and the superconducting phase of the SU(3) model appear to be very different.

It is particularly interesting that all of the physics in the SU(2) model is within reach of effective Lagrangians and chiral perturbation theory, in particular. The phase transition itself, from hadronic matter to a superfluid phase at $\mu_c = m_\pi/2$, is also predicted by these methods. This is a unique situation in which an inherently perturbative method of analysis like chiral perturbation theory can address a phase transition in full detail and with good control.

We will discuss the SU(2) transition thoroughly and not speculate further on its differences from the physics of the SU(3) model. Perhaps we will learn things relevant to color superconductivity, perhaps not. Superfluid phases have been argued for in the SU(3) model, in addition to its superconductivity, and pion condensation and kaon condensation have been modeled. There is a good chance that some of the SU(2) analysis will apply to some aspects of the physics of QCD.

10.3.1 QCD inequalities and the nature of the ground state

We begin with the fermion Lagrangian,

$$L = \sum_{f=1}^{N_f} (\bar{\psi}_f \gamma_\nu D_\nu \psi_f + \mu \bar{\psi}_f \gamma_0 \psi_f + m_q \bar{\psi}_f \psi_f) \tag{10.44}$$

where we shall assume that the quark mass is small and consider the chiral limit $m_q \to 0$ in much of this discussion.

First, we are interested in establishing general constraints on the pattern of symmetry breaking here. What quark bilinear can develop a vacuum expectation value? What are the quantum numbers of the lightest meson?

We know from ancient discussions in Euclidean QCD [119] that the pion propagator bounds all others from above. From this one can prove that the pion is a pseudo-scalar and therefore the condensate must be a scalar $\langle \bar{\psi} \psi \rangle$. This establishes that the condensate is not a pseudo-scalar, like $\langle \bar{\psi} \gamma_5 \psi \rangle$, because this would give rise to a 0^+ Goldstone boson.

The argument proceeds as follows. Consider the Dirac operator for $\mu = 0$, $D = \gamma(\partial + iA) + m_q$. which satisfies the identity

$$\gamma_5 D \gamma_5 = D^\dagger \tag{10.45}$$

Now consider the composite propagator for a meson. Generate the meson with the operator $M = \bar{\psi} \Gamma \psi$, where Γ is a generic flavor or spinor matrix. The expectation value of the composite propagator is

$$\langle M(x) M(0) \rangle_{\psi, \bar{\psi}, A} = \langle \text{tr}[S(x, 0) \Gamma S^\dagger(0, x) \Gamma] \rangle_A \tag{10.46}$$

where we have indicated which integrals in the path integral have been done, and S is the fermion propagator, $S = D^{-1}$. Taking $\Gamma = \gamma_5$, we can use Eq. (10.45) to write Eq. (10.46) as $\langle \text{tr}[S(x, 0) S(0, x)] \rangle_A$. This quantity is an upper bound on the general composite propagator, as we can see using the Schwartz inequality,

$$\text{tr}[S(x, 0) \Gamma S(0, x) \Gamma] = \text{tr}[S(x, 0) \Gamma \gamma_5 S^\dagger(x, 0) \gamma_5 \Gamma] \leq \text{tr}[S(x, 0) S^\dagger(x, 0)] \tag{10.47}$$

On taking expectation values, we obtain the desired inequality for the correlation function and, therefore, for the meson masses. Implications and consequences of these correlation-function inequalities have been discussed in the literature to shed light on chiral-symmetry breaking and spectroscopy [119].

What parts, if any, of these observations generalize to nonzero chemical potential? In the case of SU(3) color Eq. (10.45) does not hold true when the chemical potential is nonzero and $D = \gamma(\partial + iA) + \mu\gamma_0 + m_q$. However, in the case of SU(2) color there is a generalization of Eq. (10.45) that proves useful. Use the fact that SU(2) is a pseudo-real group. In particular, the second Pauli matrix satisfies $T_2 T_a T_2 = -T_a^*$. By combining this with charge conjugation, $C = i\gamma_0\gamma_2$ ($C^2 = 1$ and $C\gamma_v C = -\gamma_v^*$), we can generalize Eq. (10.45) to

$$\gamma_5 C T_2 D \gamma_5 C T_2 = D^* \tag{10.48}$$

It is particularly illuminating to consider the propagator of the SU(2) diquark, $M_{\psi\psi} = \psi^T C T_2 \gamma_5 \psi$, which is a 0^+, $I = 0$ (anti-symmetric in flavor) operator, for which

$$\langle M_{\psi\psi}(x) M_{\psi\psi}(0) \rangle_{\psi,\psi^T,A} = \langle \mathrm{tr}[S(x,0)C T_2\gamma_5 S^T(x,0)C T_2\gamma_5] \rangle_A$$
$$= \langle \mathrm{tr}[S(x,0)S^\dagger(x,0)] \rangle_A \tag{10.49}$$

This relation becomes the basis of showing that the correlator of $M_{\psi\psi} = \psi^T C T_2\gamma_5\psi$ bounds the correlator of any other meson like $\psi^T C T_2\gamma_5\Gamma\psi$. Therefore, it is the 0^+, not the 0^-, state which is the lightest. In addition, if there is condensation, it has to be that of the diquark operator $\psi^T C T_2\gamma_5\psi$ which leaves parity invariant.

10.3.2 Symmetries and Goldstone bosons in the diquark phase

Let's consider the SU(2) color version of QCD with two flavors, massless "up" and "down" quarks. We want to set up a convenient formalism for it so that we can easily read off that the theory at vanishing chemical potential, $\mu = 0$, in the chiral limit $m_q = 0$, has an SU(4) global symmetry, which, when chiral symmetry breaks dynamically, is reduced to Sp(4), producing five Goldstone bosons. The theory has two "extra" Goldstone bosons compared with SU(3) color and we want to understand their origin in the pseudo-real nature of the SU(2) color group. When the chemical potential is turned on, we will also see that the theory's global symmetry is reduced to SU(2) × SU(2) × U(1), which breaks upon diquark condensation to SU(2) × SU(2), producing one Goldstone boson corresponding to baryon-number breaking. In addition, there are four pseudo-Goldstone bosons, each with a mass of precisely 2μ, to leading order [120].

We begin by writing the fermion piece of the SU(2) color Lagrangian in terms of left- and right-handed quark fields,

$$\psi = \begin{pmatrix} q_L \\ q_R \end{pmatrix} \tag{10.50}$$

and define the extensions of the Pauli matrices,

$$\sigma_\nu = (-i, \sigma_k), \qquad \bar{\sigma}_\nu = (-i, -\sigma_k) \tag{10.51}$$

Now the Lagrangian can be written

$$L = \bar{\psi} i \not{D} \psi = q_L^\dagger i \sigma_\nu D_\nu q_L + q_R^\dagger i \bar{\sigma}_\nu D_\nu q_R \tag{10.52}$$

where $D_\nu = \partial_\nu + T_a A_\nu^a$ is the covariant derivative for SU(2) color. It will prove handy to recall that $T_a^* = -T_2 T_a T_2 = T_a^T$ for the Pauli-matrix generators of SU(2) color and verify that the covariant derivative satisfies

$$D_\nu^T = -T_2 D_\nu T_2 \tag{10.53}$$

Now we can define conjugate quarks,

$$\tilde{q} = \sigma_2 T_2 q_R^*, \qquad \tilde{q}^\dagger = q_R^T T_2 \sigma_2 \tag{10.54}$$

The Lagrangian can now be written

$$L = q^\dagger i \sigma_\nu D_\nu q + \tilde{q}^\dagger i \sigma_\nu D_\nu \tilde{q} = \Psi^\dagger i \sigma_\nu D_\nu \Psi \tag{10.55}$$

which has the global SU(4) flavor symmetry advertised above. In this expression Ψ is the Weyl spinor,

$$\Psi = \begin{pmatrix} q \\ \tilde{q} \end{pmatrix} = \begin{pmatrix} q^1 \\ q^2 \\ \tilde{q}^1 \\ \tilde{q}^2 \end{pmatrix} \tag{10.56}$$

Note that q and $\tilde{q}$ have opposite baryon numbers but identical chiral $U(1)_A$ charges.

It is useful to write various quark bilinears in this language to determine their transformation properties. For example, the chiral condensate is

$$\bar{\psi}\psi = \frac{1}{2}\Psi \sigma_2 T_2 \begin{pmatrix} 0 & 1 \\ -1 & 0 \end{pmatrix} \Psi + \text{h.c.} \tag{10.57}$$

which we identify with the sigma meson. It is a member of an SU(4) 6-plet. The remaining five members of the 6-plet are three pions, a scalar diquark, and a scalar anti-diquark.

The baryon charge reads

$$\bar{\psi}\gamma_0\psi = q_L^\dagger q_L + q_R^\dagger q_R = q^\dagger q + \tilde{q}^T \tilde{q}^{\dagger T} = q^\dagger q - \tilde{q}^\dagger \tilde{q} = \Psi^\dagger \begin{pmatrix} 1 & 0 \\ 0 & -1 \end{pmatrix} \Psi$$
(10.58)

which indicates that adding a chemical potetial to the Lagrangian breaks its algebraic symmetry down to $SU(2)_L \times SU(2)_R \times U(1)_B$. When this symmetry breaks through diquark condensation, there will be one Goldstone boson corresponding to the breaking of $U(1)_B$. We will see that the other four "pseudo-Goldstone" bosons form a $(2, 2)$ multiplet of $SU(2)_L \times SU(2)_R$ and acquire a common mass of 2μ.

10.3.3 Effective-Lagrangian construction

Now we can construct the effective Lagrangian to describe the low-energy features of this model. We will concentrate on the low-lying mesons and diquarks and restrict their interactions, guided by the symmetry considerations we have laid out above. To do this we will use Ψ because it transforms simply, $\Psi \to U\Psi$, under $SU(4)$.

We begin with the quark–quark bilinears that compose the low-energy states,

$$\Sigma \sim \Psi \Psi^T \sigma_2 T_2$$
(10.59)

which is anti-symmetric both in spin and in color indices and transforms under $SU(4)$ according to $\Sigma \to U\Sigma U^\dagger$, as a flavor $SU(4)$ 6-plet. The Lagrangian that can describe the low-energy aspects of the theory and is invariant under flavor $SU(4)$ is a nonlinear sigma model,

$$L_1 = f_\pi^2 \, \mathrm{tr}(\partial_\nu \Sigma^\dagger \, \partial_\nu \Sigma)$$
(10.60)

We want to generalize this discussion to nonzero chemical-potential. The chemical-potential term in the microscopic theory reads

$$\mu \bar{\psi}\gamma_0\psi = \mu \Psi^\dagger \begin{pmatrix} 1 & 0 \\ 0 & -1 \end{pmatrix} \Psi$$
(10.61)

which breaks the $SU(4)$ flavor symmetry because it distinguishes between quarks and conjugate quarks. However, it is useful to re-interpret this symmetry. Write this term as

$$\mu \Psi^\dagger i\sigma_\nu B_\nu \Psi$$
(10.62)

where B_ν is the $SU(4)$ matrix

$$B_\nu = \delta_{0\nu} \begin{pmatrix} 1 & 0 \\ 0 & -1 \end{pmatrix} = \delta_{0\nu} B$$
(10.63)

and give B_ν the SU(4) transformation property

$$B_\nu \to U B_\nu U^\dagger \qquad (10.64)$$

With this rule, the microscopic Lagrangian

$$\Psi^\dagger i\sigma_\nu D_\nu \Psi + \mu \Psi^\dagger i\sigma_\nu B_\nu \Psi \qquad (10.65)$$

becomes invariant.

Now we can turn our attention to the construction of the effective Lagrangian which has the same transformation properties as Eq. (10.65). It must also be invariant under these extended transformations and its dependence on μ must be unique and not involve any additional couplings. Since B_ν transforms nontrivially under SU(4), a term linear in B_ν cannot appear in the effective Lagrangian. The lowest-order nontrivial term would be proportional to $\mu^2 \operatorname{tr}(\Sigma B_\nu^T \Sigma^\dagger B_\nu)$. Such a term will produce pseudo-Goldstone boson masses that are linear in μ, but what should the overall coefficient of this term be? The answer must be unique, because there is no freedom in how to introduce the chemical potential into the microscopic, quark-level Lagrangian. We shall see in our discussions of the lattice version of QCD at nonzero μ that the chemical potential is introduced like a timelike imaginary Abelian gauge field. This inspires us to extend the symmetry properties of B_ν to a local, gauge symmetry in the microscopic theory: we require that the theory be invariant under the transformations

$$B_\nu \to U B_\nu U^\dagger + \frac{1}{\mu} U \, \partial_\nu U^\dagger \qquad (10.66)$$

The effective Lagrangian must be invariant under this local symmetry, so the derivatives in the effective L of Eq. (10.60) must become

$$D_\nu \Sigma = \partial_\nu \Sigma + \mu (B_\nu \Sigma + \Sigma B_\nu^T) \qquad (10.67)$$
$$D_\nu \Sigma^\dagger = \partial_\nu \Sigma^\dagger - \mu (\Sigma^\dagger B_\nu + B_\nu^T \Sigma^\dagger) \qquad (10.68)$$

The effective Lagrangian becomes

$$L_{\text{eff}} = f_\pi^2 \operatorname{tr}(D_\nu \Sigma^\dagger D_\nu \Sigma) \qquad (10.69)$$

Finally, we should incorporate the effect of a bare quark mass into the effective Lagrangian. We need this term in order to pick out the ground state and to discuss the phase transition to a diquark condensate at nonzero μ. The bare mass in the microscopic theory is

$$m\bar\psi\psi = \frac{1}{2} m \Psi^T \sigma_2 \tau_2 \begin{pmatrix} 0 & -1 \\ 1 & 0 \end{pmatrix} \Psi + \text{h.c.} = -\frac{1}{2} \Psi^T \sigma_2 \tau_2 M \Psi + \text{h.c.} \qquad (10.70)$$

where the mass matrix M is given by

$$M = m\hat{M}, \qquad \hat{M} = \begin{pmatrix} 0 & 1 \\ -1 & 0 \end{pmatrix} \tag{10.71}$$

We see that the bare-mass term breaks the SU(4) symmetry. The full SU(4) invariance can be restored if M is given the transformation property

$$\Psi \to U\Psi, \qquad M \to U^*MU^\dagger. \tag{10.72}$$

This extended symmetry must also be manifest in the effective theory,

$$\Sigma \to U\Sigma U^T, \qquad M \to U^*MU^\dagger \tag{10.73}$$

The lowest-order term induced by the quark mass must therefore have the form

$$L_{\text{qmass}} = -G\,\text{Re}[\text{tr}(M\Sigma)] = -mG\,\text{Re}[\text{tr}(\hat{M}\Sigma)]. \tag{10.74}$$

One can view $\text{Re}\,\text{tr}(\hat{M}\Sigma)$ as a generalized cosine of the angle between unitary matrices Σ and $\hat{M}^\dagger$. It is maximal when Σ is aligned with $\hat{M}^\dagger$. Therefore the direction of Σ minimizing Eq. (10.74) is given by

$$\Sigma_c = \hat{M}^\dagger = \begin{pmatrix} 0 & 1 \\ -1 & 0 \end{pmatrix} \tag{10.75}$$

The mass term comes with a phenomenological coefficient, which we denote by G. It is given by the derivative of the vacuum energy with respect to m and is, therefore, proportional to the chiral condensate in the chiral limit $m \to 0$ at $\mu = 0$,

$$G = \langle \bar{\psi}\psi \rangle_0 / 4 \tag{10.76}$$

The resulting Lagrangian with the mass term is

$$L_{\text{eff}} = f_\pi^2\,\text{tr}(D_\nu\Sigma D_\nu\Sigma^\dagger) - mG\,\text{Re}\,\text{tr}(\hat{M}\Sigma) \tag{10.77}$$

Expanded to second order in the pion fields, it yields a spectrum with five degenerate pseudo-Goldstone bosons with masses given by the Gell-Mann–Oakes–Renner relation

$$m_\pi^2 = mG/(2f_\pi^2) \tag{10.78}$$

We can use this relation to trade G for another parameter, m_π, and write

$$L_{\text{eff}} = f_\pi^2\big[\text{tr}(D_\nu\Sigma D_\nu\Sigma^\dagger) - 2m_\pi^2\,\text{Re}\,\text{tr}\,(\hat{M}\Sigma)\big] \tag{10.79}$$

which can be expanded, using the property $\Sigma^T = -\Sigma$, to read

$$L_{\text{eff}} = f_\pi^2\,\text{tr}(\partial_\nu\Sigma^\dagger\,\partial_\nu\Sigma) - 4\mu f_\pi^2\,\text{tr}(\partial_\nu\Sigma\,\Sigma^\dagger B_\nu)$$
$$- 2f_\pi^2\mu^2\,\text{tr}\big[(\Sigma^\dagger B_\nu + B_\nu^T\Sigma^\dagger)(B_\nu\Sigma + \Sigma B_\nu^T)\big] - 2f_\pi^2 m_\pi^2\,\text{Re}\,\text{tr}(\hat{M}\Sigma) \tag{10.80}$$

10.3.4 Vacuum alignment, diquark condensation, the phase diagram, and scaling laws

Our next task is the determination of the ground state and its symmetries as a function of μ [121]. Take the static pieces in L_{eff} and write them in the form

$$L_{\text{static}}(\Sigma) = \frac{f_\pi^2 m_\pi^2}{2}\left(-\frac{x^2}{2}\,\text{tr}\big(\Sigma B_\nu^T \Sigma^\dagger B_\nu + B_\nu B_\nu\big) - 2\,\text{Re}\,\text{tr}(\hat{M}\Sigma) \right) \quad (10.81)$$

where we introduced the dimensionless variable $x = 2\mu/m_\pi$. We need to find those fields Σ which minimize the static part of the effective Lagrangian. We have already observed that, when the chemical potential vanishes, the condensate direction is $\Sigma_c = \hat{M}^\dagger$. However, when the chemical potential becomes very large, $x \to \infty$, the best Σ becomes

$$\Sigma_d = \begin{pmatrix} iI & 0 \\ 0 & iI \end{pmatrix}, \qquad I = \begin{pmatrix} 0 & -1 \\ 1 & 0 \end{pmatrix} \quad (10.82)$$

This choice of the condensate direction is not unique. It is clear from the effective Lagrangian that we can rotate the condensate by the generator B and the effective Lagrangian remains unchanged. This degeneracy corresponds to the existence of a massless Goldstone boson.

Now we need to determine the condensate for intermediate values of the chemical potential. The simplicity of this model produces a simple result: the condensate at intermediate values of x, call it Σ_α, is just a rotation from Σ_c to Σ_d,

$$\Sigma_\alpha = \Sigma_c \cos\alpha + \Sigma_d \sin\alpha \quad (10.83)$$

where the "rotation" angle α varies form 0 at $x = 0$ to $\pi/2$ at $x = \infty$. The proof of this fact consists of some algebra, which will be done at the end of this chapter. Let's concentrate on the physics first.

On substituting Σ_α into the static piece of the effective Lagrangian, we find

$$L_{\text{static}}(\Sigma_\alpha) = 2 f_\pi^2 m_\pi^2\left(\frac{x^2}{2}[\cos(2\alpha) - 1] - 2\cos\alpha \right) \quad (10.84)$$

which determines α as a function of x,

$$\alpha = 0, \qquad \text{when } x < 1 \quad (10.85)$$

$$\cos\alpha = \frac{1}{x^2}, \qquad \text{when } x > 1 \quad (10.86)$$

This is a major result of this model. It establishes that there is a second-order phase transition in the model at $x = 1$, which means that $\mu_c = m_\pi/2$. For

$\mu < m_\pi/2$, the vacuum does not respond to the chemical potential at all. At $\mu_c = m_\pi/2$ a transition occurs and the direction of the condensate starts rotating from Σ_c to Σ_d. Such a state breaks baryon-number symmetry spontaneously and this breaking is accompanied by a Goldstone boson, a 0^+ singlet.

To understand the character of the phase transition, we should calculate the ground-state expectation values of some local observables, such as the chiral condensate $\langle \bar\psi \psi \rangle$, the diquark condensate $\langle \psi \psi \rangle$, and the induced baryon number n_B. According to lowest-order perturbation theory,

$$\langle \bar\psi \psi \rangle = -\frac{\partial E_{\text{vac}}}{\partial m}, \qquad n_B = -\frac{\partial E_{\text{vac}}}{\partial \mu} \tag{10.87}$$

To calculate the diquark expectation value, we need to put a diquark source into the Lagrangian. In the microscopic Lagrangian, such a source reads

$$-\mathrm{i}\frac{j}{2}\psi^T C \gamma_5 T_2 I \psi + \text{h.c.} = -\frac{j}{2}\Psi^T \sigma_2 T_2 \begin{pmatrix} \mathrm{i}I & 0 \\ 0 & \mathrm{i}I \end{pmatrix} \Psi + h.c.$$

$$= -\frac{j}{2}\Psi^T \sigma_2 T_2 \hat{J} \Psi + \text{h.c.} \tag{10.88}$$

where we have defined

$$\hat{J} \equiv \begin{pmatrix} \mathrm{i}I & 0 \\ 0 & \mathrm{i}I \end{pmatrix} \tag{10.89}$$

and I is the anti-symmetric matrix introduced earlier. This source term should be added to our $m\bar\psi\psi$ source term,

$$m\bar\psi\psi - \mathrm{i}\frac{j}{2}(\psi^T C \gamma_5 T_2 I \psi + \text{h.c.}) = -\frac{1}{2}\Psi^T \sigma_2 T_2 M_\phi \Psi \tag{10.90}$$

where

$$M_\phi \equiv m\hat{M} + j\hat{J} = \sqrt{m^2 + j^2}(\hat{M}\cos\phi + \hat{J}\sin\phi)$$

$$= \sqrt{m^2 + j^2}\,\hat{M}_\phi \tag{10.91}$$

where $\tan\phi = j/m$. Finally, the effective Lagrangian with both sources is, repeating the steps above when we were just considering the source $m\bar\psi\psi$,

$$L_{\text{eff}}(\Sigma) = f_\pi^2[\text{tr}(D_\nu \Sigma D_\nu \Sigma^\dagger) - 2m_\pi^2 \,\text{Re}\,\text{tr}(\hat{M}_\phi \Sigma)] \tag{10.92}$$

Now we can complete Eq. (10.87) with the addition of the diquark condensate,

$$\langle \psi\psi \rangle = -\frac{\partial E_{\text{vac}}}{\partial j} \tag{10.93}$$

Table 10.1 The values of the chiral condensate, $\langle \bar{\psi}\psi \rangle$, the diquark condensate, $\langle \psi\psi \rangle$, and the baryon density n_B in the two phases of the theory

Phase	$\langle \bar{\psi}\psi \rangle$	$\langle \psi\psi \rangle$	n_B
$\mu < m_\pi/2$	$\langle \bar{\psi}\psi \rangle_0$	0	0
$\mu > m_\pi/2$	$\langle \bar{\psi}\psi \rangle_0 \left(\frac{m_\pi}{2\mu}\right)^2$	$\langle \bar{\psi}\psi \rangle_0 \sqrt{1 - \left(\frac{m_\pi}{2\mu}\right)^4}$	$32\mu f_\pi^2 \left[1 - \left(\frac{m_\pi}{2\mu}\right)^4\right]$

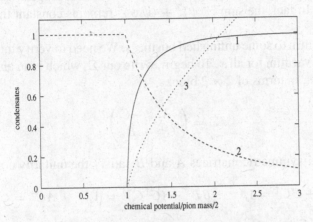

Figure 10.7. The effective-Lagrangian prediction for the diquark condensation transition, showing magnitudes of the diquark condensate $\langle \psi\psi \rangle$ (1) and the chiral condensate $\langle \bar{\psi}\psi \rangle$ (2) in units of $\langle \bar{\psi}\psi \rangle_0 = 4G$ as a function of $2\mu/m_\pi$ for zero diquark source. The density of the baryon charge (3) is shown in units of $32 f_\pi^2 m_\pi$.

Our last task is to write the effective Lagrangian at the minimum,

$$
E_{\text{vac}} = L_{\text{eff}}(\Sigma_\alpha) = f_\pi^2 \left[-2\mu^2 \, \text{tr}\left(\Sigma_\alpha B_\nu^{\text{T}} \Sigma_\alpha^\dagger B_\nu\right) - 2m_\pi^2 \, \text{Re} \, \text{tr}(\Sigma_\phi^\dagger \Sigma_\alpha)\right]
$$
$$
= -16 f_\pi^2 \mu^2 \sin^2\alpha - 4G(m\cos\alpha + j\sin\alpha) \qquad (10.94)
$$

On carrying out the differentiations, we find

$$
\langle \bar{\psi}\psi \rangle = 4G\cos\alpha, \qquad \langle \psi\psi \rangle = 4G\sin\alpha, \qquad n_B = 32 f_\pi^2 \mu \sin^2\alpha \quad (10.95)
$$

The results for these vacuum expectation values are listed in Table 10.1 and are plotted in Fig. 10.7. Note a few interesting, elementary features. First, the observables are constants in the $\mu < m_\pi/2$ phase. This was assured by the fact that the vacuum does not change its direction for $\mu < m_\pi/2$. Therefore, the chiral condensate retains its $\mu = 0$ value exactly until $\mu_c = m_\pi/2$. In addition, both the diquark condensate and the induced baryon number vanish exactly in the low-μ phase. However, at $\mu_c = m_\pi/2$ all of the observables develop nonanalytic dependences on x. In particular, the diquark condensate rises from zero with a square-root singularity characteristic of a mean-field transition, $\mu^{-2}\sqrt{\mu^4 - (m_\pi/2)^4} \sim 2\sqrt{\mu - m_\pi/2}$, for μ greater than but very near $m_\pi/2$. In addition, the induced baryon number rises linearly with μ in the critical region. Finally, in the high-μ phase the chiral condensate falls as x^{-2}. We have already observed that, in the mean-field approximation, the ground-state vacuum rotates from Σ_c to Σ_d as x grows beyond unity and we see that property reflected in the condensates. In fact, the sum $\langle \bar{\psi}\psi \rangle^2 + \langle \psi\psi \rangle^2$ remains constant throughout the transition.

Now we return to some unfinished business. We need to verify that Eq. (10.83) describes the vacuum for all x. To begin, write out Σ, which is an anti-symmetric unitary matrix, in terms of 2×2 blocks,

$$\Sigma = \begin{pmatrix} A & -C \\ C^{\mathrm{T}} & B \end{pmatrix} \tag{10.96}$$

where the anti-symmetric matrices A and B satisfy the unitarity constraints

$$AA^\dagger + CC^\dagger = 1, \qquad BB^\dagger + (C^\dagger C)^{\mathrm{T}} = 1, \qquad AC^* = CB^\dagger \tag{10.97}$$

Using these relations the static piece of the effective Lagrangian can be written just in terms of C,

$$L_{\mathrm{eff}}(\Sigma) = 2f_\pi^2 m_\pi^2 \left[x^2 \operatorname{tr}\left(C - \frac{1}{x^2} \right)\left(C^\dagger - \frac{1}{x^2} \right) - 2\left(x^2 + \frac{1}{x^2} \right) \right] \tag{10.98}$$

To minimize $L_{\mathrm{eff}}(\Sigma)$, we should take $C = 1/x^2$, and to satisfy the constraints Eq. (10.97), we can take $A = B = iI\sqrt{1 - 1/x^2}$. Then, if we use the notation $\cos\alpha = 1/x^2$, Σ_α will have the form Eq. (10.83). Note that, when $x < 1$, we have a special situation. The solution $C = 1/x^2$ is then inconsistent with the constraints Eq. (10.97) and we are forced to the solution $C = 1$, $A = B = 0$. So, for all $x < 1$, for any chemical potential below the critical value $\mu_c = m_\pi/2$, Σ_α reduces to Σ_c, the minimum with no chemical potential.

There are other interesting features of this model that are worth mentioning. In particular, the spectroscopy of the low-lying pseudo-Goldstone states is constrained by chiral symmetry and the simplicity of the coupling of the model to the chemical potential. For example, in the absence of bare-quark masses, the curvature of the potential which determines these masses is just

$$L_{\mathrm{curv}} = 2\mu^2 \operatorname{tr} B_\nu^2 - 2\mu^2 \operatorname{tr}\left(\Sigma B_\nu^{\mathrm{T}} \Sigma^\dagger B_\nu \right) \tag{10.99}$$

and to find the mass matrix of the pseudo-Goldstone bosons we should expand Σ in small, smooth fluctuations around the vacuum Σ_d,

$$\Sigma = U\Sigma_d U^T \tag{10.100}$$

It is sensible to write U as an exponent of the various generators of SU(4) and consider first those generators which do not change Σ_d, call them T_a, and those that do, call them X_b. The details are given in the literature [120] and show that there are ten T_a's and five X_b's. One of the broken generators, called X_5 in the literature, is the baryon charge B and the fluctuations it generates in the ground state leave the ground-state energy unchanged. Therefore, its Goldstone particle is massless. This is an exact result for all quark bare masses and chemical potentials, because the baryon number is always a perfect symmetry of the local Lagrangian density. The other four X generators can be identified and their common mass is found to be exactly 2μ, unadorned by strong interaction. This final result is a consequence of the local symmetry which dictated how the chemical potential can enter the effective Lagrangian.

10.4 QCD at nonzero isospin chemical potential

We have seen in Section 10.1 that the integrand of the partition function in QCD lacks positivity, which is the major obstacle to current attempts to simulate QCD in this regime using Monte Carlo methods. We have also seen in Section 10.2.3 that the quenched theory is a limit of a theory that has a positive measure/integrand in the partition function. This is due to the presence of conjugate quarks. Although unphysical, as they appear at first, they turn out to be present in two-color QCD, due to the pseudo-reality of the SU(2) group. In this section we discuss a regime of actual three-color QCD that contains conjugate quarks explicitly, and therefore can be simulated on the lattice. It also turns out that this theory can be analysed by the effective-Lagrangian methods of the previous section. This regime is QCD at finite *isospin* density [122].

Why should this regime be interesting? One of the questions we would like to address by studying QCD at finite baryon-number density is how the transition between hadronic and underlying quark–gluon degrees of freedom occurs as a function of a *conserved* number density, unlike the temperature transition. In QCD, isospin is such a conserved number, like the baryon charge, and it is natural to inquire what happens in QCD as a function of the isospin density n_I, or the isospin chemical potential μ_I.

Let us comment on the relevance of this regime to the real world. Nature provides us with nonzero-μ_I systems in the form of isospin-asymmetric matter (e.g. inside neutron stars); however, the latter contains both isospin density *and* baryon-number density. In contrast, the idealized system considered in this paper does not carry baryon number: the chemical potentials of the two light quarks, u and d, are equal in magnitude, $|\mu_I|/2$, and opposite in sign. Such a system, strictly speaking, is unstable with respect to weak decays that do not conserve isospin, and, as we shall see, is also not electrically neutral and thus does not exist in the thermodynamic limit. However, since we are interested in the dynamics of the strong interaction alone, one can imagine that all relatively unimportant electromagnetic and weak effects are turned off. Once this is done, we have a nontrivial regime that, as we shall see, is accessible to present-day lattice Monte Carlo methods, while being analytically tractable in various interesting limits. As a result, the system we consider has the potential to improve substantially our understanding of cold dense QCD. This regime carries many attractive traits of two-color QCD, but is realized in a physically relevant theory – QCD with three colors.

Our analysis here will closely parallel that of the previous section.

10.4.1 Positivity and QCD inequalities

Since the fermion determinant of our theory is real and positive in Euclidean space, some rigorous results on its low-energy behavior can be obtained from QCD inequalities. At finite isospin density, $\mu_I \neq 0$, $\mathcal{D} = \gamma(\partial + iA) + \frac{1}{2}\mu_I\gamma_0\tau_3 + m$, and Eq. (10.45) no longer holds, since the operation on the right-hand side of Eq. (10.45) changes the relative sign of μ_I. However, provided that $m_u = m_d$, interchanging up and down quarks compensates for this change of sign (the u and d quarks play the role of mutually conjugate quarks), i.e.

$$\tau_1\gamma_5\mathcal{D}\gamma_5\tau_1 = \mathcal{D}^\dagger \tag{10.101}$$

Instead of isospin τ_1 in Eq. (10.101) one can also use τ_2 (but not τ_3). Equation (10.101) replaces the now invalid Eq. (10.45) and ensures that $\det \mathcal{D} \geq 0$. On repeating the derivation of the QCD inequalities using Eq. (10.101), we find that the lightest meson, or the condensate, must be in channels $\bar{\psi}i\gamma_5\tau_{1,2}\psi$, i.e. a linear combination of $\pi^- \sim \bar{u}\gamma_5 d$ and $\pi^+ \sim \bar{d}\gamma_5 u$ states. Indeed, as shown below, in both analytically tractable regimes of small and large μ_I the lightest mode is a massless Goldstone mode that is a linear combination of $\bar{u}\gamma_5 d$ and $\bar{d}\gamma_5 u$.

10.4.2 Small isospin densities: the pion condensate

When μ_I is small, chiral perturbation theory can be used to treat the problem. To have a rough sense of how small μ_I should be, we require that no particles other than pions are excited due to the chemical potential. This gives $\mu_I \simeq m_\rho$ as the upper limit of applicability of chiral perturbation theory.

For zero quark mass and zero μ_I, the pions are the massless Goldstone bosons of spontaneously broken $SU(2)_L \times SU(2)_R$ chiral symmetry. In reality, quarks have small masses, which break this symmetry explicitly. Assuming that we have equal quark masses, the symmetry of the Lagrangian is $SU(2)_{L+R}$. The low-energy dynamics is governed by the familiar chiral Lagrangian, which is written in terms of the matrix pion field $\Sigma \in SU(2)$:

$$\mathcal{L} = \tfrac{1}{4} f_\pi^2 \, \mathrm{tr}(\partial_\mu \Sigma \, \partial_\mu \Sigma^\dagger - 2m_\pi^2 \, \mathrm{Re}\,\Sigma)$$

This Lagrangian contains only two phenomenological parameters: the pion decay constant, f_π, and the pion mass in the vacuum, m_π. We will see that interesting physics occurs at $\mu_I > m_\pi$, and, since $m_\pi \ll m_\rho$, there is a nontrivial range of μ_I for which the chiral Lagrangian is reliable and useful.

The isospin chemical potential further breaks $SU(2)_{L+R}$ down to $U(1)_{L+R}$. Its effect can be incorporated into the effective Lagrangian to leading order in μ_I, without introducing additional phenomenological parameters. Indeed, μ_I enters the QCD Lagrangian in the same way as the the zeroth component of a gauge potential. Thus the finite-μ_I chiral Lagrangian is obtained by promoting the global $SU(2)_L \times SU(2)_R$ symmetry to a local gauge symmetry: gauge invariance completely fixes the way μ_I enters the chiral Lagrangian:

$$\mathcal{L}_{\mathrm{eff}} = \frac{f_\pi^2}{4} \, \mathrm{tr}(\nabla_\nu \Sigma \, \nabla_\nu \Sigma^\dagger) - \frac{m_\pi^2 f_\pi^2}{2} \, \mathrm{Re}\,\mathrm{tr}\,\Sigma \qquad (10.102)$$

The covariant derivative is defined as

$$\nabla_0 \Sigma = \partial_0 \Sigma - \frac{\mu_I}{2}(\tau_3 \Sigma - \Sigma \tau_3), \qquad \nabla_i \Sigma = \partial_i \Sigma \qquad (10.103)$$

which follows from the transformation property of Σ under rotations by the isospin generator $I_3 = \tau_3/2$.

Using Eq. (10.102) it is straightforward to determine the vacuum alignment of Σ as a function of μ_I and the spectrum of excitations around the vacuum. We will be interested in negative μ_I, which favors neutrons over protons, as in neutron stars. The results are very similar to two-color QCD at finite baryon density, Section 10.3. From Eq. (10.102), one finds the potential energy for Σ,

$$V_{\mathrm{eff}}(\Sigma) = \frac{f_\pi^2 \mu_I^2}{8} \, \mathrm{tr}(\tau_3 \Sigma \tau_3 \Sigma^\dagger - 1) - \frac{f_\pi^2 m_\pi^2}{2} \, \mathrm{Re}\,\mathrm{tr}\,\Sigma \qquad (10.104)$$

The first term in Eq. (10.104) favors directions of Σ which anti-commute with τ_3, i.e. τ_1 and τ_2, while the second term prefers the vacuum direction $\Sigma = 1$. It turns out that the minima of Eq. (10.104) at all μ_I are captured by the following *Ansatz*:

$$\overline{\Sigma} = \cos\alpha + i(\tau_1 \cos\phi + \tau_2 \sin\phi)\sin\alpha \qquad (10.105)$$

On substituting Eq. (10.105) into Eq. (10.104), one sees that the potential energy depends only on α, not on ϕ:

$$V_{\mathrm{eff}}(\alpha) = \frac{f_\pi^2 \mu_I^2}{4}[\cos(2\alpha) - 1] - f_\pi^2 m_\pi^2 \cos\alpha \qquad (10.106)$$

On minimizing $V_{\text{eff}}(\alpha)$ with respect to α, one sees that the behavior of the system is different in two distinct regimes.

(1) For $|\mu_I| < m_\pi$, the system is in the same ground state as it is at $\mu_I = 0$: $\alpha = 0$, or $\overline{\Sigma} = 1$.

This result is easy to understand. The lowest-lying pion state costs a positive energy $m_\pi - |\mu_I|$ to excite, thus at zero temperature no pion is excited. The ground state of the Hamiltonian at such μ_I coincides with the normal vacuum of QCD. The isospin density is zero in this case.

(2) When $|\mu_I|$ exceeds m_π the minimum of Eq. (10.106) occurs at

$$\cos\alpha = m_\pi^2/\mu_I^2 \qquad (10.107)$$

In this regime the energy needed to excite a π^- quantum, $m_\pi - |\mu_I|$, is negative, thus it is energetically favorable to excite a large number of these quanta. Since pions are bosons, the result is a Bose condensate of π^-. If the pions did not interact, the density of the condensate would be infinite. However, the repulsion between pions stabilizes the system at a finite value of the isospin density. This value can be found by differentiating the ground-state energy with respect to μ_I:

$$n_I = -\frac{\partial \mathcal{L}_{\text{eff}}}{\partial \mu_I} = f_\pi^2 \mu_I \sin^2\alpha = f_\pi^2 \mu_I \left(1 - \frac{m_\pi^4}{\mu_I^4}\right) \qquad (10.108)$$

For $|\mu_I|$ just above the condensation threshold, $|\mu_I| - m_\pi \ll m_\pi$, Eq. (10.108) reproduces the equation of state of the dilute nonrelativistic pion gas (cf. Eq. (8.57)),

$$n_I = 4 f_\pi^2 (\mu_I - m_\pi)$$

At larger μ_I, $|\mu_I| \gg m_\pi$, the isospin density is linear in μ_I,

$$n_I = f_\pi^2 \mu_I, \qquad |\mu_I| \gg m_\pi$$

From Eq. (10.108) one can find the pressure and the energy density as functions of μ_I.

The fact that the minimum of the potential (10.104) is degenerate with respect to the angle ϕ corresponds to the spontaneous breaking of the $U(1)_{L+R}$ symmetry generated by I_3 in the Lagrangian (10.102). This is not unexpected since the ground state is, in essence, a pion superfluid, with one massless Goldstone mode. Since we start from a theory with three pions, there are, in addition to the massless mode, two massive modes in the superfluid phase. One can be identified with the π^0. The other is a linear combination of π^+ and π^-, which we denote as $\tilde{\pi}^+$, since it coincides with π^+ at the condensation threshold. The

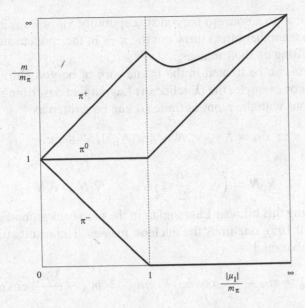

Figure 10.8. A schematic plot of masses (rest energies) of lowest-lying excitations in QCD at finite (negative) μ_I, in the regime of applicability of chiral perturbation theory: $m_\pi, \mu_I \ll m_\rho$.

mass (defined as the rest energy) of these modes can be obtained by expanding the Lagrangian (10.102) around the minimum. The result reads

$$m_{\pi^0} = |\mu_I|, \qquad m_{\tilde\pi^+} = |\mu_I|\sqrt{1 + 3(m_\pi/\mu_I)^4} \qquad (10.109)$$

At the condensation threshold, $m_{\pi^0} = m_\pi$ and $m_{\tilde\pi^+} = 2m_\pi$, while for $|\mu_I| \gg m_\pi$ both masses approach $|\mu_I|$ (see Fig. 10.8).

The values of the chiral condensate, $\langle \bar u u + \bar d d \rangle$, and the pion condensate, $\langle \bar u \gamma_5 d \rangle$, follow from (10.107):

$$\langle \bar u u + \bar d d \rangle = 2\langle \bar\psi\psi \rangle_{\rm vac} \cos\alpha \qquad \text{and} \qquad \langle \bar u \gamma_5 d \rangle + \text{h.c.} = 2\langle \bar\psi\psi \rangle_{\rm vac} \sin\alpha$$
$$(10.110)$$

i.e. the chiral condensate "rotates" into the pion condensate as a function of $|\mu_I|$.

It is also possible to find baryon masses, i.e. the energy cost of introducing a single baryon into the system. The most interesting baryons are those with the lowest energy and highest isospin, i.e. the neutron n and Δ^- isobar. There are two effects of μ_I on the baryon masses. The first comes from the isospin of the baryons, which effectively reduces the neutron mass by $\frac{1}{2}|\mu_I|$, and the Δ^- mass by $\frac{3}{2}|\mu_I|$. If this were the only effect, the effective Δ^- mass would vanish at $|\mu_I| = \frac{2}{3}m_\Delta$. For larger μ_I, baryon or anti-baryon Fermi surfaces would form, which would lead to a nonzero baryon susceptibility $\chi_B \equiv \partial n_B / \partial \mu_B$. However, long before that another effect turns on: the π^-'s in the condensate tend to repel the baryons, lifting up their masses.

These effects can be treated in the framework of baryon chiral perturbation theory [123]. For example, the (Euclidean) Lagrangian describing nucleons and their interactions with the pions at finite μ_I can be written as

$$\mathcal{L}_N = \bar{N}\gamma_\mu \nabla_\mu N + m_N(\bar{N}_L \Sigma N_R + \text{h.c.}) \tag{10.111}$$

where

$$\nabla_0 N = \left(\partial_0 - \frac{\mu_I}{2}\tau_3\right)N, \qquad \nabla_i N = \partial_i N$$

On diagonalizing this bilinear Lagrangian in the pion background given by $\Sigma = \bar{\Sigma}$ from Eq. (10.105), one finds the nucleon masses. The results for the neutron and the Δ^- isobar read

$$m_n = m_N - \frac{|\mu_I|}{2}\cos\alpha, \qquad m_{\Delta^-} = m_\Delta - \frac{3|\mu_I|}{2}\cos\alpha \tag{10.112}$$

in the approximation of nonrelativistic baryons. Equation (10.112) can be interpreted as follows: as a result of the rotation (10.105) of the chiral condensate, the nucleon-mass eigenstate becomes a superposition of vacuum n and p states. The expectation value of the isospin in this state is proportional to $\cos\alpha$ appearing in Eq. (10.112). With $\cos\alpha$ given in Eq. (10.107), we see that the two effects mentioned cancel each other out when $m_\pi \ll |\mu_I| \ll m_\rho$. Thus the baryon mass never drops to zero, and $\chi_B = 0$ at zero temperature in the region of applicability of the chiral Lagrangian.

As one forces more pions into the condensate, the pions are packed closer together and their interaction becomes stronger. When $\mu_I \sim m_\rho$, chiral perturbation theory breaks down. To find the equation of state in this regime, full QCD has to be employed. As we have seen, this can be done using present-day lattice techniques since the fermion sign problem is not present at finite μ_I, similarly to two-color QCD.

10.4.3 Asymptotically high isospin densities: the quark–antiquark condensate

In the opposite limit of very large isospin densities, or $|\mu_I| \gg m_\rho$, the description in terms of quark degrees of freedom applies since the latter are weakly interacting due to asymptotic freedom. In our case of large negative μ_I, or n_I, the ground state contains an equal number of d quarks and $\bar{u}$ anti-quarks per unit volume. If one neglects the interaction, the quarks fill two Fermi spheres with equal radii $|\mu_I|/2$. Turning on the interaction between the fermions leads to the instability with respect to the formation and condensation of Cooper pairs, which is similar to the BCS instability in metals and the diquark pairing at high baryon density in Chapter 9. To leading order of perturbation theory, quarks interact via one-gluon exchange. It is easy to see that the attraction is strongest in the color-singlet channel, thus the Cooper pair consists of an ū and a d. The ground state, hence, is a fermionic superfluid.

The perturbative one-gluon exchange, however, does not discriminate between the scalar, $\bar{u}d$, channel, and the pseudo-scalar, $\bar{u}\gamma_5 d$, channel: the attraction is the same in both cases. However, one expects that the instanton-induced interaction, however small, will favor the $\bar{u}\gamma_5 d$ channel over the $\bar{u}d$ one. The condensate is therefore a pseudo-scalar and breaks parity,

$$\langle \bar{u}\gamma^5 d \rangle \neq 0 \qquad (10.113)$$

This is consistent with our earlier observation that QCD inequalities constrain the $I = 1$ condensate to be a pseudo-scalar at any μ_I. Note that the order parameter in Eq. (10.113) has *the same quantum numbers* as the pion condensate at lower densities. We shall discuss this coincidence later.

As a consequence of Cooper pairing, the fermion spectrum acquires a gap Δ at the Fermi surface, where

$$\Delta = b|\mu_I|g^{-5}e^{-c/g}, \qquad c = 3\pi^2/2, \qquad (10.114)$$

where g should be evaluated at the scale $|\mu_I|$. The peculiar $e^{-c/g}$ behavior comes from the long-range magnetic interaction, as in the superconducting gap at large μ_B in Section 9.2.3. The constant c is smaller by a factor of $\sqrt{2}$ compared with the latter case due to the stronger one-gluon attraction in the singlet $q\bar{q}$ channel compared with the $\bar{3}$ diquark channel. Consequently, the gap (10.114) is exponentially larger than the diquark gap at comparable baryon chemical potentials. As in the BCS theory, the critical temperature at which the superfluid state is destroyed is of order Δ.

10.4.4 Quark–hadron continuity and confinement

Since the order parameter (10.113) has the same quantum numbers and breaks the same symmetry as the pion condensate in the low-density regime, it is plausible that there is no phase transition along the μ_I axis. In this case, as one increases the density, the Bose condensate of weakly interacting pions smoothly transforms into the superfluid state of ūd Cooper pairs. [5] Of course, this conjecture needs to be (and can be, due to the positivity) verified by lattice calculations.

At first sight, this conjecture seems to contradict common wisdom, which states that there is a "deconfinement" phase transition from the hadron phase to

---

[5] The situation is very similar to that of strongly coupled superconductors with a "pseudogap" [124], and possibly to that of high-temperature superconductors [125]. This also parallels the continuity between nuclear and quark matter in three-flavor QCD as conjectured by Schäfer and Wilczek (see Section 9.3.4).

the quark–gluon-plasma phase. It is logically possible that there exists a first-order phase transition at an intermediate value of μ_I. However, there are several nontrivial arguments that make the hypothesis of continuity highly plausible.

The first argument arises from considering baryons. One notices that all fermions have a gap at large $|\mu_I|$, which means that all excitations carrying baryon number are massive. In particular, at zero temperature, the baryon-number susceptibility χ_B vanishes. This is also true at small μ_I. It is thus natural to expect that all excitations with nonzero baryon number are massive at any value of μ_I, and χ_B remains zero at $T = 0$ for all μ_I. This also suggests one way to verify continuity on the lattice.

Another argument comes from considering the limit of a large number of colors N_c. Recall that, in finite-temperature QCD, there is a mismatch, at large N_c, between the number of gluon degrees of freedom, which is $\mathcal{O}(N_c^2)$, and that of hadrons, which is $\mathcal{O}(N_c^0)$. This fact is a strong hint of a first-order confinement–deconfinement phase transition, at which the effective number of degrees of freedom jumps from $\mathcal{O}(N_c^0)$ to $\mathcal{O}(N_c^2)$. It is easy to see, however, that the N_c behavior of thermodynamic quantities is the same in the "hadronic" phase (low μ_I) and the "quark" phase (large μ_I). Indeed, at very large μ_I the isospin density n_I is proportional to the number of quarks, which is $\mathcal{O}(N_c)$:

$$n_I = \frac{N_c}{3} \frac{\mu_I^3}{8\pi^2} \tag{10.115}$$

In the small-μ_I region the isospin density is given by Eq. (10.108). In the large-N_c limit, the pion decay constant scales as $f_\pi^2 = \mathcal{O}(N_c)$, and thus the isospin density in the pion gas is also proportional to N_c.[6] What happens is that the repulsion between pions becomes weaker as N_c increases, so more pions can be stacked at a given chemical potential. As a result, the N_c dependence of thermodynamic quantities is the same in the quark and the hadronic regimes, although for seemingly very different reasons.

Now let us return to the question of confinement. Naively, one would think that, at asymptotically large μ_I, the $\bar{u}$ and d quarks are packed at a very high density, and the system should become deconfined. At finite temperature, there is no rigorous way to distinguish between the confined and deconfined phases in QCD with quarks in the fundamental representation. However, at zero temperature (and finite μ_I), a sharp distinction between the two phases can be made. In the confined phase, all particle excitations carry integer baryon number; the deconfined phase can be defined as the phase in which there exist finite-energy excitations carrying fractional baryon charge. The pion superfluid at small μ_I

---

[6] With physical values of N_c, f_π and m_π, the values of n_I given by Eqs. (10.108) and (10.115), naively continued into the regime of intermediate μ_I, cross at $\mu_I \approx 800$ MeV. This agrees with the value of $\mu_I \sim m_\rho$ at which one would expect the crossover between the quark and hadron regimes to occur. This is a quantitative indication that a phase transition is not necessary.

clearly is in the confined phase. The question to ask is the following: is quark matter at large μ_I confined or deconfined?

It might seem that, at very large μ_I, there exist excitations with fractional baryon number. These are the fermionic quasi-particles near the Fermi surface, which are related to the original quarks and anti-quarks by a Bogoliubov–Valatin transformation. The opening of a BCS gap makes the energy of these excitations larger than Δ, but still finite.

To see that the logic above has a flaw and there are no such excitations, one needs to consider the dynamics of very soft gluons. The crucial observation is that, at large μ_I, gluons softer than Δ are not screened, either by the Meissner effect or by the Debye effect. The Meissner effect is absent because the condensate does not break gauge symmetry (in contrast to the color-superconducting condensate in Chapter 9). Debye screening is also absent, because on scales lower than Δ there are no charge excitations in the medium: the Cooper pairs are neutral, while the fermions are too heavy to be excited. Thus, the gluon sector below the Δ scale is described by pure SU(3) gluodynamics, *which is a confining theory*. This means that there are no quark excitations above the ground state: all particles and holes must be confined in colorless objects, namely mesons and baryons, just like in QCD in an ordinary vacuum.

If there is no transition along the μ_I axis, we expect confinement at all values of μ_I. At large μ_I, since the running strong coupling α_s at the scale of Δ is small, the confinement scale Λ'_{QCD} is much less than Δ. In more detail, let us imagine following the running of the strong coupling from the ultraviolet to the infrared. First, α_s increases until the scale $g\mu_I$ is reached, whereupon it "freezes" due to Debye screening and Landau damping. The freezing continues until we reach the scale Δ, after which the coupling runs again as in pure gluodynamics. Since the coupling is still small at the scale Δ, it can become large only at some scale Λ'_{QCD} much lower than Δ. Thus, at large $|\mu_I|$ there are three different scales separated by large exponential factors, $\mu_I \gg \Delta \gg \Lambda'_{QCD}$.

That the scale of confinement is much smaller than the gap at large μ_I has an important consequence for finite temperature. One can actually predict a temperature-driven deconfinement phase transition at a temperature T'_c of the order of Λ'_{QCD}. Indeed, at such low temperatures, quarks are unimportant, so the transition must be of first order as in pure gluodynamics. In particular, one expects the temperature dependence of the baryon-number susceptibility to change from $e^{-3\Delta/T}$ to $e^{-\Delta/T}$ around T'_c due to deconfinement.

The smallness of the confinement scale Λ'_{QCD} compared with the BCS gap Δ allows one to conclude that the binding energy of quarks and anti-quarks is small and the hadronic spectrum follows the pattern of the constituent-quark model, with Δ playing the role of the constituent-quark mass. This means that mesons weigh 2Δ and baryons weigh 3Δ, approximately. A good analog of the large-μ_I regime is vacuum QCD with only heavy quarks. As in the latter case, the string tension and string breaking are determined by parametrically different energy scales (Λ'_{QCD} and Δ, respectively). Hence the area law should work up to some distance much larger than Λ_{QCD}^{-1}, even when fundamental quarks are present. For the same reason one also expects the high-spin excited states of hadrons to be narrow at large μ_I.

10.5 Pion propagation near and below T_c

Some nonperturbative features of QCD, such as the equation of state, can be determined from numerical Monte Carlo techniques. However, many important issues related to *real-time* behavior and the response of high-temperature strongly interacting matter cannot be systematically studied by such methods. This is because lattice techniques rely on the formulation of quantum-field theory in *imaginary time*. As a result, changes in the hadron spectrum of QCD caused by the environment created in a heavy-ion collision cannot be obtained in a reliable fashion by computer simulations. This is a pity since one needs the spectrum of QCD in these extreme environments in order to calculate signals for the experimental existence of the quark–gluon plasma, such as the dilepton spectrum (see Section 7.6). In general, an analytic continuation $\tau \rightarrow -it$ is necessary in order to obtain real-time correlation functions from Euclidean correlators. Such a continuation from a numerically known Euclidean time function may be reliable for small τ, or large frequencies, $\omega \gg T$, but becomes problematic if the behavior for long real time, or low frequencies, is required. In simple terms, this is a consequence of the fact that, on the imaginary-frequency axis, the smallest nonzero Matsubara frequency is $2\pi T$. There is no direct information about the correlation function for smaller frequencies (except exactly at $\omega = 0$). New numerical methods to simulate the real-time behavior of QCD are sorely needed.

Fortunately, many quantities characterizing the real-time behavior of finite-temperature systems can be related, by exact identities, to static (thermodynamic) functions. The most familiar case is the relation among the velocity of sound u, the pressure p, and the energy density ϵ: $u = (\partial p/\partial \epsilon)^{1/2}$. A less trivial example is that of spin waves in antiferromagnets: it has long been known [126] that, at long enough wavelengths and at any temperature below the phase transition, there exist low-frequency spin waves that have a linear dispersion curve, whose slope is given exactly in terms of static quantities.

In this section, following [127], we point out that, in thermal QCD, the dispersion relation of soft pions can be determined entirely using static quantities. Such quantities can, in principle, be measured on the lattice. Using this observation, we shall see that the pion pole mass, which characterizes the propagation of the collective pion modes, must decrease as one approaches the critical temperature, despite the well-known fact that the pion screening mass increases in the same limit.

10.5.1 Pion dispersion from static quantities (summary)

From the point of view of symmetry properties, QCD at temperatures T below or just above the temperature of the chiral phase transition T_c is similar to a Heisenberg antiferromagnet [128, 129]. With two light quarks (u and d), QCD possesses an approximate chiral $SU(2)_V \times SU(2)_A \simeq O(4)$ symmetry, which is broken spontaneously to $SU(2)_V \simeq O(3)$ by the chiral condensate. This is similar to the $O(3) \to O(2)$ symmetry breaking in antiferromagnets. Moreover, the order parameter of QCD, the chiral condensate $\langle \bar{\psi}\psi \rangle$, is distinct from the conserved charges (the vector- and axial-isospin charges), which makes the real-time behavior of QCD similar to that of antiferromagnets (but not of ferromagnets, in which the order parameter – magnetization – is a conserved quantity.)

By analogy with spin waves in antiferromagnets [126], one can show that, at any T below T_c, the real part of the dispersion relation of soft pions is given by

$$\omega^2 = u^2(\vec{p}^2 + m^2) \tag{10.116}$$

provided that the quark masses are small enough. We shall use the following terminology: u is the pion *velocity* (although it is the velocity only when $m = 0$), m is the pion *screening mass*, and the energy of a pion at $\vec{p} = 0$, $m_p = um$, is the pion *pole mass*. At zero temperature, $u = 1$, and the pole mass coincides with the screening mass. At nonzero temperature, Lorentz invariance is lost, and u generally differs from 1 [130, 131]. Such pion modes with their modified dispersion relation are termed "quasipions" in [130].

The parameters u and m can be determined by measuring only static (zero-frequency) Euclidean correlators. In particular, m can be extracted from the long-distance behavior of the correlation function of the operator $\pi^a \equiv i\bar{\psi}\gamma^5\tau^a\psi$,

$$\int d\tau \, dV \, e^{-i\vec{q}\cdot\vec{x}} \frac{\langle \pi^a(x)\pi^b(0) \rangle}{\langle \bar{\psi}\psi \rangle^2} = \frac{1}{f^2} \frac{\delta^{ab}}{\vec{q}^2 + m^2} \tag{10.117}$$

where $x = (\tau, \vec{x})$, ψ is the quark field, $a, b = 1, 2, 3$, τ^a are isospin Pauli matrices, $\mathrm{tr}(\tau^a\tau^b) = 2\delta^{ab}$, and $\langle \cdots \rangle$ denotes thermal averaging. The integration over the Euclidean time variable τ is taken in the interval $(0, 1/T)$. In Eq. (10.117) $\langle \bar{\psi}\psi \rangle$ is the chiral condensate at zero quark masses. Equation (10.117) also provides the *definition* of the temperature-dependent pion *decay constant* f.

Less trivial is the relation of u to static correlators. The pion velocity u is equal to the ratio of the above-defined pion decay constant f and the axial-isospin susceptibility χ_{I5}:

$$u^2 = f^2/\chi_{I5} \tag{10.118}$$

This is a close analog of the equation $c^2 = \rho_s/\chi_m$ [126] for the velocity of spin waves in antiferromagnets. The axial-isospin susceptibility χ_{I5} can be defined as the second derivative of the pressure with respect to the axial-isospin chemical potential (see Eq. (10.120) below), or, equivalently, via the static correlator of the axial isospin charge densities,

$$\delta^{ab}\chi_{I5} = \int d\tau \, dV \, \langle A_0^a(x) A_0^b(0) \rangle, \qquad A_0^a \equiv \bar\psi \gamma^0 \gamma^5 \frac{\tau^a}{2} \psi \qquad (10.119)$$

The right-hand side of Eq. (10.119) is free of short-distance divergences in the limit of zero quark masses, when A_0^a are densities of conserved charges.

The derivation of Eqs. (10.116)–(10.119) at nonzero temperature requires an analysis of the hydrodynamic theory as developed in [132]. Here we shall use an intuitively simpler (but less rigorous) derivation based on the effective Lagrangian approach. This approach does not allow a correct treatment of dissipative effects, but will be sufficient for our purpose.

10.5.2 Derivation

Our strategy is to first write down the most general form of the effective Lagrangian of pions, and then relate its free parameters to the correlation functions of QCD by matching the partition function $\mathcal{Z} = e^{PV/T}$ and its derivatives in the effective and microscopic theories. The quark part of the QCD Lagrangian at finite axial-isospin chemical potential μ_{I5} is given by

$$\mathcal{L}_{\text{quark}} = i\bar\psi \gamma^\mu D_\mu \psi - (\bar\psi_L M \psi_R + \text{h.c.}) + \mu_{I5} A_0^3 \qquad (10.120)$$

where $M = \text{diag}(m_u, m_d)$ is the quark-mass matrix. The chemical potential μ_{I5} is coupled to the axial-isospin charge A_0^3 defined in Eq. (10.119). For simplicity, we set $m_u = m_d = m_q$.

We shall assume that, in the infrared, the pion thermal width is negligible compared with its energy. This has been seen in explicit calculations at low T [133] and is related to the fact that pions are Goldstone bosons. The dynamics of the pions is described, in this case, by an effective Lagrangian $\mathcal{L}_{\text{eff}}$, which we assume to be local, allowing its expansion in powers of momenta. This is equivalent to the assumption that the correlation functions have only pole singularities, as in hydrodynamics. To lowest order, the Lagrangian is fixed by symmetries up to three coefficients, f_t, f_s, and f_m,

$$\mathcal{L}_{\text{eff}} = \frac{f_t^2}{4} \text{tr}(\nabla_0 \Sigma \, \nabla_0 \Sigma^\dagger) - \frac{f_s^2}{4} \text{tr}(\partial_i \Sigma \, \partial_i \Sigma^\dagger) + \frac{f_m^2}{2} \text{Re} \, \text{tr}(M\Sigma) \qquad (10.121)$$

where Σ is an SU(2) matrix whose phases describe the pions. Owing to the lack of Lorentz invariance, f_t^2 and f_s^2 are independent parameters.

The chemical potential μ_{I5} enters the lowest-order effective Lagrangian (10.121) through the covariant derivative ∇_0, which is completely determined by symmetries. This can be seen by promoting the SU(2)$_A$ symmetry in

Eq. (10.120) to a local symmetry and treating μ_{15} as the time component of the SU(2)$_A$ vector potential, as we discussed earlier in this chapter. The covariant derivative ∇_0 necessarily has the form

$$\nabla_0 \Sigma \equiv \partial_0 \Sigma - \frac{i}{2}\mu_{15}(\tau_3 \Sigma + \Sigma \tau_3) \tag{10.122}$$

The structure of the Lagrangian (10.121) is analogous to that of the effective Lagrangian at finite (vector) isospin chemical potential μ_I in Section 10.4. A significant difference between the two cases is that the QCD vacuum breaks the SU(2)$_A$ (axial-isospin) symmetry *spontaneously*. It is important to note, however, that SU(2)$_A$ *is* a symmetry of the Lagrangian (at $m_q = 0$), as good as the SU(2)$_V$ symmetry. The conservation of the axial-isospin current A_μ^a in the chiral limit puts the consideration of finite μ_{15} on solid theoretical ground.

The pion dispersion relation following from Eq. (10.121) is given by Eq. (10.116) with

$$u^2 = \frac{f_s^2}{f_t^2} \quad \text{and} \quad m^2 = \frac{m_q f_m^2}{f_s^2} \tag{10.123}$$

By matching the second derivative of the pressure $\mathcal{P}$ with respect to μ_{15} in QCD and in the effective theory, we find the relation between f_t and χ_{15}:

$$\chi_{15} = \frac{\partial^2 \mathcal{P}}{\partial \mu_{15}^2} = f_t^2 \tag{10.124}$$

Together with the first of Eqs. (10.123) and $f = f_s$ (see below), this implies Eq. (10.118). The first derivative with respect to m_q gives

$$-\langle \bar{\psi}\psi \rangle = \frac{\partial \mathcal{P}}{\partial m_q} = f_m^2 \tag{10.125}$$

By combining this with the second of Eqs. (10.123), we derive the generalization of the famous Gell-Mann–Oakes–Renner (GOR) relation to finite temperature:

$$f_s^2 m^2 = -m_q \langle \bar{\psi}\psi \rangle \tag{10.126}$$

Finally, we need to show that $f = f_s$. We achieve this by treating M as an external field, which we parametrize as $M(x) = m_q e^{i\alpha^a(x)\tau^a}$. By matching derivatives of $\ln \mathcal{Z}$, we find

$$\langle \pi^a(x)\pi^b(0) \rangle = \frac{\delta^2 \ln \mathcal{Z}}{m_q^2\, \delta\alpha^a(x)\, \delta\alpha^b(0)} = f_m^4 \langle \phi^a(x)\phi^b(0) \rangle$$

$$\pi^a \equiv i\bar{\psi}\gamma_5\tau^a\psi, \qquad \phi^a(x) \equiv \operatorname{Re}\operatorname{tr}[i\tau^a\Sigma(x)/2] \tag{10.127}$$

Note that π^a is defined in the microscopic theory (QCD), while ϕ^a is a field of the effective theory.

The correlation function of $\phi^a(x)$ can be calculated by expanding the effective Lagrangian in Eq. (10.121) to second order in ϕ^a. We expect the result to match the correlator of π^a only for small momenta, which means that we have to limit ourselves to zero Matsubara frequency and small spatial momenta, e.g. smaller than the screening mass m_σ of the order parameter $\sigma = \bar\psi\psi$. For the static correlator of π^a, by integrating Eq. (10.127) over τ and using Eq. (10.125); we obtain

$$
\int d\tau \, \frac{\langle \pi^a(x)\pi^b(0)\rangle}{\langle \bar\psi\psi \rangle^2} = \int d\tau \, \langle \phi^a(x)\phi^b(0)\rangle
$$

$$
= \frac{1}{f_s^2}\int \frac{d^3\vec{q}}{(2\pi)^3}\, \frac{e^{i\vec{q}\cdot\vec{x}}\delta^{ab}}{\vec{q}^2 + m^2} = \frac{1}{f_s^2}\, \frac{e^{-m|\vec{x}|}}{4\pi|\vec{x}|}\delta^{ab} \qquad (10.128)
$$

We see that, by measuring the large-distance ($|\vec{x}| \gg m_\sigma^{-1}$) static correlation function of the operator $\pi^a = i\bar\psi\gamma^5\tau^a\psi$, we can extract two parameters of the effective Lagrangian (10.121): the screening mass m and f_s, which coincides with the decay constant f defined by Eq. (10.117). The third parameter, f_t^2, coincides with the susceptibility χ_{I5}, which can also be expressed in terms of the static correlation function in Eq. (10.119). From Eq. (10.118) we completely determine the dispersion relation of soft pions.

10.5.3 Critical behavior

For the above results to be valid, pions must be the lightest modes. In particular, this requires $m \ll m_\sigma$. If m_q is very small, this condition is satisfied everywhere below T_c, except for a region very close to T_c. As $T \to T_c$ from below and $m_\sigma \to 0$, one can ask the following question: what is the critical behavior of the parameters u and m when T remains sufficiently far from T_c that the hierarchy $m \ll m_\sigma \ll T$ is maintained?

Since u and m can be related to static correlation functions, one should expect their critical behavior to be governed by the same static critical exponents as those known from the theory of critical phenomena. We begin by considering the critical scaling of the decay constant $f = f_s$. It is defined via the behavior of a static correlator (10.128) at distances larger than m_σ^{-1}. In the range of momenta $m \ll |\vec{q}| \ll m_\sigma$ we have (see Eq. (10.117))

$$
\int d\tau \, dV \, e^{-i\vec{q}\cdot\vec{x}}\langle \pi^a(x)\pi^b(0)\rangle = \delta^{ab}\frac{\langle\bar\psi\psi\rangle^2}{f^2}\frac{1}{\vec{q}^2} \qquad (10.129)
$$

On the other hand, at distances short compared with the correlation length, i.e. for momenta such that $m_\sigma \ll |\vec{q}| \ll T$, the correlator of the order parameter

$\bar{\psi}\psi$ has the following scaling behavior:

$$\int d\tau \, dV \, e^{-i\vec{q}\cdot\vec{x}} \langle \bar{\psi}\psi(x)\bar{\psi}\psi(0) \rangle \sim \frac{1}{|\vec{q}|^{2-\eta}} \tag{10.130}$$

We also know that, in this regime, the correlators of $\sigma = \bar{\psi}\psi$ and $\pi^a = i\bar{\psi}\gamma^5\tau^a\psi$ are degenerate, since they are related by the SU(2)$_A$ symmetry, which is restored at T_c. Thus the correlator (10.130) must match with the correlator (10.129) at the scale $|\vec{q}| \sim m_\sigma$. This requires

$$f^2 = A m_\sigma^{-\eta} \langle \bar{\psi}\psi \rangle^2 \tag{10.131}$$

The coefficient A cannot be found from scaling arguments, but is finite and regular at T_c. The exponent η is in the universality class of the O(4) sigma model in $d = 3$ dimensions (to which two-flavor QCD at T_c belongs, as discussed in Section 7.4) and is known: $\eta \approx 0.03$.

At $T \to T_c$ the scaling laws for the inverse correlation length m_σ and the order parameter $\langle \bar{\psi}\psi \rangle$ are also known from universality,

$$m_\sigma \sim t^\nu \tag{10.132}$$

$$\langle \bar{\psi}\psi \rangle \sim t^\beta \tag{10.133}$$

where $t = (T_c - T)/T_c$. Thus we find

$$f^2 \sim t^{2\beta - \nu\eta} = t^{(d-2)\nu} \tag{10.134}$$

where in the last equation the relation

$$2\beta = \nu(d - 2 + \eta) \tag{10.135}$$

is used (recall that all scaling exponents can be expressed in terms of two independent ones, e.g. η and ν, assuming hyperscaling). From this point on, we set $d = 3$, so $f \sim t^{\nu/2}$. This scaling law is the same as Josephson scaling for the superfluid density in helium [134]. Contrary to naive expectations, f scales differently from the order parameter $\langle \bar{\psi}\psi \rangle \sim t^\beta$. The difference, however, is numerically small, due to the smallness of η. In the O(4) universality class in $d = 3$, $\nu \approx 0.73$ and $\beta \approx 0.38$ [135].

Next, we point out that χ_{I5} is finite at $T = T_c$, where it is degenerate with the vector-isospin susceptibility. The singular behavior of χ_{I5} is dominated by the mixing of A_0^a with operators linear or quadratic in σ or π^a in the dimensionally reduced theory describing infrared modes $|\vec{q}| \ll T$. Such mixing, however, is forbidden by the O(4) chiral symmetry, as well as by charge conjugation. This is consistent with the lattice result that the vector-isospin susceptibility is finite at T_c [136]. Since $f_t^2 = \chi_{I5}$, finiteness of χ_{I5} invalidates the common assumption that $f_t \to 0$ at T_c. Note that, above T_c, there are no propagating soft pion modes, so the parameters of the effective Lagrangian (such as f_t), and hence Eq. (10.124), lose their meaning, even as χ_{I5} remains well-defined and finite. This is not surprising if one recalls that, as $T \to T_c$ from below, the domain of validity of the Lagrangian (10.121) ($|\vec{q}| \ll m_\sigma$) shrinks away and disappears at T_c.

Now we are ready to find the scaling of u. Using Eq. (10.118), the scaling of f in Eq. (10.134), and the fact that χ_{I5} is finite at T_c, we find

$$u^2 \sim f^2 \sim t^\nu \tag{10.136}$$

This means that the velocity of pions vanishes at T_c.

The scaling of the screening mass m can be found from the GOR relation (10.126) (recall that m_q is assumed to be small):

$$m^2 = -\frac{m_q \langle \bar{\psi}\psi \rangle}{f^2} \sim m_q t^{\beta-\nu} \tag{10.137}$$

In the O(4) universality class $\beta < \nu$, which implies that the static screening pion mass grows (at fixed $m_q \neq 0$) as $T \to T_c$. This fact agrees with lattice simulations of QCD.

The *pole* mass of the pion, m_p, scales differently:

$$m_p^2 \equiv u^2 m^2 = -\frac{m_q \langle \bar{\psi}\psi \rangle}{\chi_{I5}} \sim m_q t^\beta \tag{10.138}$$

This means that the pole mass of the pion drops as $T \to T_c$.

For the formulas (10.136)–(10.138) to be valid it is necessary that $t \ll 1$. However, for any $m_q \neq 0$, these formulas break down when t is so small that the condition $m \ll m_\sigma$ is violated. Using Eqs. (10.132) and (10.137), we see that this happens when $t \sim m_q^{1/(\beta\delta)}$ or smaller. In the regime $t \ll m_q^{1/(\beta\delta)}$ the "distance" from the critical point ($T = T_c$, $m_q = 0$) is controlled by m_q, but not by t. The scaling with m_q of all quantities can be obtained starting from

$$\langle \bar{\psi}\psi \rangle \sim m_q^{1/\delta} \qquad \text{at} \qquad t = 0 \tag{10.139}$$

On comparing this with Eq. (10.133), we see that t and $m_q^{1/(\beta\delta)}$ have the same scaling dimension. Using the scaling hypothesis we can easily obtain the scaling with m_q by replacing t by $m_q^{1/(\beta\delta)}$. For example,

$$m^2 \sim m_q^{1-(\nu-\beta)/(\beta\delta)}, \qquad m_p^2 \sim m_q^{1+1/\delta} \tag{10.140}$$

(m_p now has the meaning of the typical frequency of the pion mode with zero momentum. This mode may be overdamped in this regime.) Both masses vanish as $m_q \to 0$ at $T = T_c$; however, for the screening mass $m^2 \gg m_q$, whereas for the pole mass $m_p^2 \ll m_q$. In particular, near the phase transition $m_p \ll m$.

The decrease of the pion pole mass may have interesting consequences for heavy-ion collisions. It is the pole mass of a hadron, rather than its static screening mass, that affects the observed spectrum. Within statistical models for hadron production, the drop in the pion pole mass would lead to an over-population of pions at low momenta, provided that the chemical freezeout temperature T_{ch}, at which the hadron abundances are fixed, is close to T_c. For a crude estimate of this effect we use $\Lambda_{QCD} \sim 200\,\text{MeV}$ as the typical QCD scale in Eq. (10.140), and $T_{ch} \sim 170\,\text{MeV}$ as an estimate for the freezeout temperature [137], which is indeed very close to T_c. The shift of the pole mass near T_c is approximately $\Delta m \equiv m_p - m_\pi \approx m_\pi[(m_q/\Lambda_{QCD})^{1/(2\delta)} - 1] \approx -0.3 m_\pi$, where $\delta \approx 5$, and the pion multiplicity at small momenta is enhanced by roughly $\exp(-\Delta m/T_{ch}) \approx 1.3$. This is a noticeable effect, although it is smaller than the known contribution to pion overpopulation due to the feed-down from the decays of resonances [137]. This enhancement is comparable to the effect of the pion chemical potential $\mu_\pi \sim 50\,\text{MeV}$ induced by pion kinetics after the chemical freezeout [138].

Another potential consequence of the fact that the pion velocity decreases at T_c is the possibility of Čerenkov radiation of pions by a hard probe moving through the hot medium created in a heavy-ion collision.

11
Lattice-gauge theory at nonzero chemical potential

11.1 Propagators and formulating the chemical potential on a Euclidean lattice

Now that we know how lattice-gauge theory is formulated and know a few tricks of the trade, we need to formulate it in the presence of a chemical potential. To gain an appreciation of the idea and to invent an elegant formalism, let's review some simple problems involving μ [139]. Consider a nonrelativistic field that describes particles that can propagate only forward in time. The finite-T partition function reads

$$Z = \int d[\phi] \, d[\phi^*] \exp\left(-\int_0^\beta [\phi^*(-\partial^2 - \mu)\phi + \phi^*\partial_\tau\phi] \, d\tau \, d^3x\right) \quad (11.1)$$

Here ϕ could represent either a boson field or a fermion field. Periodic boundary conditions in the temporal direction apply in the Bose case and anti-periodic in the Fermi case, in which ϕ would be a Grassmann variable. Letting a $-$ sign apply to the Bose case and a $+$ sign to the Fermi case, we can evaluate Z,

$$Z = [\det(\partial^2 - \mu + \partial_\tau)]^{\mp 1} \quad (11.2)$$

We can evaluate the Bose or the Fermi determinant using the thermal Green function, G^β, which is the familiar Green function defined on all of space but a temporal strip ranging from 0 to β. Differentiate the partition function Z with respect to the chemical potential,

$$\frac{dZ}{d\mu} = \pm \int d^3x \, d\tau \, G^\beta(x, x, 0) \quad (11.3)$$

where

$$G^\beta(x, y, \tau) = (-\partial^2 - \mu + \partial_\tau)^{-1}_{x,y,\tau} \quad (11.4)$$

324

It is particularly illuminating to express the thermal Green function, G^β, in terms of the ordinary Green function, G, which lives in unbounded space and time $R^3 \times R$. G satisfies the "heat" equation

$$(\partial^2 + \mu)G(x, y, \tau) = \frac{\partial G}{\partial \tau}$$
$$G = 0 \quad \text{if} \quad \tau < 0$$
$$G(x, y, 0) = \delta^4(x - y) \tag{11.5}$$

We can use the method of images to implement the boundary conditions and find the series of images for the thermal Green function,

$$\pm \int d^3x \, d\tau \, G^\beta(x, x, 0) = \pm \text{tr}(-\partial^2 - \mu + \partial_\tau)^{-1}$$
$$= \pm \beta \int d^3x \, [G(x, x, 0) \pm G(x, x, \beta)$$
$$+ G(x, x, 2\beta) \pm G(x, x, 3\beta) \cdots] \tag{11.6}$$

This equation has the usual interpretation associated with the method of images: the "heat" arriving at $\tau = 0$ consists of that put in at $\tau = 0$ ($G(x, x, 0)$), plus the heat that has traveled once around the cylinder of circumference β ($G(x, x, \beta)$), etc. The $\pm$ signs in this equation record the fact that a fermion picks up a minus sign each time it crosses anti-periodic boundary conditions.

We can make the μ dependence in this equation explicit using the following observations. First, since the combination $-\mu + \partial_\tau$ enters the Green function, we have

$$G(x, y, \tau)_\mu = e^{\mu\tau} G(x, y, \tau)_{\mu=0} \tag{11.7}$$

In addition,

$$G(x, y, \tau_1 + \tau_2) = \int d^3z \, G(x, z, \tau_1)G(z, y, \tau_2) \tag{11.8}$$

This allows us to simplify the equation above,

$$\pm \text{tr}(-\partial^2 - \mu + \partial_\tau)^{-1} = \pm \beta \int d^3x \, [G(x, x, 0) \pm e^{\mu\beta} G(x, x, \beta)]$$
$$+ e^{2\mu\beta} \int d^3x' \, G(x, x', \beta)G(x', x, \beta) \cdots \tag{11.9}$$

where these Green functions now refer to $\mu = 0$ and are solutions to the ordinary heat equation. The partition function can now be obtained by integrating over μ. For Bose fields,

$$\ln Z = \pm \beta\mu G(x, x, 0) - \text{tr} \ln(1 - e^{\beta\mu} G(\beta)) \tag{11.10}$$

For fermions,

$$\ln Z = \pm \beta \mu G(x, x, 0) + \text{tr} \ln(1 + e^{\beta \mu} G(\beta)) \tag{11.11}$$

The first term in these equations arises because of a normal ordering ambiguity in the density and is usually set to zero. If we evaluate the traces in momentum space, we obtain more familiar answers. For Bose fields,

$$\ln Z = -\int \frac{d^3 k}{(2\pi)^3} \ln(1 - e^{\beta \mu} e^{-\beta k^2}) \tag{11.12}$$

and for Fermi fields,

$$\ln Z = +\int \frac{d^3 k}{(2\pi)^3} \ln(1 + e^{\beta \mu} e^{-\beta k^2}) \tag{11.13}$$

The point we want to emphasize here is the appearance of the chemical potential in the exponential factors, recording each excursion that the field makes around the cylinder $[0, \beta]$ in the temporal dimension. When we sum over all worldlines, the factor $e^{\beta \mu}$ "encourages" the particle to move forward in "time" and, if μ exceeds the rest mass so that $e^{\beta \mu} e^{-\beta k^2} > 1$, or $e^{\beta \mu} e^{-\beta E} > 1$ in more generality, the worldline can wrap around the cylinder an unlimited number of times. For Bose fields, this leads to a catastrophic divergence, but for Fermi fields, it simply fills the Fermi sea.

This formalism illustrates how one can fill a Fermi sea with just one fermion. In other words, in a quenched lattice-gauge simulation one can fill a Fermi sea and generate an appreciable baryon number. Unfortunately, the quenched approximation, which is such a useful starting point for lattice-gauge investigations in other fields, suffers from a serious disease at nonzero chemical potential, as we have discussed in Section 10.2.3.

11.2 Naive fermions at finite density

We want to discuss lattice-gauge theory of QCD at nonzero chemical potential. We will begin with naive fermions because they are easy to deal with and because the lessons learned here will carry over to staggered fermions.

It is convenient to normalize the fermion fields so the fermion piece of the lattice QCD action reads

$$S_f = \sum_x \left(\bar{\psi}(x) \psi(x) \right.$$

$$\left. + K \bar{\psi}(x) \sum_{v=1}^{4} \gamma_v [U(x, \hat{n}_v) \psi(x + \hat{n}_v) - U^\dagger(x - \hat{n}_v, \hat{n}_v) \psi(x - \hat{n}_v)] \right) \tag{11.14}$$

where the Euclidean γ matrices satisfy

$$[\gamma_\kappa, \gamma_\nu]_+ = 2\delta_{\kappa\nu}, \qquad \gamma_\nu^\dagger = \gamma_\nu \qquad (11.15)$$

In this expression it was convenient to define K in terms of the bare lattice mass m_0,

$$m_0 = \frac{1}{2K} \qquad (11.16)$$

and to normalize lattice fields, ψ, in terms of their continuum relatives as

$$\psi = \sqrt{\frac{1}{2K}}\, \psi_{\text{continuum}} \qquad (11.17)$$

The fermion propagator for this action can be written as a sum over worldlines between points x and y,

$$G_{\alpha\beta}^{ab}(x, y) = \sum_{\text{paths}\,\Gamma} K^{|\Gamma|} \left(\prod_\Gamma \gamma_\nu\right)_{\alpha\beta} \left(\prod_\Gamma U\right)_{ab} \qquad (11.18)$$

This form makes it clear how to introduce the chemical potential into the lattice theory so that it has all the desired properties, allowing us to count fermions appropriately. We just replace the gauge-field factors U on timelike links in the $+$ direction with $e^\mu U$ and replace the gauge-field factors $U^\dagger$ on timelike links in the $-$ direction with $e^{-\mu}U^\dagger$ [139, 140]. Since the circumference of the cylinder in the temporal direction is β, this prescription encourages forward propagation with a factor $e^{\beta\mu}$ and inhibits backward propagation with a factor $e^{-\beta\mu}$. Notice that only those fermions which actually wind around the cylinder pick up the factors $e^{\pm\beta\mu}$. Other fermions that "double back" do not contribute any chemical-potential dependence.

Now we can write the grand canonical partition function for the lattice theory,

$$Z = \int d[\bar\psi]\, d[\psi]\, d[U] \exp(-S_f - S_{\text{gauge}}) = \text{tr}(e^{\beta\mu N}e^{-\beta H}) \qquad (11.19)$$

We can now differentiate the action with respect to the chemical potential to obtain the number operator N,

$$Q(\tau) = \sum_{\tau=\text{constant}} J_0(x) = \sum_{\tau=\text{constant}} K[\bar\psi(x)\gamma_0 U(x, \hat{n}_0)e^\mu \psi(x + \hat{n}_0)$$
$$- \bar\psi(x + \hat{n}_0)\gamma_0 e^{-\mu} U^\dagger(x, \hat{n}_0)\psi(x)] \qquad (11.20)$$

We see that J_0 counts paths passing through the link $x + \hat{n}_0$ by assigning a $+K$ for upward paths and $-K$ for downward paths. So $\langle J_0(x)\rangle$ is literally the expectation value of the number of paths through the link $x + \hat{n}_0$. The spatial components of the vector

$$J_a(x) = K[\bar{\psi}(x)\gamma_a U(x, \hat{n}_0)\psi(x + \hat{n}_a) - \bar{\psi}(x + \hat{n}_a)\gamma_a U^\dagger(x, \hat{n}_a)\psi(x)]$$

$$(11.21)$$

together with J_0 form a conserved vector current that counts the flux of world-lines.

We can obtain many useful, informative observables from various derivatives of the partition function. For example,

$$\langle J_0 \rangle = \frac{1}{\beta V} \frac{\partial \ln Z}{\partial \mu} \tag{11.22}$$

and the chiral condensate is

$$\langle \bar{\psi}\psi \rangle = \frac{1}{\beta V} \frac{\partial \ln Z}{\partial m} \tag{11.23}$$

With these general relations, the Maxwell relation follows,

$$\left(\frac{\langle \partial J_0 \rangle}{\partial m} \right)_\mu = \left(\frac{\langle \partial(\bar{\psi}\psi) \rangle}{\partial \mu} \right)_m \tag{11.24}$$

It is also informative to observe that

$$\frac{1}{\beta V} \ln Z = \frac{\Omega}{V} = P_f \tag{11.25}$$

where P_f is the fermionic contribution to the pressure and Ω is the grand canonical potential.

By collecting everything, we can relate the chiral condensate to the mass derivative of the pressure,

$$\langle \bar{\psi}\psi \rangle = -\frac{\partial P_f}{\partial m} \tag{11.26}$$

So, in principle, we could obtain the pressure by integrating the area under the chiral-condensate curve and obtain the equation of state. Care is needed here because one must subtract the chiral condensate at vanishing chemical potential in order to remove the vacuum contribution.

It is interesting to illustrate these results for the free-fermion field. At zero T, one can compute $\langle \bar{\psi}\psi \rangle$ and $\langle J_0 \rangle$ and consider their μ dependences. In particular,

$$\langle \bar{\psi}\psi \rangle = \int_{-\pi}^{+\pi} \frac{d^4 p}{(2\pi)^4} \frac{m}{m^2 + \sum_{i=1}^{4} \sin^2 p_i} \tag{11.27}$$

and

$$\langle J_0 \rangle = \int_{-\pi}^{+\pi} \frac{\mathrm{d}^4 p}{(2\pi)^4} \frac{i \sin p_0 \cos p_0}{m^2 + \sum_{i=1}^{4} \sin p_i^{\,2}} \tag{11.28}$$

The μ dependence enters these momentum expansions through the substitution $p_0 \to p_0 + i\mu$. It is convenient to change integration variables to $z = e^{ip_0}$, so after the substitution we have

$$\langle \bar{\psi} \psi \rangle = \frac{1}{2\pi i} \int_{|z|=e^{-\mu}} \frac{\mathrm{d}z}{z} f(z) \tag{11.29}$$

and the charge becomes

$$\langle \bar{\psi} \gamma_0 \psi \rangle = \frac{1}{2\pi i} \int_{|z|=e^{-\mu}} \frac{\mathrm{d}z}{z} \frac{1}{4} \left(z^2 - \frac{1}{z^2} \right)^2 f(z) \tag{11.30}$$

where the function $f(z)$ is

$$f(z) = \int_{-\pi}^{+\pi} \frac{1}{m^2 + \sum_{i=1}^{3} \sin^3 p_i - \frac{1}{4}(z - 1/z)^2} \frac{\mathrm{d}^3 p}{(2\pi)^3} \tag{11.31}$$

Using the fact that the integrand in the expression for $f(z)$ behaves as $4z^2$ as $z \to 0$, and behaves as $4/z^2$ as $z \to \infty$, we find the asymptotic behavior of the chiral condensate and the baryon charge density as $\mu \to \pm\infty$,

$$\langle \bar{\psi} \psi \rangle \to 0 \tag{11.32}$$

and

$$\langle \bar{\psi} \gamma_0 \psi \rangle \to \int_{-\pi}^{+\pi} \frac{\mathrm{d}^3 k}{(2\pi)^3} = \pm 1 \tag{11.33}$$

It is interesting to understand these simple results in more physical terms that will help us later to interpret lattice-gauge-theory simulation data. The reason why $\langle \bar{\psi} \psi \rangle$ vanishes in these limits is that

$$\langle \bar{\psi} \psi \rangle = -\frac{\partial(\Omega/V)}{\partial m} \tag{11.34}$$

where Ω/V is essentially the energy of the filled states,

$$\frac{\Omega}{V} = -\frac{1}{V\beta} \ln Z = \frac{1}{V} \sum_{\text{filled states}} (E(k) - \mu) \tag{11.35}$$

Only the states near $E = 0$ can be affected by variations in m, so, if $\mu \gg m$, as many states rise in energy as fall when m increases, so the chiral condensate vanishes. Similarly, when μ is large and negative no states move when m is varied and $\langle \bar{\psi}\psi \rangle = 0$ again.

This completes our introductory remarks about incorporating the chemical potential into lattice-gauge theory. Later we will consider simulations of the model and aspects of the fermion determinant at nonzero μ, and find some challenging subtleties.

11.3 The three-dimensional four-Fermi model at nonzero T and μ

Before discussing QCD it will prove worthwhile to consider a simpler model in some detail. We choose the four-Fermi model in three dimensions for several reasons. First, it is renormalizable in a $1/N$ expansion and displays chiral-symmetry breaking. The theory is, of course, nonrenormalizable in a coupling-constant expansion, but that is a problem with perturbation theory, not with the physics, as emphasized by K. G. Wilson in his early work on the ϵ expansion [141]. The model's continuum Lagrangian reads

$$L = \sum_{j=1}^{N} \left(\bar{\psi}^{(j)} \partial\!\!\!/ \psi^{(j)} - \frac{g^2}{2N} (\bar{\psi}^{(j)} \psi^{(j)})^2 \right) \tag{11.36}$$

and $\psi^{(j)}$ is a four-component spinor even though we are working in three space-time dimensions. In three dimensions one could also consider a model with two-component spinors, but we shall see that the four-component version studied here is more relevant to our ultimate interests in four dimensions. N is the number of fermion species which is the basis for the $1/N$ expansion and its soluble $N \to \infty$ limit. The theory has a discrete Z_2 chiral symmetry, $\psi \to \gamma_5 \psi$, $\bar{\psi} \to -\bar{\psi}\gamma_5$, which is broken dynamically for sufficiently strong coupling $g^2 > g_c^2$. The critical coupling g_c^2 serves as a fixed point of the renormalization group and its existence underlies the good features of the $1/N$ expansion about this critical coupling. The theory's critical indices are not those of mean-field theory. To $O(1/N)$ they are given by [142]

$$\nu = 1 + \frac{8}{3N\pi^2}, \qquad \delta = 2 + \frac{8}{N\pi^2}, \qquad \beta_{\text{mag}} = 1 \tag{11.37}$$

$$\gamma = 1 + \frac{8}{N\pi^2}, \qquad \eta = 1 - \frac{16}{3N\pi^2} \tag{11.38}$$

These indices satisfy the hyperscaling relations expected of a critical system with one dynamical mass scale. In addition, the theory has an interesting finite-T and finite-μ transition that can be studied analytically in the $1/N$ expansion [143] as well as through lattice simulations. Both approaches will be discussed here.

In order to analyze this model by conventional methods, we need to introduce an auxiliary Bose field, σ, so that the Lagrangian can be written as a quadratic form in the Fermi field,

$$L = \sum_{j=1}^{N} (\bar{\psi}^{(j)} \partial\!\!\!/ \psi^{(j)} + \sigma \bar{\psi}^{(j)} \psi^{(j)}) - \frac{N}{2g^2} \sigma^2 \tag{11.39}$$

Clearly, $\langle \sigma \rangle$ can serve as an order parameter for chiral-symmetry breaking. $\langle \sigma \rangle$ acts as the dynamical mass of the fermion and will prove particularly important in lattice simulations as well as in analytic analyses of the model.

Begin with the large-N solution of the model. We wish to calculate the order parameter $\Sigma = \langle \sigma \rangle$ as a function of the coupling g, chemical potential μ, and temperature $T \equiv 1/\beta$. Since the large-N limit suppresses fluctuations around the saddle-point solution of the path integral, this is equivalent to a mean-field treatment. However, the Fermi nature of the underlying degrees of freedom will produce the critical indices listed above, not the mean-field exponents so familiar from commuting classical spin models of statistical mechanics.

At large N, the leading contribution to Σ comes from the simple fermion-loop tadpole. The gap equation determines Σ self-consistently. It states that the tadpole graph generates the dynamical mass which determines the propagator in the loop graph that produced it,

$$\Sigma = -g^2 \langle \bar{\psi}\psi \rangle = \frac{g^2}{V} \, \mathrm{tr} \, S_\mathrm{f}(\mu, T, \Sigma) \tag{11.40}$$

where the fermion propagator in Euclidean space reads

$$S_\mathrm{f}^{-1}(k; \mu, T, \Sigma) = i\gamma_0(k_0 - i\mu) + \sum_{\nu=1,2} ik_\nu\gamma_\nu + \Sigma \tag{11.41}$$

where the chemical potential has already been introduced and the temperature appears implicitly through the fact that the allowed values of k_0 are given by the Matsubara frequencies for fermions, enforcing anti-periodic boundary conditions in the temporal direction,

$$k_0 = (2n - 1)\pi T \tag{11.42}$$

where n is an integer. On collecting these features into the gap equation, we find

$$\frac{1}{g^2} = 4T \sum_{n=-\infty}^{\infty} \int \frac{\mathrm{d}^2 p}{(2\pi)^2} \frac{1}{[(2n-1)\pi T - i\mu]^2 + p^2 + \Sigma^2} \tag{11.43}$$

The sum over n can be done by first rewriting it using the Poisson summation formula,

$$\frac{1}{g^2} = 4T \sum_{m=-\infty}^{\infty} \int \frac{\mathrm{d}^2 p}{(2\pi)^2}$$

$$\times \int_{-\infty}^{\infty} \mathrm{d}\phi \, \frac{e^{2\pi i m\phi}}{[(2\phi - 1)\pi T - i\mu + iE][(2\phi - 1)\pi T - i\mu - iE]} \tag{11.44}$$

Lattice-gauge theory

where $E = \sqrt{p^2 + \Sigma^2}$. The integral of ϕ can be done easily if $\mu < \Sigma$ because the poles in the integrand lie on opposite sides of the integration contour for all values of p,

$$\frac{1}{g^2} = \int_{\Sigma}^{\infty} \frac{dE}{\pi} \left(1 - \frac{1}{e^{\beta(E-\mu)} + 1} - \frac{1}{e^{\beta(E+\mu)} + 1} \right) \tag{11.45}$$

If $\mu > \Sigma$, then the position of the poles, above or below the contour, depends on p and the integral must be redone, giving

$$\frac{1}{g^2} = \int_{\mu}^{\infty} \frac{dE}{\pi} \left[1 - \frac{1}{e^{\beta(E-\mu)} + 1} \right] + \int_{\Sigma}^{\mu} \frac{dE}{\pi} \frac{1}{e^{\beta(\mu-E)} + 1}$$
$$- \int_{\Sigma}^{\infty} \frac{dE}{\pi} \frac{1}{e^{\beta(\mu+E)} + 1} \tag{11.46}$$

Finally, we can eliminate the coupling constant from this equation in favor of Σ_0, the order parameter at vanishing T and μ. Replace the right-hand side of this equation with the equation before it after setting $T = \mu = 0$. This produces an implicit equation for Σ in terms of the physical mass scale Σ_0, with no reference to any ultraviolet-dependent quantity. For $\mu < \Sigma$ we have

$$\Sigma_0 - \Sigma = T \left[\ln \left(1 + e^{-\beta(\Sigma-\mu)} \right) + \ln \left(1 + e^{-\beta(\Sigma+\mu)} \right) \right] \tag{11.47}$$

and for $\mu > \Sigma$ we have

$$\Sigma_0 - \mu = T \left[\ln \left(1 + e^{-\beta(\mu-\Sigma)} \right) + \ln \left(1 + e^{-\beta(\mu+\Sigma)} \right) \right] \tag{11.48}$$

Actually, these two solutions are identical for $\Sigma(\mu, T)$, and demonstrate that curves of $\Sigma(\mu, T)$ at constant T are symmetric under reflection across the line $\Sigma = \mu$.

The last two equations produce the complete solution for $\Sigma(\mu, T)$ in terms of Σ_0. Since Σ approaches zero smoothly from above, except precisely at $\mu = 0$, we have a line of second-order phase transitions throughout the T–μ plane. To obtain the critical line we set $\Sigma = 0$ in the last equation and find

$$1 - \frac{\mu}{\Sigma_0} = 2 \frac{T}{\Sigma_0} \ln(1 + e^{-\beta\mu}) \tag{11.49}$$

At vanishing chemical potential, we find a chiral-symmetry-restoring transition at the critical temperature,

$$T_c = \frac{\Sigma_0}{2 \ln 2} \approx 0.72 \Sigma_0 \tag{11.50}$$

For vanishing μ the gap equation in the broken phase becomes

$$\Sigma_0 - \Sigma = 2T \ln(1 + e^{-\beta\Sigma}) \tag{11.51}$$

Finally, at vanishing temperature, we find that $\Sigma = \Sigma_0$ independently of μ up to a critical value

$$\mu_c = \Sigma_0 \tag{11.52}$$

where the transition becomes of first order at this isolated point. For small excursions into the μ–T plane, the gap equation predicts

$$\left.\frac{\partial\Sigma}{\partial\mu}\right|_{T\to 0} = \lim_{T\to 0}\left(-\frac{\sinh(\beta\mu)}{\sinh(\beta\Sigma)}\right) \simeq -e^{\beta(\mu-\Sigma)} \tag{11.53}$$

So the slope of the surface $\Sigma(\mu, T)$ diverges in an essentially singular fashion as $\mu \to \mu_c$ and $T \to 0$. This means that the large-N limit predicts a first-order transition for $T = 0$, which becomes of second order as soon as $T > 0$.

The fact that the critical chemical potential at vanishing temperature is exactly the fermion dynamical mass indicates that interactions other than those that create the dynamical mass itself are relatively negligible at large N.

Using similar calculational methods we can predict other features of this model. In particular, the fermion number density $\langle n \rangle$ in the broken phase, $\mu < \Sigma$, reads

$$\langle n \rangle = \frac{\Sigma T}{\pi} \ln\left(\frac{1 + e^{-\beta(\Sigma-\mu)}}{1 + e^{-\beta(\Sigma+\mu)}}\right) - \frac{2T^2}{\pi}\sum_{k=1}^{\infty}(-1)^k \frac{e^{-\beta k\Sigma}\sinh(\beta k\mu)}{k^2} \tag{11.54}$$

whereas for $\mu > \Sigma$,

$$\langle n \rangle = \frac{\mu^2 - \Sigma^2}{2\pi} + \frac{\Sigma T}{\pi} \ln\left(\frac{1 + e^{-\beta(\mu-\Sigma)}}{1 + e^{-\beta(\mu+\Sigma)}}\right) - \frac{2T^2}{\pi}\sum_{k=1}^{\infty}(-1)^k$$
$$\times \frac{1 - e^{-\beta k\mu}\cosh(\beta k\Sigma)}{k^2} \tag{11.55}$$

This expression reduces to the two-dimensional free relativistic Fermi-gas result in the symmetric phase,

$$\langle n \rangle = \frac{\mu^2}{2\pi} - \frac{2T^2}{\pi}\sum_{k=1}^{\infty}(-1)^k \frac{1 - e^{-\beta k\mu}}{k^2} \tag{11.56}$$

This free Fermi behavior is another manifestation of the simplicity of the large-N limit and is not expected at finite N or in richer models.

We see that the fermion density is strongly suppressed for low temperatures in the symmetric phase, $\mu < \Sigma$, jumps discontinuously as soon as μ exceeds Σ, and then continues to rise quadratically in the phase with chirality restored.

These results have been verified in lattice simulations. In fact, the theory's critical behavior and the indices listed above have all been reproduced in computer simulations. With that in mind, let's consider the lattice formulation of this model following our discussion of lattice QCD at nonzero T and μ given above.

To place this four-Fermi model on the lattice we begin with the Bose form of the Lagrangian and use the staggered-lattice-fermion method,

$$S = \sum_{i=1}^{N/2} \left(\sum_{x,y} \bar{\chi}_i(x)\tilde{M}_{x,y}\chi_i(y) + \frac{1}{8}\sum_x \bar{\chi}_i(x)\chi_i(x) \sum_{\langle \tilde{x},y\rangle} \sigma(\tilde{x}) \right)$$
$$+ \frac{N}{4g^2}\sum_{\tilde{x}} \sigma^2(\tilde{x}) \tag{11.57}$$

where χ_i and $\bar{\chi}_i$ are Grassmann-valued staggered-fermion fields defined on the lattice sites x, the auxiliary scalar field σ is defined on the dual lattice sites $\tilde{x}$, and $\langle \tilde{x}, y\rangle$ denotes the set of eight dual lattice sites $\tilde{x}$ surrounding the site y. In this case the dual lattice is just a copy of the original lattice but translated by half a link in each lattice direction so that each dual site lies at the center of each cube of the original lattice. The auxiliary fields are placed on the dual lattice because the symmetry of the construction guarantees that the lattice action better approximates the continuum action for nonzero lattice spacing a [144]. The fermion lattice Dirac operator in the presence of a chemical potential reads

$$\tilde{M}_{x,y} = \frac{1}{2}(e^\mu \delta_{y,x+\hat{0}} - e^{-\mu}\delta_{y,x-\hat{0}}) + \frac{1}{2}\sum_{v=1,2} \eta_v(x)(\delta_{y,x+\hat{v}} - \delta_{y,x-\hat{v}}) \tag{11.58}$$

where $\eta_v(x)$ are the staggered-fermion phases $(-1)^{x_0+\cdots+x_{v-1}}$. The lattice spacing a has been set to unity everywhere.

One can simulate this lattice action by using the hybrid Monte Carlo method introduced earlier. The Grassmann fields representing the fermions are replaced by real Bose "pseudo-fermion" fields $\phi(x)$, governed by the action

$$S = \sum_{x,y} \sum_{i,j=1}^{N/2} \frac{1}{2}\phi_i(x)(M^\mathrm{T}M)^{-1}_{xyij}\phi_j(y) + \frac{N}{4g^2}\sum_{\tilde{x}} \sigma^2(\tilde{x}) \tag{11.59}$$

where

$$M_{xyij} = \tilde{M}_{xy}\delta_{ij} + \delta_{xy}\delta_{ij}\frac{1}{8}\sum_{\langle \tilde{x},x\rangle} \sigma(\tilde{x}) \tag{11.60}$$

The sum over N extends only up to $N/2$ because of fermion doubling: for each pseudo-fermion field on the lattice, two will emerge in the continuum limit. The fact that the pseudo-fermions are real (not complex) fields puts the ratio here at 2 to 1 rather than 4 to 1, as one might have expected. Note that the hopping matrix M is strictly real. When the field ϕ is integrated over in the path integral one obtains the measure $\sqrt{\det(M^{\mathrm{T}}M)} \equiv \det M$ if the determinant of M is positive semi-definite. This condition is true as long as $N/2$ is an even integer, even if the chemical potential $\mu \neq 0$. This simplicity relies on the fact that there are no weights on the links. For example in SU(3) lattice-gauge theory there are SU(3) matrices on each link, the fermion determinant will not be real, there will be no positive-semi-definite measure in the lattice action, and therefore no probabilistic interpretation of the Euclidean path integral. This will leave us without a reliable simulation method in lattice QCD at nonzero chemical potential, a point we will ponder in more detail in later sections.

In the case of the three-dimensional four-Fermi model, we can calculate observables from simulations of the pseudo-fermion action without difficulty. The expectation value of the auxiliary field $\langle \sigma \rangle$ is immediately accessible, as are its correlation functions. The chiral condensate can be measured through

$$-\langle \bar{\psi} \psi \rangle = \frac{1}{V} \operatorname{tr} S_{\mathrm{f}} = \frac{1}{V} \langle \operatorname{tr} M^{-1} \rangle \tag{11.61}$$

the expectation of the energy density is

$$\langle \epsilon \rangle = -\frac{1}{V_{\mathrm{s}}} \frac{\partial \ln Z}{\partial \beta} = \frac{1}{V} \operatorname{tr}(\partial_0 \gamma_0 \, S_{\mathrm{f}}) = \frac{1}{2V} \left\langle \sum_x e^\mu M^{-1}_{x,x+\hat{0}} - e^{-\mu} M^{-1}_{x,x-\hat{0}} \right\rangle \tag{11.62}$$

and the expectation of the number density is

$$\langle n \rangle = -\frac{1}{V_{\mathrm{s}} \beta} \frac{\partial \ln Z}{\partial \mu} = \frac{1}{V} \operatorname{tr}(\gamma_0 S_{\mathrm{f}}) = \frac{1}{2V} \left\langle \sum_x e^\mu M^{-1}_{x,x+\hat{0}} + e^{-\mu} M^{-1}_{x,x-\hat{0}} \right\rangle \tag{11.63}$$

where V_{s} is the spatial volume, β is the inverse temperature, and $V = V_{\mathrm{s}} \beta$, the overall volume of spacetime.

11.4 Four-flavor SU(2) lattice-gauge theory at nonzero μ and T

SU(2) lattice-gauge theory is of considerable interest at nonzero chemical potential because it has a transition to a superfluid state complete with a diquark condensate predicted by effective-Lagrangian methods [145]. The equivalence of quarks and anti-quarks in the two-color model makes mesons and baryons equivalent physically and leads to a critical chemical potential of $\mu_c = m_\pi/2$. Diquarks are color singlets in this model and, when they condense at μ_c, a superfluid phase, rather than a superconducting phase, is formed. Many of the predictions of the effective-Lagrangian analysis have been confirmed by lattice simulations and are worth understanding [120]. We will display and discuss some simulation results here, although much more accurate and precise simulations are constantly updating these findings.

We will present the lattice simulations of this model as a prototype of such studies in QCD-like models in extreme environments. However, first we will review some facts about the model's fermion determinant and compare it with SU(3) lattice-gauge theory with a baryon-number chemical potential.

Consider the Dirac operator in QCD at vanishing chemical potential, $D = \gamma(\partial + A) + m_q$. In Euclidean space where the γ matrices are Hermitian and the matrix A is anti-Hermitian, the Dirac operator satisfies the identity

$$\gamma_5 D \gamma_5 = D^\dagger \tag{11.64}$$

which guarantees that its determinant is positive. This fact allows us to interpret the Euclidean path integral of QCD as a probabilistic weight and apply algorithms such as the hybrid molecular-dynamics algorithm and the hybrid Monte Carlo algorithm to its simulation.

When the chemical potential μ is turned on, the fermion determinant remains positive for SU(2) color, but not, alas, for the physical case of SU(3) color. In the special case of the SU(2) color group, we can use the pseudo-reality of its representations to find a useful generalization of Eq. (11.64). In particular, recall the charge-conjugation matrix, $C = i\gamma_0\gamma_2$ ($C^2 = 1$, $C\gamma_\mu C = -\gamma_\mu^*$) and the second Pauli-matrix generator, T_2 ($T_2 T_a T_2 = -T_a^*$), of SU(2) color. Then defining the product $K = \gamma_5 C T_2$, and considering the SU(2) Dirac operator at nonzero chemical potential, $D(\mu) = \gamma(\partial + A) + \mu\gamma_0 + m_q$, we can easily show that

$$K D(\mu) K = D^*(\mu) \tag{11.65}$$

which allows us to show that the fermion determinant is again positive for any μ.

It is easy to check that the fermion determinant for SU(3) color is not real when $\mu_B \neq 0$. An example of an explicit demonstration of this will be given for heavy quarks below. The pseudo-real feature of SU(2) color was essential in establishing the reality of its determinant when $\mu \neq 0$.

We shall see that there is considerable interesting physics at nonzero chemical potential in the SU(2) color model. Some of it may be relevant to the real world. However, there are gross qualitative shortcomings in the model that must be appreciated. First, SU(2) color does not have baryons. Because of its pseudo-reality, its two-quark "baryons" are degenerate with its quark–anti-quark mesons and there are no spin-$\frac{1}{2}$ quark-model states in the theory. Second, the critical chemical potential per quark must be half the mass of the lightest "baryon," which must be degenerate with the lightest meson, the pion. So, $\mu_c = m_\pi/2$ and the threshold for restoration of chiral symmetry and diquark condensation vanishes in the chiral limit where the mass of the quark goes to zero and the pion becomes massless. This peculiarity is the reason why chiral perturbation theory has precise predictions for the phase transition here. The chiral-perturbative approach and effective Lagrangians were discussed above in Chapter 10. Of course, in SU(3) QCD, the lightest baryon, the nucleon multiplet, remains massive in the chiral limit, so chiral perturbation theory does not describe its transition at $\mu_B = M_N/3$.

In addition, the diquark-condensate phases in the two models are qualitatively different. In the SU(2) theory, the diquarks are colorless, scalar mesons and their condensation is expected to lead to a superfluid state, not so different from a Bose–Einstein condensate. By contrast, the diquarks in the SU(3) QCD theory are color $\bar{3}$ states and their condensation breaks the global color symmetry dynamically, not unlike the Higgs mechanism. A condensate of such diquarks should have color-superconducting properties that should be qualitatively different from the properties expected of SU(2) both in their spectroscopies and in their bulk thermodynamics. Color superconductivity was discussed above in Chapter 9. At this time its properties have been "established" by symmetry considerations and weak-coupling estimates. It has not been possible to study this phase in terms of lattice-gauge theory because the theory has no reliable algorithms for phases in which the fermion determinant is complex. It is hoped that this shortcoming will be conquered in the near future.

Now we return to the SU(2) model at nonzero temperature and chemical potential and we consider the system's phase diagram. For fixed quark mass, which insures that the pion has a nonzero mass and chiral symmetry is explicitly broken, simulations give a line of transitions separating a phase with no diquark condensation from one with a diquark condensate. Along this line there is a tricritical point where the transition switches from being of second order (and well described by mean-field theory) at relatively low μ to a first-order transition at an intermediate value of μ. At large μ the line of first-order transitions separates the diquark-condensate phase at low T from a quark–gluon phase at high T. The transition is clear both in diquark observables and in the Wilson line. A schematic phase diagram is shown in Fig. 11.1. (Note that we have not included the line of first-order transitions starting from the $\mu = 0$, finite-temperature transition, which exists for m small enough, and the line of crossovers that should exist for larger m. We are just concentrating on the transition to the diquark condensate instead.) The existence of a tricritical point has a natural explanation in the context of chiral Lagrangians. In fact, since μ plays the role of a second "temperature" in this theory, in that it is a parameter that controls diquark

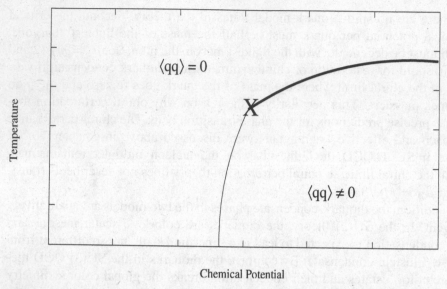

Figure 11.1. A schematic phase diagram in the $T-\mu$ plane for the two-color model. The thin (thick) line consists of second (first)-order transitions. X labels the tricritical point.

condensation, but does not explicitly break the (quark-number) symmetry, tricritical behaviour is expected.

We begin with a discussion of the simulation method and the numerical results.

The lattice action of the staggered-fermion version of this theory is

$$S_{\mathrm{f}} = \sum_{\mathrm{sites}} \left(\bar{\chi}[\slashed{D}(\mu) + m]\chi + \frac{1}{2}\lambda[\chi^{\mathrm{T}}\tau_2\chi + \bar{\chi}\tau_2\bar{\chi}^{\mathrm{T}}] \right) \qquad (11.66)$$

where the chemical potential μ is introduced by multiplying links in the $+t$ direction by e^μ and those in the $-t$ direction by $\mathrm{e}^{-\mu}$ [139]. The diquark source term (Majorana-mass term) is added in order to allow observation of the spontaneous breakdown of quark number on a finite lattice. The parameter λ and the usual mass term m control the amount of explicit symmetry breaking in the lattice action.

Integrating out the fermion fields in Eq. (11.66) gives

$$\mathrm{pfaffian}\begin{bmatrix} \lambda\tau_2 & \mathcal{A} \\ -\mathcal{A}^{\mathrm{T}} & \lambda\tau_2 \end{bmatrix} = \sqrt{\det(\mathcal{A}^\dagger\mathcal{A} + \lambda^2)} \qquad (11.67)$$

where

$$\mathcal{A} \equiv \slashed{D}(\mu) + m \qquad (11.68)$$

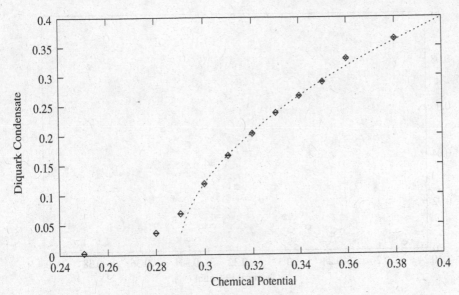

Figure 11.2. The diquark condensate versus μ, on a 16^4 lattice with $\beta = 1.85$, $m = 0.05$, and $\lambda = 0.00$ (spline fit).

Note that the Pfaffian is strictly positive, so the hybrid molecular-dynamics [38] method can be used to simulate this theory using "noisy" fermions to take the square root, giving $N_f = 4$.

For $\lambda = 0, m \neq 0$, we expect no spontaneous symmetry breaking for small μ. For μ large enough ($\mu > m_\pi/2$) we expect spontaneous breakdown of quark number and one Goldstone boson – a scalar diquark.

Now consider the simulation results for the $N_f = 4$ theory measuring the chiral and diquark condensates ($\langle\chi^T\tau_2\chi\rangle$), the fermion number density, the Wilson/Polyakov line, etc.

First consider measurements on a 16^4, "zero-temperature" lattice. The SU(2) model was simulated at $\beta = 1.85$, within the theory's scaling window. The quark mass was $m = 0.05$ and simulations were done at $\lambda = 0.0025, 0.005$, and 0.01 so that the results could be extrapolated to vanishing diquark source, $\lambda = 0$. For second-order phase transitions, only in the limit of vanishing λ do we expect real diquark phase transitions.

In Fig. 11.2 the diquark condensate, extrapolated to $\lambda = 0$, is plotted against the chemical potential μ.

There is good evidence for a quark-number-violating second-order phase transition in Fig. 11.2. The dashed line is a power-law fit, confidence level 24%, which picks out the critical chemical potential of $\mu_c = 0.2749(2)$ and a critical index of $\beta_{\text{mag}} = 0.54(5)$, which is compatible with the mean-field result

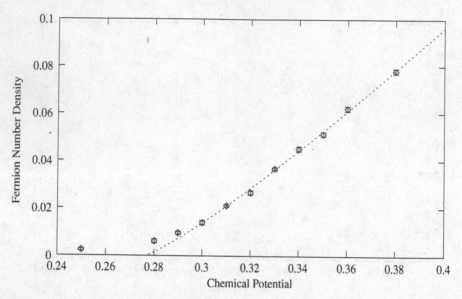

Figure 11.3. The fermion number density versus μ, on a 16^4 lattice, with $\beta = 1.85$, $m = 0.05$, and $\lambda = 0.0025$.

$\beta_{\text{mag}} = \frac{1}{2}$ predicted by loop-improved chiral perturbation theory [121], [146]. Note that the quark mass is fixed at $m = 0.05$ throughout this simulation, so chiral symmetry is explicitly broken and this transition is due to quark-number breaking alone. The fermion number density, shown in Fig. 11.3, also shows the diquark continuous phase transition. The approximate linear dependence of the fermion number density with μ is the expected scaling behavior above the transition in lowest-order chiral perturbation theory [121], [146].

Now consider the phase diagram in the temperature–chemical-potential plane. To begin, we review some small ($8^3 \times 4$) lattice results.

Consider a slice through the phase diagram at a fixed small temperature, for variable μ. In Fig. 11.4 we show the diquark condensate for $\beta = 1.3$, $m = 0.05$, with the data extrapolated to $\lambda = 0$ as discussed above. As μ increases we find that a second-order phase transition to a diquark condensate appears at $\mu_c = 0.2919(4)$. The dashed-line fit in Fig. 11.4 has the critical index $\beta_{\text{mag}} = 0.50(15)$, in good agreement with mean-field theory. The fit has a confidence level of 21%. The diquark transition is also clear in a plot of the fermion number density, as expected. However, the chiral condensate varies smoothly over the diquark critical region, with $\langle \bar{\chi} \chi \rangle$ ranging from 0.80 to 0.50. The Wilson line is also smooth, varying from 0.10 to 0.20, with no obvious sign of a transition.

Simulations at $\beta = 1.1$ and 1.5 gave similar conclusions. At $\beta = 1.1$, there was a continuous diquark transition at $\mu = 0.2544(3)$ and at $\beta = 1.5$ there was

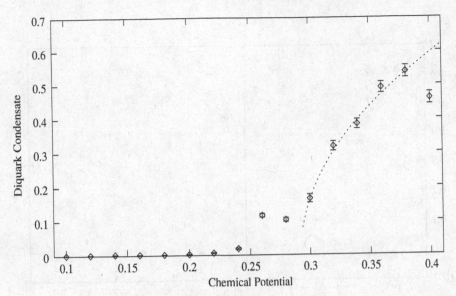

Figure 11.4. The diquark condensate versus μ, on an $8^3 \times 4$ lattice, with $\beta = 1.30$.

a continuous diquark transition at $\mu = 0.2950(3)$. So, as the system is heated (β increases), a larger chemical potential is needed to order the system into a diquark condensate.

Consider simulation results at fixed μ and variable β. The diquark condensate was measured as a function of β for $\mu = 0.40$ in Fig. 11.5. There is a clear jump in the condensate at $\beta = 1.55(5)$, suggesting a first-order phase transition.

A discontinuous transition is also strongly suggested by the behavior of the Wilson line shown in Fig. 11.6.

If this transition at relatively large μ and T is really of first order, then there should be a tricritical point separating the region of continuous transitions at smaller μ and T.

The evidence for a first-order transition at $\mu = 0.40$ was even stronger on a larger ($12^3 \times 6$) lattice. In Figs. 11.7, 11.8, and 11.9 we show the diquark condensate, the Wilson line, and the chiral condensate on the $12^3 \times 6$ lattice. The computer-time evolution of the diquark condensate at $\mu = 0.40$ reveals a two-state signal in Fig. 11.10, giving excellent evidence for a first-order transition.

$\beta = 1.97(2)$ lies in the scaling window of the SU(2) lattice-gauge theory with four species of dynamical quarks, so we can compare our $\mu_c = 0.40$ with the theory's spectroscopic mass scales [147]. In fact, $m_\pi/2 = 0.31(1)$, so finite-T effects have apparently raised μ_c somewhat from its zero-T value, $m_\pi/2$. Such effects were discussed in Chapter 10 in the context of effective Lagrangians.

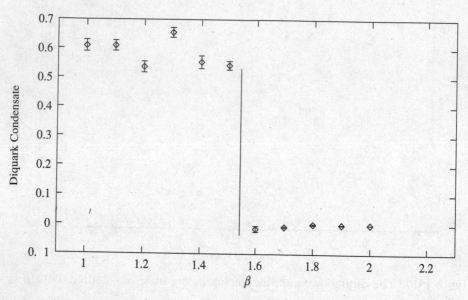

Figure 11.5. The diquark condensate versus β for $\mu = 0.40$, on an $8^3 \times 4$ lattice, with $m = 0.05$, and $\lambda = 0.000$.

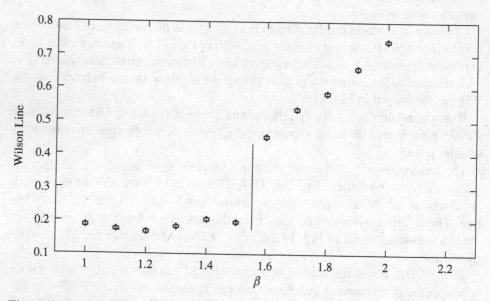

Figure 11.6. The Wilson line versus β for $\mu = 0.40$, on an $8^3 \times 4$ lattice, with $m = 0.05$, and $\lambda = 0.000$.

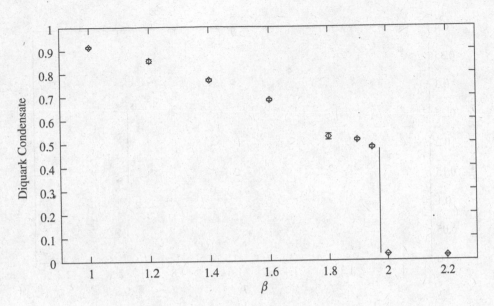

Figure 11.7. The diquark condensate versus β, on a $12^3 \times 6$ lattice, with $\mu = 0.40$, $m = 0.05$, and $\lambda = 0.005$.

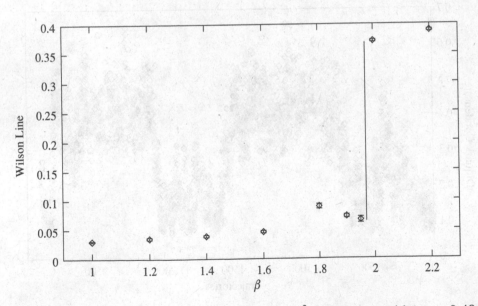

Figure 11.8. The Wilson line versus β, on a $12^3 \times 6$ lattice, with $\mu = 0.40$, $m = 0.05$, and $\lambda = 0.005$.

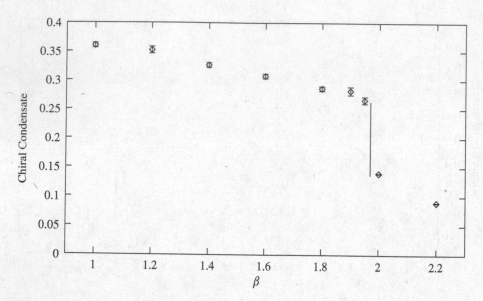

Figure 11.9. The chiral condensate versus β, on an $12^3 \times 6$ lattice, with $\mu = 0.40$, $m = 0.05$, and $\lambda = 0.005$.

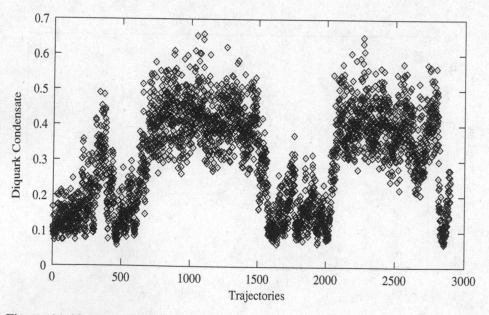

Figure 11.10. The diquark condensate versus computer time, for $\beta = 1.97$, $\mu = 0.40$, $m = 0.05$, and $\lambda = 0.005$.

The exact location of the tricritical point could be found in larger-scale simulations. On the small $8^3 \times 4$ lattices there is some evidence that it lies between $\mu = 0.30$ and 0.40 since results of simulations at $\mu = 0.30$ were compatible with mean-field theory.

11.5 High-density QCD and static quarks

We will discuss algorithms that may develop into systematic simulation methods of SU(3) lattice-gauge theory at nonzero baryon chemical potential below, but, since that is a difficult and longstanding barrier in the field, we begin with a less-ambitious but instructive topic – simulating heavy-quark SU(3) QCD at nonzero baryon chemical potential. The idea is that, if we take the limit of heavy quarks and large chemical potentials, the complex fermion determinant becomes tractable but nontrivial [148]. Of course, this limiting case is far from the interesting physical situation of light (or massless) quarks, spontaneous chiral-symmetry breaking, and chemical potentials of a few hundred MeV, so the lessons learned here might not be germane.

First consider how the lattice Dirac operator simplifies in this limit,

$$M(x, y) = 2am_q\delta_{x,y} + \sum_{v=1,2,3} [U_v(x)\eta_v(x)\delta_{x+\hat{v},y} - U_v^\dagger(y)\eta_v(y)\delta_{x-\hat{v},y}]$$

$$+ [e^{\mu a}U_t(x)\eta_t(x)\delta_{x+\hat{t},y} - e^{-\mu a}U_t^\dagger(y)\eta_t(y)\delta_{x-\hat{t},y}] \tag{11.69}$$

To evaluate the determinant of the hopping matrix, organize it according to time slices,

$$M(x, y) = \begin{pmatrix} S_0 & e^{\mu a}W_0 & 0 & \cdots & e^{-\mu a}W_{N_t}^\dagger \\ -e^{-\mu a}W_0^\dagger & S_1 & e^{\mu a}W_1 & \cdots & \\ 0 & -e^{-\mu a}W_1^\dagger & S_2 & & \\ \vdots & \vdots & & \ddots & \\ -e^{\mu a}W_{N_t} & & & & \end{pmatrix} \tag{11.70}$$

S_i contains the mass term and the spatial hopping terms in the ith time slice and W_i consists of the hopping terms for passage from the ith to the $(i+1)$th slice.

When we take the limit in which $m \to \infty$ and $\mu \to \infty$, we are left with just the $2ma$ terms along the diagonal and the forward hopping terms $e^{\mu a}W_i$ connecting adjoining time slices. Spatial points are now decoupled and the fermion determinant becomes a product of determinants on each static worldline,

$$\det M = \prod_{\vec{x}} e^{\mu a N_c N_t} \det(L(\vec{x}) + C) \tag{11.71}$$

where $L(\vec{x})$ is the Wilson (Polyakov) line at spatial site $\vec{x}$, N_c is the number of colors, and N_t is the number of time slices. The parameter C is just $(2ma/e^{\mu a})^{N_t}$.

Now the determinant can be evaluated for any $SU(N)$ gauge theory. For $SU(2)$,

$$\det(L(\vec{x}) + C) = C^2 + C \operatorname{tr} L(\vec{x}) + 1 \tag{11.72}$$

and for $SU(3)$,

$$\det(L(\vec{x}) + C) = C^3 + C^2 \operatorname{tr} L(\vec{x}) + C \operatorname{tr} L^\dagger(\vec{x}) + 1 \tag{11.73}$$

We read off that the $SU(2)$ determinant is explicitly real because the trace of an $SU(2)$ matrix is real. However, the $SU(3)$ determinant is explicitly complex because of the different weightings of the $\operatorname{tr} L(\vec{x})$ and the $\operatorname{tr} L^\dagger(\vec{x})$.

Even though the $SU(3)$ determinant is complex, we can evaluate its partition function using standard methods. The partition function reads

$$Z = \int [DU] \prod_{\vec{x}} e^{\mu a N_c N_t} \det(L(\vec{x}) + C)\, e^{-S_{\text{gauge}}} \tag{11.74}$$

$$= \int [DU] e^{-S_{\text{gauge}} + \sum_{\vec{x}} \{\mu a N_c N_t + \ln[\det(L(\vec{x}) + C)]\}} \tag{11.75}$$

So we can update the spatial links with a Metropolis algorithm borrowed from quenched QCD simulations. The temporal links require the trick of using the magnitude of the determinant and the gauge action as an effective action and then incorporating the phase of the determinant θ into the expectation value of interest,

$$\langle A \rangle = \frac{\langle A e^{i\theta} \rangle}{\langle e^{i\theta} \rangle} \tag{11.76}$$

where the expectation values are taken with respect to the ensemble of configurations weighted by the magnitude of the determinant. Of course, this method will not be practical if the phase oscillates wildly and/or averages to zero. In this case, as opposed to examples with light fermions known in statistical mechanics and to QCD itself with light quarks, the average of $e^{i\theta}$ is free of such problems and useful estimates can be made. One calculates all the usual thermodynamic quantities, including the average quark density, and investigates the fate of the $\mu = 0.0$ finite-T transition, which is of first order for three light quarks in $SU(3)$ QCD, as one turns the chemical potential up. One expects that the first-order transition will be stable against perturbations and that the first-order transition will persist to some intermediate value of $1/C$. Instead, plots of the Wilson line against coupling for small values of $1/C$ indicate that the first-order transition becomes a crossover immediately.

This peculiar result is not understood. Perhaps it is related to the limiting procedure $\mu \to \infty$ and $m \to \infty$. Certainly it can be tested with light quarks in the case of SU(2) lattice-gauge theory and developments we shall review below suggest that this limited question can also be investigated in three-flavor SU(3) lattice-gauge theory for temperatures near the quark–gluon-plasma transition T_c at $\mu = 0.0$.

11.6 The Glasgow algorithm

The original Glasgow algorithm was put forward, tested, and develped by Ian Barbour [149]. The idea is, very roughly, the following. Since we do not know how to simulate lattice QCD at nonzero μ for the baryon number because the action becomes complex, we will isolate the action's μ dependence in a fugacity expansion. The coefficients of the fugacity expansion will then be computed by using an ensemble of configurations generated at the coupling of interest, but at vanishing μ.

This naive idea faces some serious hurdles. Forgetting about the details of the fugacity expansion, the method would be like inferring that the Ising model can exist at low temperature in a state of spontaneous magnetization by accumulating Ising configurations at high temperature. One would have to have an astronomically large ensemble of configurations and hope to find every now and then a statistical fluctuation so that a state with nonvanishing mean magnetization would appear. These configurations might be singled out and used to make up a Boltzmann distribution of configurations at low temperatures that possess a nonvanishing mean magnetization.

How efficient could such a "reweighting" procedure be? Since we start in the wrong phase where the spins are fluctuating almost independently of each other, an initial estimate of the probability of finding a configuration having a nonzero mean magnetization would be exponentially small in the number of degrees of freedom of the system, i.e. $\exp(-cV)$, where V is the volume of the system and c is a finite constant.

Actually, prospects for this method working at zero T might be much worse than these estimates which are based on analogies with similar strategies intended to work for transitions along the temperature axis. We know from the studies of soluble models and general arguments that many observables are completely unaffected by the chemical potential for $\mu < \mu_c$ if our physical system has vanishing temperature. How then could one use a fugacity expansion starting in the low-μ phase and find the existence of a phase transition at μ_c and a completely different phase at high μ? The nonanalyticities involved in the e^μ plane suggest that this is not possible.

As simple-minded as this assessment of the method is, quantitative studies of the method generally support its conclusion. Why then is there any interest in such an approach? The hope is that there are generalizations of the method that do not have such serious limitations. We will review below a physical situation in which a generalization of the method does, in fact, look promising. We will see that the prospects for a multi-variable generalization of the Glasgow method are much better at a substantial T, near the quark–gluon phase transition at vanishing μ.

Glasgow-style methods begin with a fugacity expansion of the fermion determinant. The fermion matrix, which is the discrete form of the Dirac operator on the lattice using staggered fermions, reads

$$2iM = G + 2im + \hat{V}e^\mu + \hat{V}^\dagger e^{-\mu} \tag{11.77}$$

Then the determinant of M reads

$$\det(2iM) = e^{-3\mu V} \det(P - e^\mu) \tag{11.78}$$

where V is the lattice size $N^3 \times N_t$ and P is the propagator matrix, which is independent of μ,

$$P = \begin{pmatrix} -(G + 2im)\hat{V} & \hat{V} \\ -\hat{V} & 0 \end{pmatrix} \tag{11.79}$$

A basis in which the propagator matrix is diagonal can be chosen,

$$\det(2iM) = e^{-3\mu V} \prod_{k=1,6V} (\lambda_k - e^\mu) \tag{11.80}$$

This representation makes it clear that the zeros of the determinant in the e^μ plane are the eigenvalues of the propagator matrix. Since the eigenvalues λ_k of the propagator matrix satisfy $\lambda_{k+j} = e^{i2\pi j/N_t}\lambda_k$, $j = 0, 1, \ldots, N_t - 1$, we can simplify this expression using the polynomial decomposition

$$\prod_{j=0,N_t-1} (e^{i2\pi j/N_t}\beta - e^\mu) = \beta^{N_t} - e^{\mu N_t} \tag{11.81}$$

and find

$$\det(2iM) = e^{-3\mu V} \prod_{k=1,6N^3} (\lambda_k^{N_t} - e^{\mu N_t}) \tag{11.82}$$

which produces the desired fugacity expansion,

$$\det(2iM) = \sum_{k=-3N^3,3N^3} b_k f^k \tag{11.83}$$

where $f = e^{\mu N_t}$ is the fugacity.

This final expression produces a polynomial expansion of the partition function $Z(\mu)$,

$$Z(\mu) = \sum_{k=-3N^3,3N^3} \langle b_k \rangle e^{k\mu T} \tag{11.84}$$

where the expectation value is taken in a configuration having $\mu = 0.0$.

Tests of this method on Gross–Neveu models, which we reviewed above and can be simulated at nonzero chemical potential by conventional algorithms, indicate that the overlap problem does invalidate it. In the case of the Gross–Neveu model one could test whether the Glasgow method could find the chirality-restoring phase transition at the known μ_c by generating configurations in either phase, calculating the fugacity expansion as explained above, and searching for the Lee–Yang zeros in the result. In all cases the method failed to find the known transition, indicating that the coefficients of the fugacity expansion for the partition function cannot be estimated accurately starting from an ensemble of limited, but large, statistics in the wrong phase.

It is generally believed that the Glasgow method needs some additional ingredients if it is to succeed. Some interesting suggestions have been made and we will review a particularly promising and relevant one next.

11.7 The Fodor–Katz method for high T, low μ

The Glasgow reweighting method appears to be more effective at nonzero T. Suppose that you are interested in the region of the QCD phase diagram where T is near the transition from the hadronic phase to the quark–gluon-plasma phase and the baryon chemical potential is relatively small. This region of phase space is of considerable interest because of its relevance to the heavy-ion experimental program which produces nuclear matter under these conditions, very hot with a small excess of baryon number. Using conventional algorithms one can generate field configurations all along the T axis. For moderate quark masses in three-flavor SU(3) lattice-gauge theory, the transition on the T axis should be smoothed to a sharp crossover. However, if the chemical potential μ were turned up, the crossover would be expected to move to slightly smaller T. Eventually, if μ were increased to μ_T, a critical point where the crossover becomes a line of first-order transitions would be found. It would be particularly interesting if this critical point were experimentally accessible, as we have discussed in previous chapters.

A multi-variable Glasgow algorithm has been suggested and tested on small lattices, which suggests that μ_T can be found by lattice methods available at present. One begins with the reweighting expression for the partition function [150],

$$Z[\alpha] = \int D[U] \exp(-S_{\text{gauge}}(U; \alpha_0)) \det M(U; \alpha_0)$$

$$\times \exp\left(-\Delta S_{\text{gauge}}(U; \alpha, \alpha_0) \frac{\det M(U; \alpha)}{\det M(U; \alpha_0)}\right) \qquad (11.85)$$

which means that configurations are generated at the parameter setting $\alpha_0 = (\beta_0, m_0, \mu_0)$, but, by evaluating the matrix element in large parentheses, we

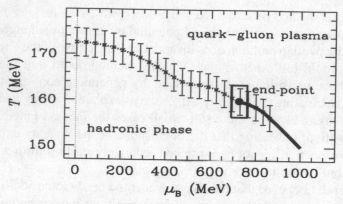

Figure 11.11. The crossover, critical point, and first-order line predicted by Fodor and Katz.

effectively attempt to reweight those configurations to generate an ensemble of configurations at the new parameter setting $\alpha = (\beta, m, \mu)$. One expects the overlap between the two ensembles to be exponentially small in the volume of the system, but large enough on small lattice volumes to simulate successfully in particular region of the $T-\mu$ plane.

An intuitive and compelling proposal that the overlap is large along the line of crossovers in the $T-\mu$ plane has been suggested [110]. It is necessary to vary both β and μ in order to stay on the line. The idea is that configurations along the line are strongly fluctuating between hadronic behavior and plasma behavior. So, if we just move along the crossover line, the character of the configurations does not change dramatically and the overlap between the configurations at $\alpha_0 = (\beta_0, m_0, \mu_0)$ and $\alpha = (\beta, m, \mu)$ is large.

In practice one identifies $\beta_c(\mu)$ by calculating the lowest Yang–Lee zero as one steps from smaller to larger μ. At the tricritcal point the Yang–Lee zero would pinch the real axis in the thermodynamic limit $N \to \infty$. In practice one uses finite-size scaling analysis to infer this behavior. The result of this exercise is shown in Fig. 11.11, which is very encouraging but not yet quantitative because the sizes of lattices were small in the actual simulations and the quark masses were rather large.

11.8 QCD at complex chemical potential

Although the QCD partition function becomes complex in the presence of a real, physical baryon chemical potential, the partition function remains real for an imaginary chemical potential. This brings up the possibility that the way to study QCD in a baryon-rich environment is to simulate it with an imaginary chemical potential and then analytically continue observables to real chemical potential.

In the early days of lattice-gauge theory this approach was tried and abandoned because, at low T and variable imaginary μ, the chemical potential had little effect on observables. However, if one is less ambitious and more focused, the approach appears very promising for high temperatures, near the quark–gluon-plasma transition, and for small chemical potentials [151].

Write the chemical potential as

$$\mu = \mu_R + i\mu_I \tag{11.86}$$

and note that, for purely imaginary chemical potential, the effect on the theory is the same as a purely temporal Abelian vector potential coupled to the quark number. Such a theory clearly has a real fermion determinant. The imaginary chemical potential affects a forward-hopping fermion with a pure constant phase, $e^{i\mu_I}$, and affects a backward-hopping fermion with a phase, $e^{-i\mu_I}$. There is a maximum μ_I that can be studied by this method. Consider a fermion path that goes once around the anti-periodic temporal lattice. The fermion worldline picks up a total phase of $e^{i\mu_I N_\tau}$. If this phase lies in the center of the gauge group SU(3), i.e. Z(3), then it has no physical effect because it can be removed by a gauge transformation. This restricts the physical region of μ_I,

$$\mu_I < \frac{2\pi}{3N_\tau} = \frac{2\pi}{3}T \tag{11.87}$$

In other words, the physics of the theory is periodic in the variable μ_I/T.

Let's write this as a symmetry of the grand canonical partition function,

$$Z(V, \mu, T) = \text{tr}\left(e^{(-\hat{H}-\mu\hat{Q})/T}\right) \tag{11.88}$$

where $\hat{H}$ is the Hamiltonian operator and $\hat{Q}$ is the quark-number operator. (In terms of the physical baryon number B and the baryon chemical potential μ_B, we must remember the factors of three, $B = Q/3$ and $\mu = \mu_B/3$.) At nonzero temperature Z depends on μ/T, so it is often more convenient to use the variable

$$\bar{\mu} \equiv \mu/T \tag{11.89}$$

and express Z as $Z(V, \bar{\mu}, T)$. Our exercise above, which concluded that the physics is periodic in μ_I/T, reads

$$Z(\bar{\mu}_R, \bar{\mu}_I) = Z(\bar{\mu}_R, \bar{\mu}_I + 2\pi/3) \tag{11.90}$$

In addition Z is an even function of $\bar{\mu}$ because switching the sign of the chemical potential can always be compensated by a time reflection,

$$Z(\bar{\mu}) = Z(-\bar{\mu}) \tag{11.91}$$

On combining these two symmetry relations, we see that the partition function is symmetric about every multiple of $\bar{\mu}_I = \pi/3$. In particular,

$$Z(\bar{\mu}_R = 0, \bar{\mu}_I = \pi/3 + \Delta\bar{\mu}_I) = Z(\bar{\mu}_R = 0, \bar{\mu}_I = \pi/3 - \Delta\bar{\mu}_I) \qquad (11.92)$$

and the free energy and other observables must inherit this symmetry.

Simulations can certainly be made for small imaginary chemical potential. Since the theory has a very clear transition from its hadronic phase to the quark–gluon plasma at $T = T_c$ and $\mu = 0$, there must be a clear line of transitions (crossovers, actually, for intermediate quark masses) inside the phase diagram for $T \approx T_c$ and $\bar{\mu}$ small. In order to use simulation data at imaginary chemical potential, we must analytically continue them to real chemical potential. We need to know the radius of convergence of the free energy in the complex-$\bar{\mu}$ plane in order to do this.

The key to doing this lies in the Z(3) structure of the theory's gauge group. Consider again fermion loops that wind once around the temporal lattice. As we discussed before, this quantity is best monitored with the free energy of a heavy quark, or the expectation of the Wilson line, $\langle L \rangle$. In the pure gauge theory, the phase of the Wilson line is one of the members of the center of the gauge group, 0, $\pm 2\pi/3$. These three states have identical free energies F in the pure gauge theory. As a function of the phase of the Wilson line, ϕ, $F(\phi)$ must have barriers between its three minima. The physics of the barriers can be studied by simulating interfaces between the different Z(3) vacua. The first-order transitions between them have been studied in great detail because of their intrinsic interest and their potential applications to the physics of the early universe, etc.

When dynamical fermions are restored into the theory, the Z(3) symmetry is explicitly broken and, therefore, the degeneracy of the three minima is lifted. The $\phi = 0$ vacuum becomes the unique minimum and the other two states share a common, higher free energy. At high T one can calculate the $\bar{\mu}_I$ dependence of the free energy of the $\phi = 0$ vacuum [152],

$$F^{(\phi=0)}(\bar{\mu}_I) = -\frac{1}{4}\pi^2 T^4\left(1 - \frac{\bar{\mu}_I^2}{\pi^2}\right) \qquad (11.93)$$

where $-\pi < \bar{\mu}_I < \pi$, and $F^{(\phi=0)}(\bar{\mu}_I)$ is periodic, $F^{(\phi=0)}(\bar{\mu}_I \pm 2\pi k) = F^{(\phi=0)}(\bar{\mu}_I)$. To determine the analogous results for the other Z(3) vacua, we consider Z(3) transformations as illustrated above, leading us to conclude that

$$F^{(\phi=+2\pi/3)}(\bar{\mu}_I) = F^{(\phi=0)}(\bar{\mu}_I - 2\pi/3),$$
$$F^{(\phi=-2\pi/3)}(\bar{\mu}_I) = F^{(\phi=0)}(\bar{\mu}_I + 2\pi/3) \qquad (11.94)$$

These three free energies are shown in Fig. 11.12. The three curves are copies of

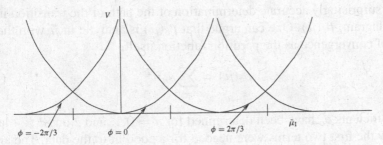

Figure 11.12. The ground-state energy for the Z(3) vacua.

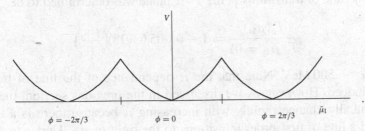

Figure 11.13. The ground-state energy for the physical vacuum at high T.

each other, just translated by $\Delta\bar{\mu}_I = 2\pi/3$. The physical branch of these curves is found by choosing the minimum free energy at each $\bar{\mu}_I$. This produces the free-energy curve in Fig. 11.13, where we see that the vacuum with $\phi = 0$ is preferred for $-\pi/3 < \bar{\mu}_I < +\pi/3$, the vacuum with $\phi = +2\pi/3$ is preferred for $+\pi/3 < \bar{\mu}_I < +3\pi/3$, and the vacuum with $\phi = -2\pi/3$ is preferred for $+3\pi/3 < \bar{\mu}_I < +5\pi/3$, etc. Since the physical free energy has cusps at $\bar{\mu}_I = 2\pi(k + \frac{1}{2})/3$, the transitions between the different vacua have latent heats and are of first order.

Now we can return to the question of the radius of convergence of the partition function in the complex-chemical-potential plane. Since there is a phase transition at $\bar{\mu}_I = \pi/3$, this is an upper bound to the radius of convergence. So the method is restricted to values of the real part of the chemical potential for $\mu_r < \pi T/3$. In practice, the method can be used to follow the finite-temperature transition out into the nonzero-chemical-potential portion of the phase diagram. Since $T_c \approx 170\,\text{MeV}$, this method is resticted to $\mu_B \leq 500\,\text{MeV}$.

Simulations to date using this method have been restricted to small lattices, so much remains to be demonstrated. However, the results are encouraging [151]. The transition between hadronic matter and the quark–gluon plasma has been tracked into the β–$\bar{\mu}$ plane by measuring susceptibilities and following their peaks. It turns out that the peaks in various susceptibilities are quite sharp and allow a surprisingly accurate determination of the path of the transition into the phase diagram, $\beta_c(\bar{\mu})$. One can argue that $\beta_c(\bar{\mu})$ is analytic in $\bar{\mu}$ with the same radius of convergence as the partition function itself,

$$\beta_c(\bar{\mu}) = \sum_{n=0} a_n \bar{\mu}^{2n} \tag{11.95}$$

The coefficients a_n have been determined for $n = 0, 1$, and 2 on $8^3 \times 4$ lattices, and only the first two terms were needed for a good fit to the data. The analytic continuation from imaginary μ, where all the simulations and fits were done, to real μ is done by trivial substitution. Assuming the applicability of asymptotic scaling, the line of transitions in the $T-\mu$ plane was determined to be

$$\frac{T_c(\mu_B)}{T_c(\mu_B = 0)} = 1 - 0.005\,63(38)\left(\frac{\mu_B}{T}\right)^2 \tag{11.96}$$

up to $\mu_B \sim 500\,\text{MeV}$. Note that the μ dependence of the line of transitions is weak indeed. However, as we discussed in the previous section, the physics will eventually change rapidly with increasing μ because there is a tricritical point and a line of first-order transitions to the right of μ_T. Understanding this quantitatively through simulations is an ongoing theme in this young field.

The Z(3) first-order transitions for high temperatures were also found numerically. The calculations appear to be under good control, with solid theoretical underpinnings.

Immediate tasks for this method are the following.

1. Check the feasibility of imaginary-chemical-potential simulations for large lattices.

2. Carry out finite-size-scaling studies in the vicinity of the line of crossovers, $\beta_c(\bar{\mu})$.

3. Determine the order of the hadronic-matter–quark–gluon-plasma transition as a function of μ.

4. Find the critical point μ_T, assuming that it is within the radius of convergence of the method, and determine whether it is relevant to the RHIC.

5. Map out the equation of state for all T and μ within the radius of convergence of the method.

12
Epilogue

The problems of quark confinement, chiral-symmetry breaking, and strongly interacting matter in extreme environments are converging and much is being learned, both experimentally and theoretically. We hope that this book helps researchers find common ground in this developing field.

The breadth of the field is quite remarkable. Concepts from critical behavior, effective-field theory and random-matrix models, lattice-gauge theory and numerical methods, spin models and duality, equilibrium and nonequilibrium thermodynamics, and even string theory play important roles here. It is impossible to tell which approach will lead to the next major advance in the field.

The field depends on a "new culture" in theoretical physics, one in which theorists use lattice-gauge theory as a realistic laboratory to test new ideas through numerical methods and computer simulations. The theorist is no longer limited, in many but, alas, not all cases, to unrealistic models for concrete inspiration. Lattice-gauge simulations have led to the sharpening of several approaches to QCD that have influenced the field very productively. We can look forward to the development of new algorithms that can address SU(3) QCD at nonzero baryon chemical potential and find new states of matter at high densities. Once such an algorithm is in place, the space of variable quark masses (m_u, m_d, and m_s) and variable chemical potentials (μ_u, μ_d and μ_s) will become accessible, and the promise of a new "chemistry" will play itself out. Certainly new phases of condensates, crystal structures, and flavor symmetries are likely to be found. Perhaps the conventional formulation of lattice-gauge theory will not be powerful enough to make this progress and a new formulation will have to be invented. Time will tell.

One should not underestimate the accomplishments of lattice-gauge theory to date. The development of the thermodynamics of hot QCD, complete with the "best" estimates of T_c at which the quark–gluon plasma is produced, is a milestone in the field. Many more specific predictions relevant to the early universe

355

and to RHIC experiments have also been made and are being improved. Much of this work is producing insight into the working of QCD in ordinary environments, shedding light on confinement and chiral-symmetry breaking. Perhaps the lofty goal of producing an approach to QCD that can incorporate asympotic freedom at short distances, chiral-symmetry breaking and light-hadron spectroscopy at intermediate distances, and confinement at large distances is not so far off. Most probably it is, but the process of searching for it may produce surprising rewards.

As this book has emphasized, it is the convergence of numerical methods with traditional and developing analytic and theoretical methods which makes this field particularly interesting and productive, at least in the eyes of the authors. The hand-in-hand development of both approaches, theoretical approaches to suggest numerical work and numerical work to challenge theoretical approximations, is new to field theory and will, it is hoped, have a bright and long future.

The authors hope that this book will serve the reader well in new pioneering efforts from confinement to extreme environments.

References

[1] J. B. Kogut, *Rev. Mod. Phys.* **51** (1979), 659; **55** (1983), 775.

[2] J. B. Kogut and K. G. Wilson, *Phys. Rep.* **C12** (1974), 75.

[3] *Phase Transition and Critical Phenomena*, edited by C. Domb and M. S. Green (London, Academic Press, 1974).

[4] N. D. Mermin and H. Wagner, *Phys. Rev. Lett.* **17** (1966), 1133.

[5] J. M. Kosterlitz and D. J. Thouless, *J. Phys.* **C6** (1973), 118.

[6] J. M. Kosterlitz, *J. Phys.* **C7** (1974), 1046.

[7] A. M. Polyakov, *Phys. Lett.* **B59** (1975), 79.

[8] R. H. Swendsen and J.-S. Wang, *Phys. Rev. Lett.* **58**, 86 (1987); U. Wolff, *Phys. Rev. Lett.* **62** (1989), 361.

[9] C. Stzyleson and J.-M. Drouffe, *Statistical Field Theory* (Cambridge, Cambridge University Press, 1989).

[10] K. G. Wilson, *Phys. Rev.* **D14** (1974), 2455; A. M. Polyakov, *Phys. Lett.* **B59** (1975), 82.

[11] C. N. Yang and R. L. Mills, *Phys. Rev.* **96** (1954), 1605.

[12] H. D. Politzer, *Phys. Rev. Lett.* **30** (1973), 1346; D. J. Gross and F. Wilczek, *Phys. Rev. Lett.* **30** (1973), 1343.

[13] J. Schwinger, *Phys. Rev.* **128** (1962), 2425.

[14] S. Adler, *Phys. Rev.* **177** (1969), 2426; J. S. Bell and R. Jackiw, *Nuov. Cim.* **60A** (1969), 47.

[15] J. B. Kogut and L. Susskind, *Phys. Rev.* **D11** (1975), 3594.

[16] J. Lowenstein and A. Swieca, *Ann. Phys. (NY)* **68** (1971), 172.

[17] S. Coleman, R. Jackiw, and L. Susskind, *Ann. Phys. (NY)* **93** (1975), 267.

[18] B. Klaiber, in *Lectures in Theoretical Physics*, Vol. XA, edited by A. O. Barut and W. Brittin (New York, Gordon and Breach, 1968).

[19] T. Schafer and E. Shuryak, *Rev. Mod. Phys.* **70** (1998), 323.

[20] G. 't Hooft, *Nucl. Phys.* **B138** (1978), 1.

[21] B. C. Barrois, *Nucl. Phys.* **B129** (1977), 390.

[22] D. Bailin and A. Love, *Phys. Rep.* **107** (1984), 325, and references therein.

[23] M. Alford, K. Rajagopal, and F. Wilczek, *Phys. Lett.* **B422** (1998), 247.

[24] R. Rapp, T. Schäfer, E. V. Shuryak, and M. Velkovsky, *Phys. Rev. Lett.* **81** (1998), 53.

[25] H. B. Nielsen and M. Ninomiya, *Nucl. Phys.* **185** (1981), 20.

[26] W. Shockley, *Phys. Rev.* **56** (1939), 317.

[27] D. Kaplan, *Phys. Lett.* **B288** (1992), 342.

[28] C. Callan and J. Harvey, *Nucl. Phys.* **B250** (1985), 427.

[29] P. Vranas, *Nucl. Phys. Proc. Suppl.* **94** (2001), 177.

[30] P. Ginsparg and K. Wilson, *Phys. Rev.* **D25** (1982), 2649.

[31] M. Luscher, *Phys. Lett.* **B428** (1998), 342.

[32] H. Neuberger, *Phys. Lett.* **B417** (1998), 141.

[33] A. Casher, *Phys. Lett.* **B83** (1979), 395.

[34] Y. Nambu and G. Jona Lasinio, *Phys. Rev.* **122** (1961), 345.

[35] T. Banks and A. Casher, *Nucl. Phys.* **B169** (1980), 103.

[36] S. Raby, S. Dimopoulos, and L. Susskind, *Nucl. Phys.* **B169** (1980), 373.

[37] J. M. Blairon, R. Brout, F. Englert, and J. Greensite, *Nucl. Phys.* **B180** (1981), 439.

[38] S. Duane and J. B. Kogut, *Phys. Rev. Lett.* **55** (1985), 2774; S. Gottlieb, W. Liu, D. Toussaint, R. L. Renken, and R. L. Sugar, *Phys. Rev.* **D35** (1987), 2531.

[39] S. Duane, A. D. Kennedy, B. J. Pendleton, and D. Roweth, *Phys. Lett.* **B195** (1987), 216.

[40] J. B. Kogut and L. Susskind, *Phys. Rev.* **D11** (1975), 395.

[41] M. Luscher, *Nucl. Phys.* **B180** (1981), 317.

[42] R. B. Pearson, J. Shigemitsu, and J. B. Kogut, *Phys. Lett.* **98B** (1980), 63; J. B. Kogut, D. K. Sinclair, R. B. Pearson, J. L. Richardson, and J. Shigemitsu, *Phys. Rev.* **D23** (1981), 2945.

[43] L. Susskind, *Phys. Rev.* **D20** (1979), 2610.

[44] B. Svetitsky and L. G. Yaffe, *Nucl. Phys.* **B210** (1982), 423.

[45] R. D. Pisarski and F. Wilczek, *Phys. Rev.* **D29** (1984), 338.

[46] J. B. Kogut and C. Detar, *Phys. Rev. Lett.* **59** (1987), 399.

[47] F. Karsch, E. Laermann, P. Petreczky, S. Stickan, and I. Wetzorke, hep-lat/0110208.

[48] M. Asakawa, Y. Nakahara, and T. Hatsuda, *Prog. Part. Nucl. Phys.* **46** (2001), 459.

[49] T. Matsui and H. Satz, *Phys. Lett.* **B178** (1986), 416.

[50] F. Karsch, in *Proceedings of QCD@Work* (Washington, American Institute of Physics, 2001), 323.

[51] J.-P. Blaizot, hep-ph/0107131.

[52] N. Bilić, K. Demeterfi, and B. Petersson, *Nucl. Phys.* **B377** (1992), 651.

[53] S. P. Klevansky, *Rev. Mod. Phys.* 64 (1992), 649.

[54] G. E. Brown, M. Buballa, and M. Rho, *Nucl. Phys.* **A609** (1996), 519.

[55] J. Stachel, *Nucl. Phys.* **A610** (1996), 509c.

[56] A. L. Fetter and J. D. Walecka, *Quantum Theory of Many-Particle Systems* (McGraw-Hill, New York, 1971).

[57] A. B. Migdal, E. E. Saperstein, M. A. Troitsky, and D. N. Voskresensky, *Phys. Rep.* **192** (1990), 179.

[58] I. Klebanov, *Nucl. Phys.* **B262** (1985), 133.

[59] D. B. Kaplan and A. E. Nelson, *Phys. Lett.* **B175** (1986), 57.

[60] E. Witten, *Phys. Rev.* **D30** (1984), 272.

[61] E. Farhi and B. Jaffe, *Phys. Rev.* **D30** (1984), 2379.

[62] M. Alford, K. Rajagopal, and F. Wilczek, *Nucl. Phys.* **B537** (1999), 443.

[63] T. Schäfer and F. Wilczek, *Phys. Rev. Lett.* **82** (1999), 3956.

[64] L. P. Csernai and J. I. Kapusta, *Phys. Rep.* **131** (1986), 223.

[65] M. A. Stephanov, *Phys. Rev. Lett.* **76** (1996), 4472.

[66] W. Trautmann, Multifragmentation in relativistic heavy ion collisions, in *Proceedings the International Summer School on Correlations and Clustering Phenomena in Subatomic Physics*, Dronten, the Netherlands, 1996 (to be published by Plenum, New York); e-print: nucl-ex/9611002.

[67] HEMCGC and HTMCGC collaborations, *Nucl. Phys. Proc. Suppl.* **B30** (1993), 315.

[68] R. Pisarski and F. Wilczek, *Phys. Rev.* **D29** (1984), 338.

[69] C. Bernard, T. Blum, C. DeTar *et al. Phys. Rev. Lett.* **78** (1997), 598.

[70] F. R. Brown, F. P. Butler, H. Chen *et al. Phys. Rev. Lett.* **65** (1990), 2491.

[71] I. D. Lawrie and S. Sarbach, in *Phase Transitions*, Vol. 9, edited by C. Domb and J. L. Lebowitz (New York, Academic Press, 1984).

[72] E. Shuryak, hep-ph/0205031.

[73] F. Karsch, in *40th Internationale Universitätswochen für Theoretische Physik: Dense Matter* (Berlin, Springer, 2002), 209.

[74] J. B. Kogut and L. Susskind, *Phys. Rep.* **8** (1973), 75.

[75] R. P. Feynman, *Photon–Hadron Interactions* (New York, W. A. Benjamin, 1972).

[76] L. N. Lipatov, *Sov. J. Nucl. Phys.* **23** (1976), 338.

[77] J. B. Kogut and L. Susskind, *Phys. Rev.* **D9** (1974), 3391.

[78] G. Altarelli and G. Parisi, *Nucl. Phys.* **B126** (1977), 298.

[79] L. V. Gribov, E. M. Levin, and M. G. Ryskin, *Phys. Rep.* **100** (1983), 1; for a recent review, see E. Iancu, A. Leonidov, and L. McLerran, hep-ph/0202270.

[80] E. Ferreiro, E. Iancu, K. Itakura, and L. McLerran, hep-ph/0206241.

[81] D. Kharzeev, *Nucl. Phys.* **B119** (2003).

[82] J. D. Bjorken, *Phys. Rev.* **D27** (1983), 140.

[83] M. A. Stephanov, K. Rajagopal, and E. V. Shuryak, *Phys. Rev. Lett.* **81** (1998), 4816.

[84] M. A. Stephanov, K. Rajagopal, and E. V. Shuryak, *Phys. Rev.* **D60** (1999), 114 028.

[85] B. Berdnikov and K. Rajagopal, *Phys. Rev.* **D61** (2000), 105 017.

[86] D. K. Hong, *Phys. Lett.* **B473** (2000), 118 (hep-ph/9812510).

[87] D. T. Son, *Phys. Rev.* **D59** (1999), 094 019.

[88] R. D. Pisarski and D. H. Rischke, *Phys. Rev.* **D61** (2000), 051 501 (nucl-th/9907041).

[89] T. Schafer and F. Wilczek, *Phys. Rev.* **D60** (1999), 114 033 (hep-ph/9906512).

[90] W. E. Brown, J. T. Liu, and H. C. Ren, *Phys. Rev.* **D61** (2000), 114 012 (hep-ph/9908248); *Phys. Rev.* **D62** (2000), 054 016 (hep-ph/9912409).

[91] R. Rapp, T. Schafer, E. V. Shuryak, and M. Velkovsky, *Ann. Phys.* **280** (2000), 35 (hep-ph/9904353); T. Schafer, *Nucl. Phys.* **B575** (2000), 269 (hep-ph/9909574); *Phys. Rev.* **D65** (2002), 094 033 (hep-ph/0201189).

[92] D. T. Son, M. A. Stephanov, and A. R. Zhitnitsky, *Phys. Rev. Lett.* **86** (2001), 3955.

[93] R. Casalbuoni and R. Gatto, *Phys. Lett.* **B464** (1999), 111 (hep-ph/9908227).

[94] D. T. Son and M. A. Stephanov, *Phys. Rev.* **D61** (2000), 074 012 (hep-ph/9910491); *Phys. Rev.* **D62** (2000), 059 902 (hep-ph/0004095).

[95] E. Witten, *Ann. Phys.* **128** (1980), 363.

[96] N. N. Bogolyubov, V. V. Tolmachev, and D. V. Shirkov, *New Methods in the Theory of Superconductivity* (New York, Consultants Bureau, 1959); P. W. Anderson, *Phys. Rev.* **112** (1958), 1900; A. J. Leggett, *Rev. Mod. Phys.* **47** (1975), 331.

[97] T. Schafer, *Phys. Rev.* **D65** (2002), 074 006 (hep-ph/0109052).

[98] K. Rajagopal and F. Wilczek, *Phys. Rev. Lett.* **86** (2001), 3492 (arXiv:hep-ph/0012039).

[99] P. F. Bedaque and T. Schafer, *Nucl. Phys.* **A697** (2002), 802.

[100] K. Rajagopal and E. Shuster, *Phys. Rev.* **D62** (2000), 085 007.

[101] J. Berges and K. Rajagopal, *Nucl. Phys.* **B538** (1999), 215.

[102] D. B. Kaplan and S. Reddy, *Phys. Rev.* **D65** (2002), 054 042.

[103] M. G. Alford, J. Berges, and K. Rajagopal, *Nucl. Phys.* **B558** (1999), 219.

[104] M. G. Alford, J. A. Bowers, and K. Rajagopal, *Phys. Rev.* **D63** (2001), 074 016.

[105] P. Fulde and A. Ferrell, *Phys. Rev.* **135** (1964), A550; A. I. Larkin and Yu. N. Ovchinnikov, *Sov. Phys. JETP* **20** (1965), 762.

[106] J. A. Bowers and K. Rajagopal, *Phys. Rev.* **D66** (2002), 065 002.

[107] M. Alford and K. Rajagopal, *JHEP* **0206** (2002), 031.

[108] K. Rajagopal and F. Wilczek, arXiv:hep-ph/0011333.

[109] M. Alford, *Ann. Rev. Nucl. Part. Sci.* **51** (2001), 131.

[110] M. G. Alford, A. Kapustin, and F. Wilczek, *Phys. Rev.* **D59** (1999), 054 502.

[111] E. V. Shuryak and J. J. M. Verbaarschot, *Nucl. Phys.* **A560** (1993), 306.

[112] A. D. Jackson, M. K. Sener, and J. J. M. Verbaarschot, *Nucl. Phys.* **B479** (1996), 707.

[113] E. V. Shuryak, *Nucl. Phys.* **B203** (1982), 93.

[114] M. A. Halasz, A. D. Jackson, R. E. Shrock, M. A. Stephanov, and J. J. Verbaarschot, *Phys. Rev.* **D58** (1998), 096 007.

[115] A. D. Jackson and J. J. M. Verbaarschot, *Phys. Rev.* **D53** (1996), 7223.

[116] T. Wettig, A. Schäfer, and H. A. Weidenmuller, *Phys. Lett.* **B367** (1996), 28.

[117] C. N. Yang and T. D. Lee, *Phys. Rev.* **87** (1952), 104; T. D. Lee and C. N. Yang, *Phys. Rev.* **87** (1952), 410.

[118] V. L. Girko, *Spektralnaia teoriia sluchainykh matrits* (Moscow, Nauka, 1988).

[119] D. Weingarten, *Phys. Rev. Lett.* **51** (1983), 1830.

[120] J. B. Kogut, D. Toublan, and D. K. Sinclair, *Phys. Lett.* **B514**, (2001), 77.

[121] J. B. Kogut, M. A. Stephanov, D. Toublan, J. J. Verbaarschot, and A. Zhitnitsky, *Nucl. Phys.* **B582** (2000), 477.

[122] D. T. Son and M. A. Stephanov, *Phys. Rev. Lett.* **86** (2001), 592.

[123] See, e.g., H. Georgi, *Weak Interaction and Modern Particle Theory* (Menlo Park, Benjamin-Cummings, 1984).

[124] A. J. Leggett, *J. Physique* **41** (1980), C7-19; P. Nozières and S. Schmitt-Rink, *J. Low Temp. Phys.* **59** (1985), 195.

[125] See, e.g., M. Randeria, cond-mat/9710223 and references therein.

[126] B. I. Halperin and P. C. Hohenberg, *Phys. Rev.* **188** (1969), 898; P. C. Hohenberg and B. I. Halperin, *Rev. Mod. Phys.* **49** (1977), 435.

[127] D. T. Son and M. A. Stephanov, *Phys. Rev. Lett.* **88** (2002), 202 302 (arXiv:hep-ph/0111100).

[128] R. D. Pisarski and F. Wilczek, *Phys. Rev.* **D29** (1984), 338.

[129] K. Rajagopal and F. Wilczek, *Nucl. Phys.* **B399** (1993), 395.

[130] E. V. Shuryak, *Phys. Rev.* **D42** (1990), 1764.

[131] R. D. Pisarski and M. Tytgat, *Phys. Rev.* **D54** (1996), 2989.

[132] D. T. Son and M. A. Stephanov, *Phys. Rev.* **D66** (2002), 076 011.

[133] J. L. Goity and H. Leutwyler, *Phys. Lett.* **B228** (1989), 517.

[134] B. D. Josephson, *Phys. Lett.* **21** (1966), 608.

[135] G. A. Baker, B. G. Nickel, and D. I. Meiron, *Phys. Rev.* **B17** (1978), 1365.

[136] S. Gottlieb, W. Liu, D. Toussaint, R. L. Renken, and R. L. Sugar, *Phys. Rev. Lett.*
       59 (1987), 2247; S. Gottlieb, U. M. Heller, A. D. Kennedy, S. Kim, J. B. Kogut,
       C. Liu, R. L. Renken, D. K. Sinclair, R. L. Sugar, D. Toussaint, and K. C. Wang,
       Phys. Rev. **D55** (1997), 6852.

[137] P. Braun-Munzinger, J. Stachel, J. P. Wessels, and N. Xu, *Phys. Lett.* **B344**
       (1995), 43; *Phys. Lett.* **B365** (1996), 1; P. Braun-Munzinger, I. Heppe, and
       J. Stachel, *Phys. Lett.* **B465** (1999), 15.

[138] C. M. Hung and E. V. Shuryak, *Phys. Rev.* **C57** (1998), 1891.

[139] J. B. Kogut, H. Matsuoka, S. H. Shenker, J. Shigemitsu, D. K. Sinclair, M.
       Stone, and H. W. Wyld, *Nucl. Phys.* **B225** (1983), 93.

[140] F. Karsch and P. Hasenfratz, *Phys. Lett.* **B125** (1983), 308.

[141] K. G. Wilson, *Phys. Rev.* **D7** (1973), 2911.

[142] Hands, J. B. Kogut, and A. Kocic, *Ann. Phys.* **224** (1993), 29.

[143] B. Rosenstein, B. J. Warr, and S. H. Park, *Phys. Rev. Lett.* **62** (1989), 1433; *Phys.
       Rep.* **205** (1991), 59.

[144] Y. Cohen, S. Elitzur, and E. Rabinovici, *Nucl. Phys.* **B220** (1983), 102.

[145] J. B. Kogut, M. A. Stephanov, and D. Toublan, *Phys. Lett.* **B464** (1999), 183.

[146] K. Splittorff, D. Toublan, and J. J. M. Verbarschot, *Nucl. Phys.* **B620** (2002), 290.

[147] K. D. Born, E. Laermann, F. Langhammer, T. F. Walsh, and P. M. Zerwas, *Phys.
       Rev.* **D40** (1989), 1664.

[148] T. C. Blum, J. E. Hetrick, and D. Toussaint, *Phys. Rev. Lett.* **76** (1996), 1019.

[149] I. Barbour and A. J. Bell, *Nucl. Phys.* **B372** (1992), 385.

[150] Z. Fodor and S. D. Katz, *Nucl. Phys. Proc. Suppl.* 106 (2002), 441.

[151] P. De Forcrand and O. Philipsen, hep-lat/0205016.

[152] A. Roberge and N. Weiss, *Nucl. Phys.* **B275** (1986), 734.

Index

action, 14
anomalies, 9
area law, 68, 75
asymptotic freedom, 8, 35, 80, 214, 229
axial anomaly, 65, 83, 94, 117, 125, 176

Bardeen–Cooper–Schrieffer, 99, 127,
    204, 246, 305, 321
baryon number density, 6, 219
beta function, 40, 81, 174
Bose condensation, 210

canonical ensemble, 195
chemical potential, 7, 98, 198, 214, 223,
    246, 249, 289, 292, 304, 309, 327,
    333, 346
chemistry of QCD, 7, 219
chiral condensate, 97, 126, 127, 181, 220,
    229, 235, 292, 338
chiral perturbation theory, 304, 317
chiral–symmetry breaking, 5, 83, 96, 126,
    132, 206
color superconductivity, 8
color–flavor locking, 226, 248, 279
computer simulations, 43, 135, 173
condensate, 6, 29, 246
confinement, 5, 61, 69, 75, 78, 80, 155,
    168
continuum limit, 16, 40, 61, 81
correlation length, 16, 17, 19, 25, 39
critical end-point, 245
critical exponent, 17

critical point, 23
crystalline color superconductor, 288

dimensional reduction, 179
diquark, 98, 305, 308, 345
diquark condensation, 12, 247, 256, 272,
    305, 312, 347
domain-wall fermion, 10, 120
duality, 21, 23, 30, 55, 77, 162, 171

effective Lagrangian, 272, 308
effective Lagrangians, 12, 278, 304, 327
entropy, 28, 78, 196, 201, 217
extreme environments, 5, 234

Fermi surface, 99, 202, 247, 249, 267,
    335
flux tubes, 5, 66, 91, 155
free energy, 15, 177, 362

gap equation, 252, 254, 257
Ginsparg–Wilson, 11, 123
Glasgow algorithm, 357, 359
Goldstone, 72, 82, 96, 112, 132, 271, 305,
    311
grand canonical ensemble, 198

Hamiltonian, 47, 52, 100, 145
harmonic oscillator, 48
Higgs, 72, 75, 85, 207, 248, 270, 305
high temperature, 8, 168, 213, 235
high-temperature expansion, 23, 68, 154

363

hybrid algorithm, 135, 141, 346
hyperscaling, 20, 340

instantons, 42, 91, 98
Ising model, 14, 190

Kosterlitz-Thouless, 24, 30, 162

Landau damping, 247, 261
Langevin, 135, 136, 141
lattice fermions, 100
linearly confining potential, 5, 66, 75, 78, 91, 156, 172
local gauge invariance, 9, 59, 66, 69, 145
low-temperature expansion, 22

magnetic gluons, 261
magnetization, 15
Matsubara, 179, 212, 325, 341
mean field, 19, 131, 231, 297, 314
Meissner mass, 276
Metropolis algorithm, 43, 356
microcanonical ensemble, 195
modified electromagnetism, 286
monopole, 69, 77

Nambu–Jona Lasinio model, 96, 127, 254, 339
nuclear matter, 6, 224

order parameter, 14, 126, 178, 233

partition function, 14, 21, 30, 47, 70, 91, 135, 169, 194, 289, 333
path integral, 14, 47, 48, 196, 289
phase diagram, 219, 222, 247

pion condensate, 319
plaquette, 61

quark–gluon plasma, 98, 168, 184, 185, 188, 213, 229, 234, 237, 325
quenched, 7, 168, 176, 291, 300

random matrix, 7, 291, 292, 297
renormalization group, 19, 36, 81
RHIC, 11, 237
rotational symmetry, 63
roughening, 30, 72, 161

screening, 6, 177, 181, 260, 266, 325
screening lengths, 184
sign problem, 246, 289
species doubling, 10, 344
staggered fermions, 10, 104, 111, 153
SU(3), 46, 352
susceptibility, 14, 181

theta vacuum, 87, 94
topology, 29, 41, 69, 92, 125
transfer matrix, 49, 52
tricritical point, 230, 232, 233, 297, 347

universality, 19, 55, 179, 230, 330

vortices, 27, 41, 69

Wilson fermions, 104
Wilson line, 178
Wilson loop, 65

zero modes, 95, 122, 125, 127